New Methods for Polymer Synthesis

New Methods for Polymer Synthesis

Edited by

W. J. Mijs

University of Technology, Delft
Delft, The Netherlands

Plenum Press • New York and London

Library of Congress Cataloging-in-Publication Data

New methods for polymer synthesis / edited by W.J. Mijs.
p. cm.
Includes bibliographical references and index.
ISBN 0-306-43871-2
1. Polymerization. I. Mijs, W. J.
QD281.P6N48 1992
547'.28--dc20 91-44084
CIP

ISBN 0-306-43871-2

A Division of Plenum Publishing Corporation
233 Spring Street, New York, N.Y. 10013

Printed in the United States of America

Contributors

T. Aida • Department of Synthetic Chemistry, Faculty of Engineering, University of Tokyo, Hongo, Bunkyo-ku, Tokyo 113, Japan

B. C. Anderson • Central Research and Development Department, Du Pont Experimental Station, Wilmington, Delaware 19880-0328

S. Boileau • Collège de France, 11 place Marcelin Berthelot, 75231 Paris Cédex 05, France

R. R. De Clercq • Laboratory of Organic Chemistry, Polymer Division, University of Ghent, Krijgslaan 281 (S4), B-9000 Ghent, Belgium

W. Drenth • Emeritus Professor of Organic Chemistry, University at Utrecht, Ovidiuslaan 4, 3584 AW Utrecht, The Netherlands

T. Endo • Research Laboratory of Resources Utilization, Tokyo Institute of Technology, 4259 Nagatsuta, Midori-ku, Yokohama 227, Japan

E. J. Goethals • Laboratory of Organic Chemistry, Polymer Division, University of Ghent, Krijgslaan 281 (S4), B-9000 Ghent, Belgium

F. Higashi • Department of Polymer Engineering, Tokyo University of Agriculture and Technology, Koganei-shi, Tokyo 184, Japan

S. Inoue • Department of Synthetic Chemistry, Faculty of Engineering, University of Tokyo, Hongo, Bunkyo-ku, Tokyo 113, Japan

R. J. M. Nolte • Department of Organic Chemistry, University at Nijmegen, Toernooiveld, 6265 ED Nijmegen, The Netherlands

O. Nuyken • Macromolecular Chemistry I, University of Bayreuth, D-8580 Bayreuth, Germany

E. A. Ofstead • Research Division, The Goodyear Tire & Rubber Company, Akron, Ohio 44316

S. D. Pask • Bayer A.G., KA/F, D-4047 Dormagen, Germany

K. B. Wagener • Department of Chemistry and Center for Macromolecular Science and Engineering, University of Florida, Gainesville, Florida 32611

O. W. Webster • Central Research and Development Department, Du Pont Experimental Station, Wilmington, Delaware 19880-0328

T. Yokozawa • Research Laboratory of Resources Utilization, Tokyo Institute of Technology, 4259 Nagatsuta, Midori-ku, Yokohama 227, Japan

Preface

The art and science of macromolecular architecture is based on synthesis, analysis, processing, and evaluation of physical properties of polymers. The growing specificity of available synthetic methods and the increasing refinement of analytical and physical analysis are gradually providing a deeper insight into structure–property relationships of polymers, upon which many applications can be based.

This book deals with recent methods for polymer synthesis. Those that lead to specific structures have been selected especially. Background, mechanism scope and limitations, and illustrative procedures are given for each method. With this layout the editor hopes that the book will provide a practical guideline, for the synthetic polymer chemist in industry or at a university graduate school, on how to apply the methods in the design of new polymer structures.

The editor is grateful to the authors not only for their contributions containing interesting new developments in polymer synthesis, but also for the way they have fitted their text into the general framework of the book.

The elegant chemistry described in the following chapters will, it is hoped, inspire more organic chemists to apply their skills to polymer synthesis, where the beauty of organic chemistry in terms of structural control and reactivity may be even more apparent than in the low molecular field.

W. J. Mijs

Delft, The Netherlands

Contents

1. Group Transfer Polymerization
O. W. Webster and B. C. Anderson

2. *Anionic Ring-Opening Polymerization*
Shohei Inoue and Takuzo Aida

8. Polymer Synthesis via Metathesis Chemistry

E. A. Ofstead and K. B. Wagener

9. Synthesis of Polymers of Isocyanides

Roeland J. M. Nolte and Wiendelt Drenth

1

Group Transfer Polymerization

O. W. Webster and B. C. Anderson

1.1. INTRODUCTION

The synthesis of acrylic polymers by free radical initiation is low cost and operable under easily obtainable conditions. It is the method of choice if one merely needs to obtain high-molecular-weight homopolymer and random copolymers from acrylic monomers. To obtain low-molecular-weight polymers and control polymer architecture one needs a living polymerization, i.e., one in which all the polymer chains grow simultaneously with no termination or chain transfer. Anionic polymerization and the related group transfer polymerization (GTP) are two such methods. This chapter gives a description of the GTP method, how it compares with anionic polymerization and other related methods, examples of polymer architectural control, mechanistic considerations, and synthetic procedures. A glossary at the end of the chapter gives the meanings of abbreviations and codes.

1.2. DESCRIPTION OF THE METHOD

1.2.1. The Silyl Michael Addition Reaction

The addition of silyl ketene acetals to α,β-unsaturated ketones, the silyl Michael addition reaction, was first studied extensively by

O. W. Webster and B. C. Anderson • Central Research and Development Department, Du Pont Experimental Station, Wilmington, Delaware 19880-0328.

New Methods for Polymer Synthesis, edited by W. J. Mijs. Plenum Press, New York, 1992.

Mukaiyama.[1,2] The reaction of α-trimethylsilyloxystyrene (**1**) with mesityl oxide (**2**) illustrates the procedure [Eq. (1.1)].

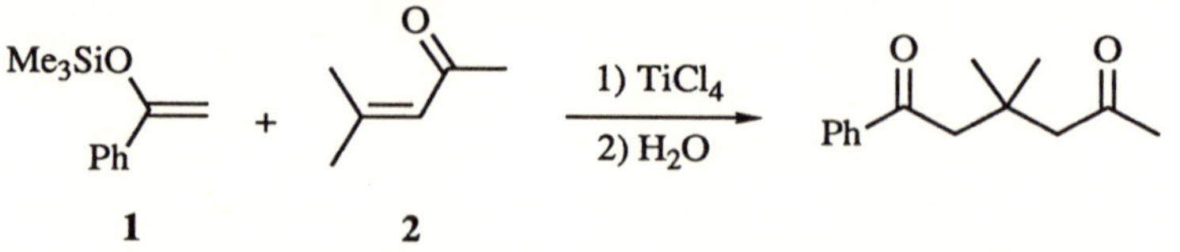

(1.1)

In the early work $TiCl_4$ is used as a catalyst and a likely first step is exchange of $TiCl_4$ with the Me_3Si group. Milder Lewis acids[3] as well as nucleophilic salts[4] are now used. The method operates with no catalyst at all if acetonitrile is used as a solvent[5] or high pressure is used.[6] In this recent work the trimethylsilyl group transfers to the oxygen of the ketone. Yields are good but not in the 99% range which one needs for high polymer synthesis.

1.2.2. Group Transfer Polymerization (GTP)

1.2.2.1. Acrylic Monomers

When the silyl Michael addition reaction is applied to an excess amount of an acrylic monomer, a polymer results.[7–9] For example, on a one-to-one basis, the silyl ketene acetal **3** and methyl methacrylate (**4**) produce adduct **5** in fair yield. The silyl ketene acetal which is generated by transfer of the silyl group undergoes repeated addition with excess monomer, each time with silyl transfer, to give polymer **6**. The process has been named group transfer polymerization (GTP) and is illustrated in Scheme 1.1. The silyl ketene acetal end group can be used for further reactions on the polymer.

1.2.2.2. Aldol GTP

Another widely used organosilicon reaction is the silyl aldol condensation[10] in which a ketone or aldehyde adds to a silyl enolate to give

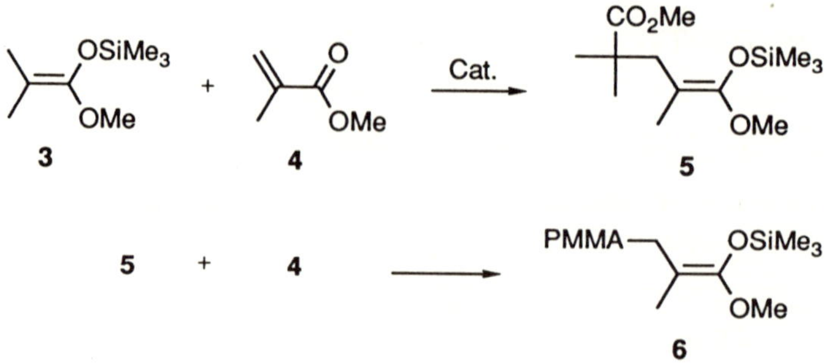

SCHEME 1.1.

the aldol product. Here again the silyl group transfers from the enolate to the carbonyl group regenerating a new carbonyl function. When this reaction is applied to excess silyl vinyl ether a silylated poly(vinyl ether) **9** results.[11] The method is illustrated for the polymerization of *t*-butyldimethylsilyl vinyl ether (**8**) initiated by benzaldehyde (**7**) [Eq. (1.2)]. A mild Lewis acid, such as $ZnBr_2$, is needed as a catalyst. Trimethylsilylvinyl ether does not work as well as *t*-butyldimethyl silyl ether. The aldehyde end group is useful for further modification of the polymer.

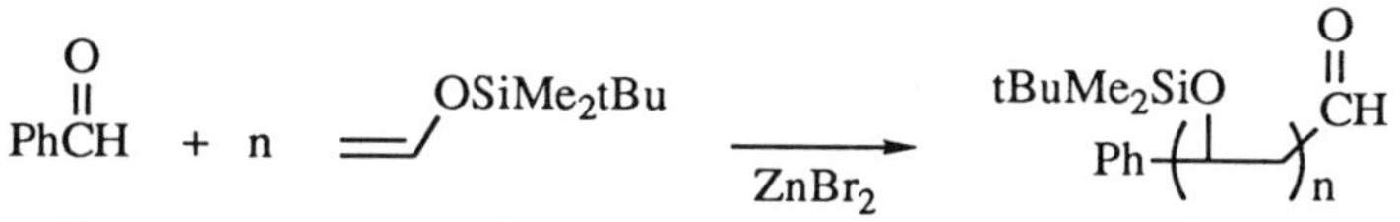

7 **8** **9** (1.2)

Trimethylsilyl dienolates **10** polymerize even more readily and give the all-*trans* 1,4-addition product **11**[12] [Eq. (1.3)].

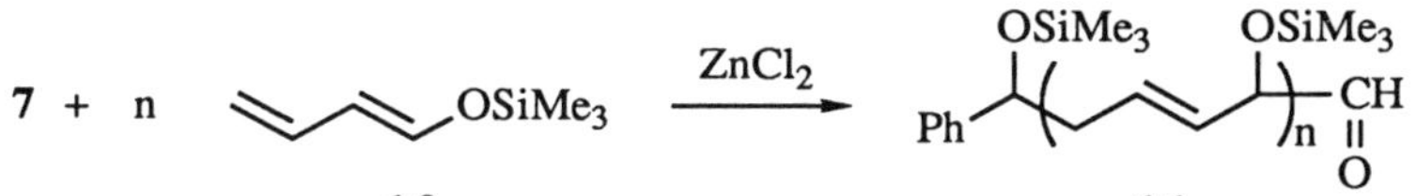

10 **11** (1.3)

1.2.2.3. *Ring-Opening GTP*

A ring-opening polymerization of propylene sulfide is initiated by trimethylsilyl thioethers. As in other GTP types, the molecular weight of the polymer is regulated by the monomer initiator ratio and M_w/M_n values in the 1.01 range result. The reaction is illustrated for propylene sulfide (**13**) initiated by trimethylsilyl 2-trimethylsiloxyethyl sulfide (**12**) (Scheme 1.2).[13] Since the trimethysilyl thioether end group on **14** initiates GTP of methacrylates (see Section 1.3.1.2) a block polymer can be formed by addition of a methacrylate after the propylene sulfide is consumed.

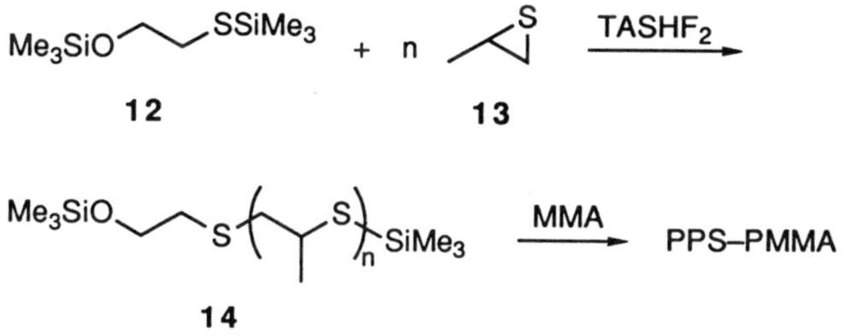

SCHEME 1.2.

1.2.3. Comparison to Other Controlled Syntheses of Acrylics

1.2.3.1. Michael Addition Polymerization

For economic synthesis of specialty resins a process is needed which operates above room temperature so that reflux condensers can be used to remove the heat of polymerization. Like GTP, Michael addition polymerization of methacrylates proceeds at above room temperature.[14] One simply refluxes the monomer in a mixture of alkoxide and alcohol [Eq. (1, 4)].

$$\text{MMA} + \xrightarrow{\text{KOMe}} \text{MeO–PMMA} \tag{1.4}$$

The molecular weight of the polymer can be controlled to some degree by adjusting the amount of alcohol used. Although DPs above 50 are mentioned, most of the examples are for polymers in the DP 30 range. Random copolymers can be made. Block polymers cannot be made, since the ester groups on the polymer exchange with the alcohol solvent. In a mixture of acrylate and methacrylate monomers only the acrylate polymerized. Hydroxy and carboxy functionality can be introduced by post-treatment of polymers with diols and by selective hydrolysis.

1.2.3.2. Anionic Polymerization

(a) Initiation by Metal Enolates. For controlled synthesis of small amounts of polymethacrylates for research purposes, a situation where cost is not a factor, one can use low-temperature anionic polymerization.[15] The best initiators are hindered alkyl lithium compounds, such as 1,1-diphenylhexyl lithium (**15**). Above about −60 °C chain termination occurs by condensation of the enolate end group with the second ester group back, to form a cyclohexanone ring **16**[16] (Scheme 1.3).

If the alkyl group on the ester is bulky, termination reactions are retarded and one can conduct a living polymerization at higher temperatures. For example, *t*-butyl methacrylate can be polymerized at 25 °C by

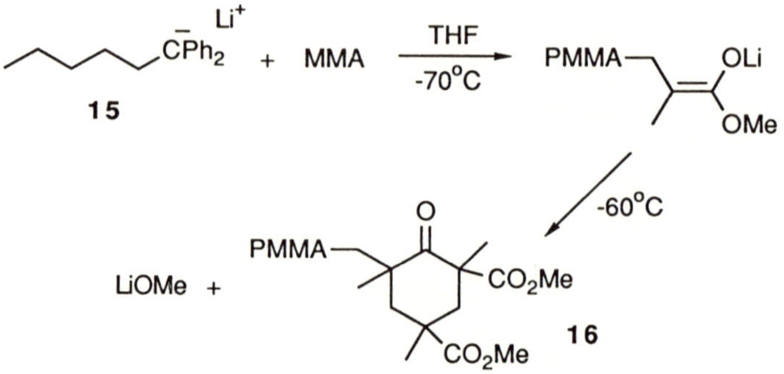

SCHEME 1.3.

initiation with 1,1-diphenylhexyl lithium.[17] A remarkable stabilizing effect of Lewis acids on living anionic polymerization of *t*-butyl acrylate has been found by Teyssie *et al.*[18] Although *t*-butyl acrylate can be polymerized at −78 °C, a polymer with broad molecular-weight distribution is obtained (M_w/M_n 3.6).[18,19] By simply adding 1–2 equivalents of LiCl to the system, the polymerization can be conducted at up to 0 °C under living conditions [M_w/M_n = 1.20–1.63, M_n (GPC)/M_n calc'd = 1.25–1.64]. Grignard initiators plus $MgCl_2$ can also be used.[20]

Living anionic polymerization has been used to prepare block polymers, functional ended polymers, macromonomers, star polymers, and polymers with groups which interfere with free radical polymerization.[17,21] If low temperature and strongly basic conditions can be tolerated, it is a viable alternative to GTP.

(b) Tetraalkylammonium Salts as Initiators. In a remarkable finding Reetz has shown that tetraalkylammonium salts with suitable nucleophilic counterions will initiate living polymerization of acrylates at room temperature.[22–24] Thus tetrabutyl ammonium thiolates and resonance-stabilized carbanions **17** gave poly(butyl acrylate) with molecular weights as high as 20,000 and M_w/M_n 1.3 [Eq. (1.5)]. On addition of a different

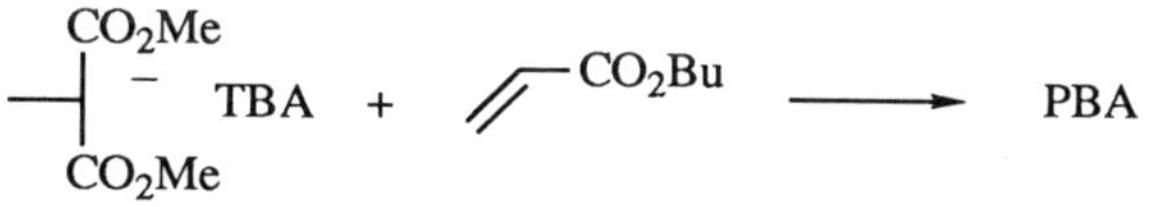

17

(1.5)

monomer to a completed polymerization, the polymerization restarts to give a block polymer. It is surprising that dispersities as low as 1.3 can be obtained since the pK_a values of the mercaptan and malonic ester initiators are in the 11 to 12 range while the pK_a of alkyl ester living end is about 24. On this basis one would predict that the rate of initiation would be much lower than the rate of propagation and that the molecular-weight dispersity would therefore be high. Surprisingly, the nitro-stabilized carbanion **18** proved to be optimal for polymerization of methacrylate. A polymer with M_n 2050, M_n (theory) 2090, M_w/M_n 1.17 was obtained in THF at room temperature. The authors suggest that, since the major termination reactions for anionic polymerization of acrylics involve elimination of alkoxides, and that, since the tetraalkylammonium alkoxides are much stronger bases than metal alkoxides, they would be less prone to elimination (Scheme 1.4).

If the Reetz method for anionic polymers proves to be a general system for acrylic monomers, it will be a major competitor to GTP.

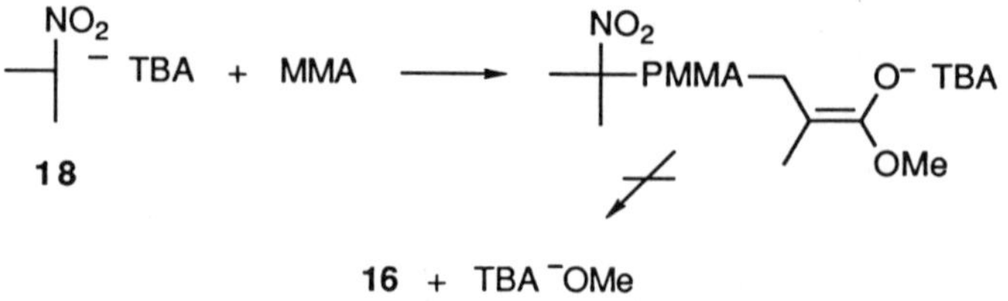

SCHEME 1.4.

1.2.3.3. GTP with Other Transfer Groups

(a) Aluminum Tetraphenylporphyrin. GTP with trialkylsilyl as the transfer group appears to be the best system. The closest competition is polymerization involving transfer of an aluminum tetraphenylporphyrin (TPP) (**19**) group.[25,26] The authors name their system "Immortal Polymerization" since it is so difficult to kill. Aluminum coordinated to the TPP group has a free covalent bond which can be attached to a nucleophilic group such as —Me or —SMe. These functional groups react with acrylates or methacrylates to produce the corresponding enolates. (Light activation is necessary for the TPPAlMe to function.) The enolates initiate living polymerization of the acrylic monomers at room temperature [Eq. (1.6)].

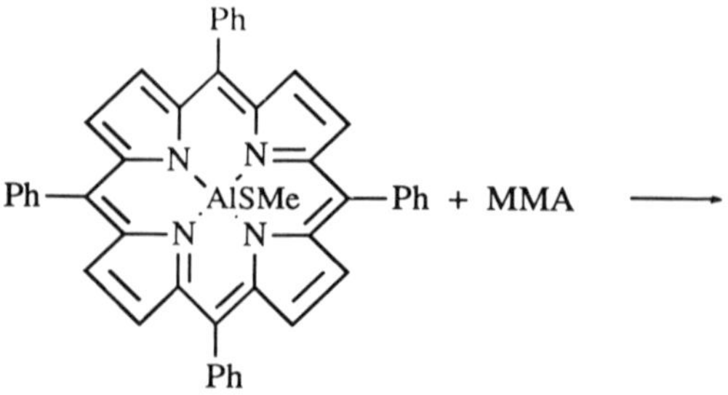

(1.6)

19

PMMA, M_w/M_n 1.06–1.20 with M_n up to 80,000, has been made. A block polymer was synthesized from methyl methacrylate and *t*-butyl acrylate (M_w/M_n 1.1–1.25). The high color of the TPP group and cost of synthesis makes commercial use unattractive; however, less color and more readily attainable aluminum complexes which function as well as TPPAl— will no doubt be found.

(b) Titanium Alkoxides. Titanium triisopropoxyenolates are reported to initiate living polymerization of methacrylates.[27] A M_w/M_n value of 1.4 was obtained. No experimental conditions were given [see Eq. (1.7)].

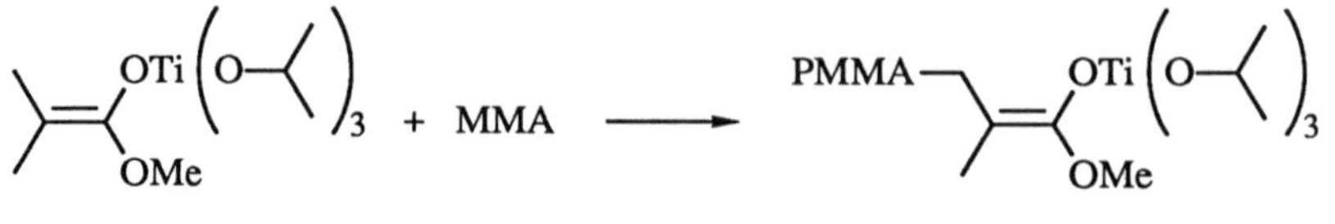

(1.7)

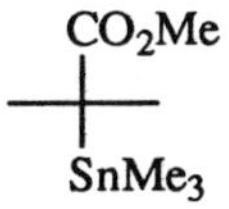

20

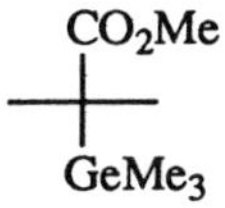

21

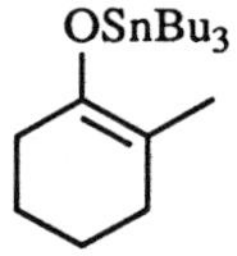

22

(c) Trialkyltin and Germanium Transfer Groups. Certain alkyl tin and alkyl germanium compounds function at room temperature as GTP initiator,[9] such as compounds **20**, **21**, and **22**. The work was not optimized to any extent since the molecular-weight dispersities of the polymers which were obtained were rather broad.

1.3. SCOPE AND LIMITATIONS OF THE METHOD

1.3.1. Initiators

1.3.1.1. Silyl Ketene Acetals

To obtain polymer with low dispersity in a living polymerization, the rate of initiation must be comparable to or faster than the rate of propagation. Otherwise the first formed chains will grow to higher molecular weight than the later formed chains. To insure that the initiation step is comparable in rate to propagation, one needs only choose an initiator that is close in structure to the growing chain ends. Thus for GTP of methyl methacrylates, one finds trialkylsilyl ketene acetals (**23**) in which R is methyl and R′ is alkyl to be ideal. Functional groups attached to R′ will be on the end of every polymer chain; for example, one can synthesize a macromonomer by using **24**.[28] If both R and R′ are large the silyl ketene acetal will not function as an initiator, for example, internally migrated silyl ketene acetal in acrylate polymerization **25**.[9] Large alkyl groups on the silicon slow the rate of polymerization[9]; hydrogen or ether groups do not interfere.[9] The R‴ group on the ether can also be used to introduce functionality on chain ends, e.g., **26** will provide a hydroxy group after deprotection and **27** will result in a carboxylic end group.[9] Bis initiators **28**[9] and **29**[29] grow chains in both directions.

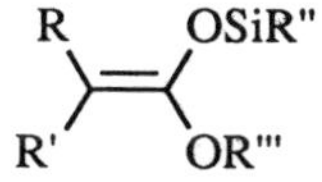

23

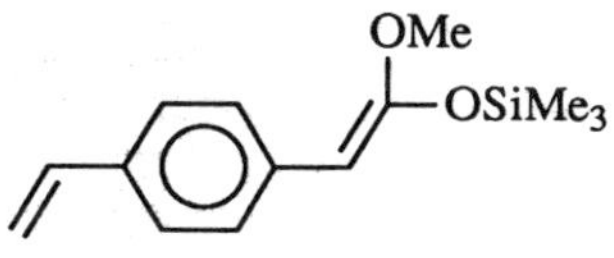

24

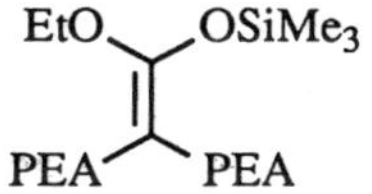

25

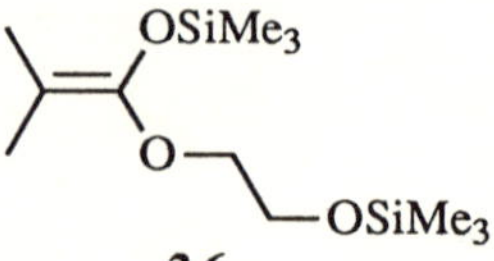

26

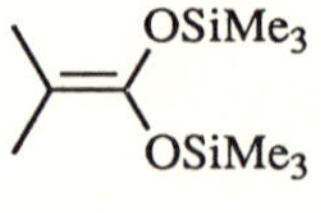

27

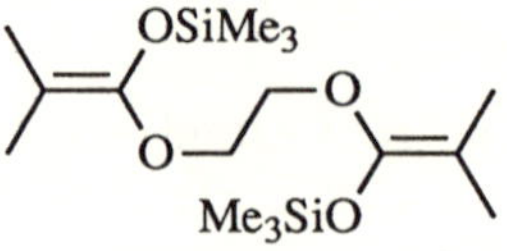

28

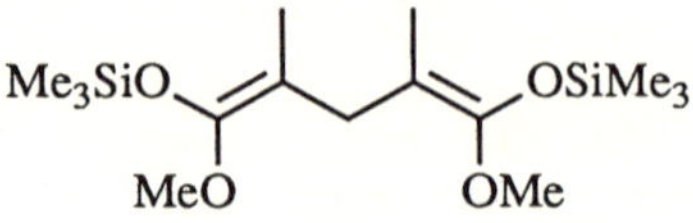

29

1.3.1.2. Other Trialkylsilyl Compounds

Trialkylsilicon compounds which add to acrylic monomers to generate ketene acetals will also operate as initiators. Trimethylsiyl cyanide **30** is unique in that its addition to methacrylate is slower than GTP propagation yet low dispersity polymer is obtained.[9] The Me_3SiCN complexes with the GTP catalyst and polymerization does not proceed until all of the Me_3SiCN is consumed [Eq. (1.8)]. Trimethylsilyl thiolates have the advantage that the alkyl group on the thiolate can accommodate other functionality, for example, one can make telechelic polymers by initiation with **31**[9,30] [Eq. (1.9)]. They have the disadvantage that acrylic polymers with an alkylthiol end group are not stable to weathering. $Me_3Si—OP(OEt)_2$ **32** gives polymer with a phosphonate end group[9] [Eq. (1.10)]. α-Trimethylsilyl esters initiate more slowly than the rate of chain propagation; thus polymers made with these initiators have broad molecular-weight distributions.

30 (1.8)

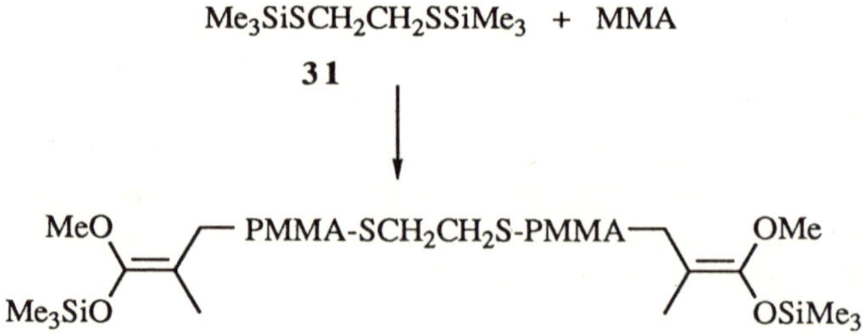

31 (1.9)

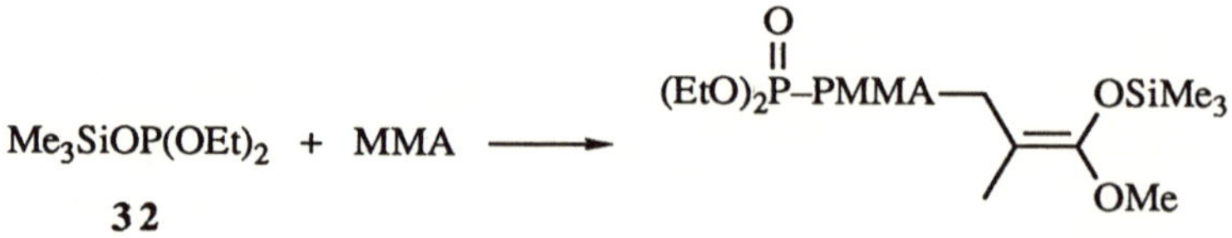

32 (1.10)

1.3.2. Catalysts

1.3.2.1. Anionic

A catalyst is necessary for GTP to function. Most of the early work used tris(dimethylamino)sulfonium bifluoride (TASHF2).[9] A surprisingly good catalyst is tetrabutylammonium fluoride trihydrate.[9] The water does not have to be removed, ionic catalysts are used at 1 to 0.1% of initiator concentration, and at these levels the water of hydration has little effect on molecular weights based on [M]/[I]. Salts of acids in the pK_a range 5 to 24 (DMSO) having large nonmetallic cations also function as catalysts.[31,32] Tetrabutylammonium carboxylates are the present preferred catalysts although about 100 times less active than bifluorides. They are easier to synthesize and results using them are more reproducible. To minimize termination reactions and to obtain polymer of low polydispersity it is best to use the smallest amount of catalyst consistent with the desired rate of polymerization, particularly when preparing polymer of Mn above 20,000. There is a difference in useful temperature range for oxyanion catalysts vs fluorides and bifluorides. Salts of the less basic oxyanion work best at above room temperature.[31] In contrast fluorides or bifluorides and the more basic oxy-anions, such as acetate, will polymerize MMA at −78 °C.[31]

Just as bifluoride is better than fluoride for molecular-weight control, a bi-oxyanion is better than the corresponding anion.[31] The reason for this may be that the bianion supplies anion in a low concentration by dissociation [Eq. (1.11)]. Under ordinary conditions RCO_2H reacts with silyl ketene acetals. However, at the low concentrations produced by the equilibrium (1.11), this reaction may be insignificant.

$$RCO_2\text{–H–}O_2CR \rightleftharpoons RCO_2^- + RCO_2H \qquad (1.11)$$

1.3.2.2. Lewis Acid

Lewis acids such as zinc halides and dialkylaluminum chloride also catalyze GTP.[8] The catalysts operate by activating the monomer (Scheme 1.5). Since the polymer itself has ester groups which compete with monomer for the catalyst, relatively large amounts of Lewis acid are required. The

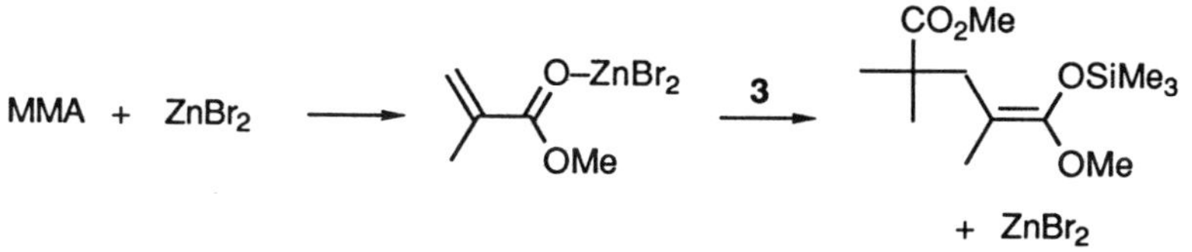

SCHEME 1.5.

excess Lewis acid complicates isolation of the resin in a large-scale process. Lewis acids, especially HgI_2, work better for acrylate polymerization.[33]

1.3.3. Enhancing Agents

Addition of a few percent of acetonitrile[34,35] or of trimethylsilyl ester[35] of the same carboxylate that is used as catalyst increases the longevity of the living polymerization. Fewer dead ends are obtained and there is a better correspondence of molecular weight with initiator concentration (see Section 1.5.1.4). These agents may be complexing with the catalyst and providing a means of supplying it continuously at a low concentration [Eq. (1.12)], or they may be scavenging trace amounts of anionic chain ends [(Eq. (1.13)]. Controlled bulk polymerization was possible in the presence of a few percent added acetonitrile.[34]

$$RCO_2SiMe_3 + RCO_2^- \rightleftharpoons (RCO_2)_2SiMe_3^- \qquad (1.12)$$

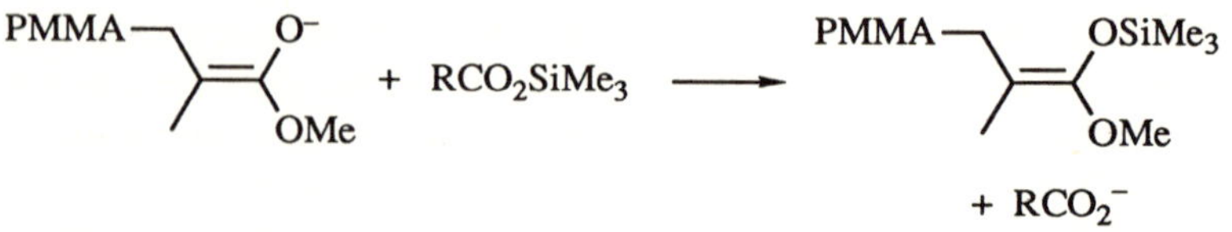

(1.13)

1.3.4. Monomers

1.3.4.1. Methacrylates

The methacrylate family of monomers are by far the best candidates for GTP. Special care must be made to remove traces of alcohols remaining in the higher alkyl members as a result of their synthesis by ester interchange with MMA. Alcohols and other protonic material deactivate the living ends in GTP. Reactivity ratios for methyl methacrylate vs butyl methacrylate, MMA = 0.44 and BMA = 0.26, have been reported.[36] In general the higher members of the series polymerize more slowly. To prepare hydroxyethyl methacrylate polymers as well as methacrylic acid polymers, nonreactive masking groups are used, such as **33**[9] (Scheme 1.6) and **34**[37] (Scheme

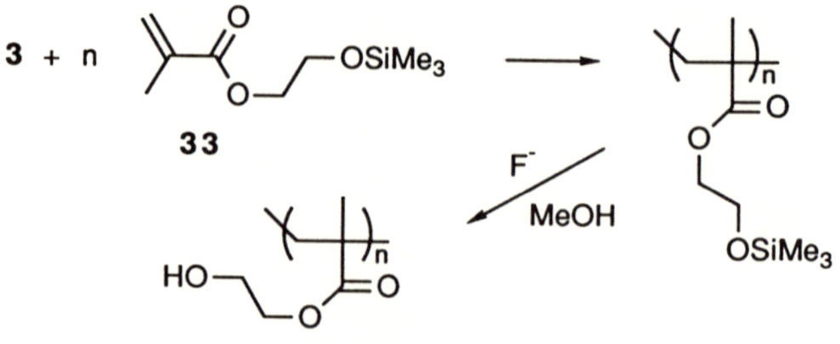

SCHEME 1.6.

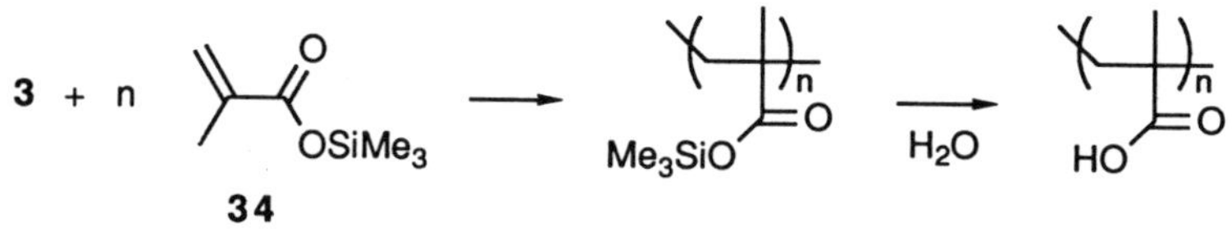

SCHEME 1.7.

1.7). Allyl methacrylate and *p*-vinylbenzyl methacrylate, with groups sensitive to free radical polymerization on the ester alkyl, polymerize without crosslinking.[9,38]

1.3.4.2. *Acrylates*

Acrylates polymerize about 100 times faster than methacrylates. Living acrylate polymer dies more easily than does living methacrylate. The silyl groups have been shown to migrate from the chain ends to produce inactive internal ketene acetals or less active α-silylesters.[9] Lewis acid catalysts, especially HgI_2, are better than anionic catalysts for acrylate polymerization. 2-(Methylacryloxy)ethyl acrylate **35** can be polymerized without crosslinking to a linear polymer with pendant methacryloxyethyl groups **36**[8] [Eq. (1.14)].

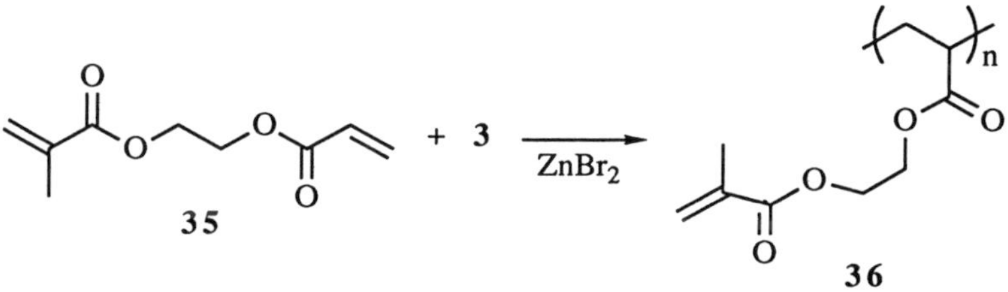

35 36 (1.14)

1.3.4.3. *Acrylonitrile and Methacrylonitrile*

These monomers polymerize extremely rapidly by GTP.[(9)] Molecular-weight control is difficult since the polymer forms before mixing can be accomplished. A polar solvent like DMF is necessary, because the polymer from these monomers is insoluble in most ordinary solvents. Although **37** can be used to initiate methacrylonitrile (Scheme 1.8), **3** is better. It has not been established whether an α-silyl nitrile **38** or a silyl imine **39** is the chain carrier.

1.3.4.4. *Dienoates and Trienoates*

Dienoates substituted with methyl groups, or unsubstituted, **40** as well as the trienolate, **42** polymerize to give pure 1,4-*trans* and 1,6-*trans,trans* polymer chains[39] [Eq. (1.15)]. These monomers are more reactive than

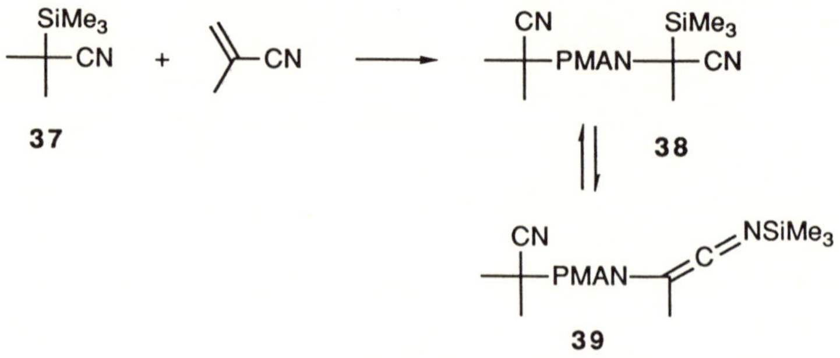

SCHEME 1.8.

methacrylates. The diensilyl ketal **41** gives polymer with a lower M_w/M_n than **3** indicating the initiation rate for **3** is less than the propagation rate. Polymerization of these monomers has a bearing on the mechanism of GTP, since associative silyl transfer for the trienoate would have to occur through a 10- or 12-membered ring (see Section 1.4.1.1).

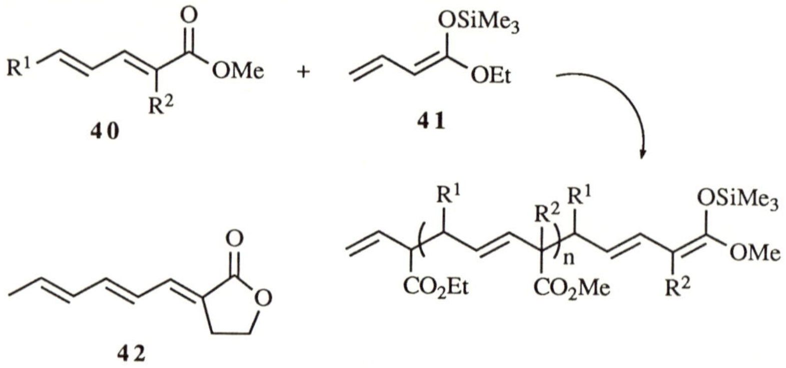

(1.15)

1.3.5. *Control of Polymer Structure*

1.3.5.1. *Molecular Weight*

Polymers with molecular weights as high as 250,000 have been obtained; however, GTP is best for preparing polymers in the 1000 to 30,000 range.[9] As in any living polymerization in GTP the number average molecular weight is determined by multiplying the molar ratio of monomer to initiator by the monomer molecular weight. This holds, provided initiation is faster than propagation and that the polymerization goes to 100% completion. Also, in a living polymerization the M_w/M_n will approach 1. In practice, for polymer in the 1000 to 20,000 molecular-weight range, experimental M_n values correspond fairly well with theoretical M_n values. At high molecular weights (over 200,000), where the impurity level may approach the initiator level, the correspondence is poor.[9] In polymerizations conducted above room temperature and aimed at moderately high molecular weights, the polydispersity M_w/M_w is as high as 1.5[9,35] The main reason for this is an inherent chain termination process during which nonreactive cyclohexanone ring **43** forms on the chain ends by a backbiting process [(Eq. (1.16)].[40] This chain termination occurs at about 1/2500 the

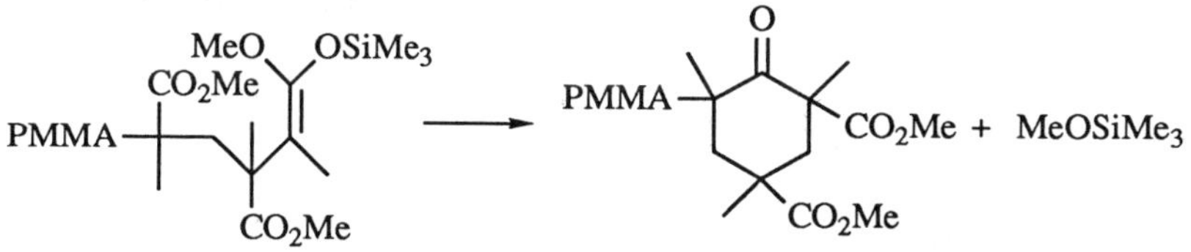

(1.16)

43

rate of polymerization.[40] Termination of polymer chains by cyclization is tenfold faster at the trimer stage than at other chain lengths. Incubation of living oligomer results in more cyclic trimer than could have been produced by the available linear trimer. This experiment demonstrates that GTP is reversible.[40]

Phenylacetic acid esters **44** and various other activated hydrogen compounds can act as chain transfer agents for GTP.[41] These substances react with the silyl ketene acetal end group to stop propagation but form a new silyl ketene acetal **44** which starts new chain growth (Scheme 1.9). This method of molecular-weight control requires less silyl initiator and can be used to introduce functional groups.

1.3.5.2. *Chain Stereochemistry*

The chain configuration for poly(methyl methacrylates) is nearly the same as that produced by free radical polymerization.[9] Polymerization conducted at 60 °C gives polymer with 48% syndiotactic, 45% heterotactic, 7% isotactic triads; at −90 °C the ratio is 80% syndiotactic and 20% heterotactic. In anionic living polymers with lithium enolate end groups the tacticity can be controlled to get mainly isotactic or mainly syndiotactic polymer by merely changing the polarity of the solvent.[42] Tacticity in GTP is not solvent-sensitive. This lack of control indicates that the configuration of each monomer as it adds is not determined by the configuration of the chain ends.

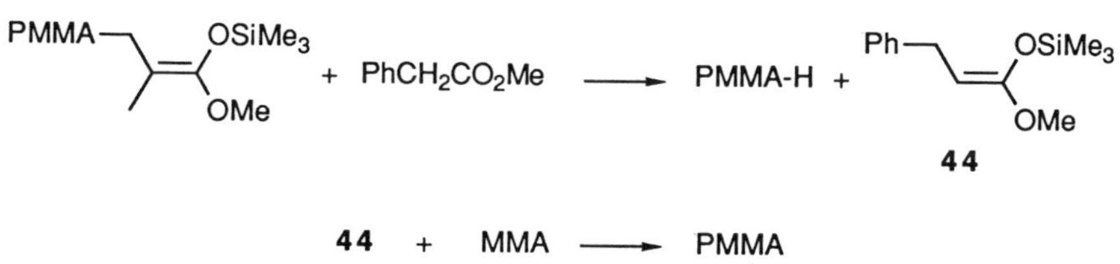

SCHEME 1.9.

1.3.5.3. End Groups

The end group on the initiator end of the polymer is determined with 100% efficiency by the structure of the initiator. Functionally, the nonsilyl portion of the initiator will be on one end of each polymer chain, dead or alive. The silyl ketene acetal function on the other end of chains which are still alive can be converted to other functional groups or can be coupled to form telechelic polymer. Based on the maximum yield of telechelic polymer one can estimate that in an average synthesis of 10,000 molecular weight polymer, 5 to 10% of the chain ends die.[43]

Examples of the use of initiator structure for synthesis of polymers with specific end groups are given below. Macromonomer, **45**, was synthesized by initiation of MMA with **24** containing a styrene moiety.[28] [Eq. (1.17)].

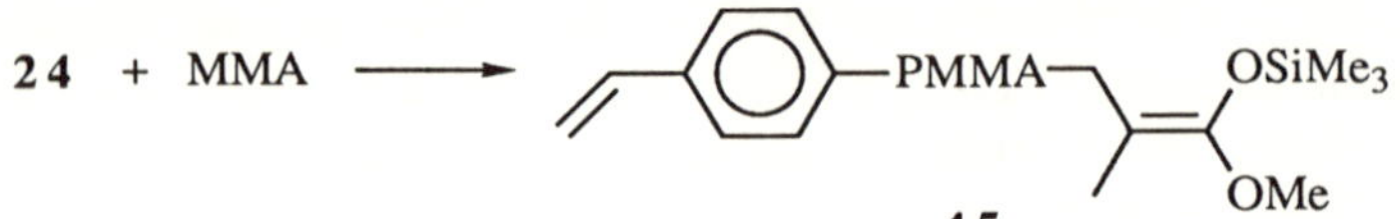

45

(1.17)

Hydroxy-ended polymer **48** was made by initiation of MMA with **26** followed by disprotection with methanolic fluoride.[9] A telechelic polymer **47** was synthesized by coupling **46** with terephthaloyl fluoride (Scheme 1.10). Telechelic hydroxy ended poly(ethyl acrylate) was obtained by a similar initiation of ethyl acrylate with **26** and then addition of a second hydroxy function by capping with benzaldehyde (Scheme 1.11).

Carboxy-ended polymer **53** results from initiation with **27** followed by hydrolysis of the trimethylsilyl ester **52** (Scheme 1.12).

1.3.5.4. Copolymers

(a) Random Copolymers. Due to the large difference in polymerization rates, random copolymer of acrylates, methacrylates, and acrylonitrile

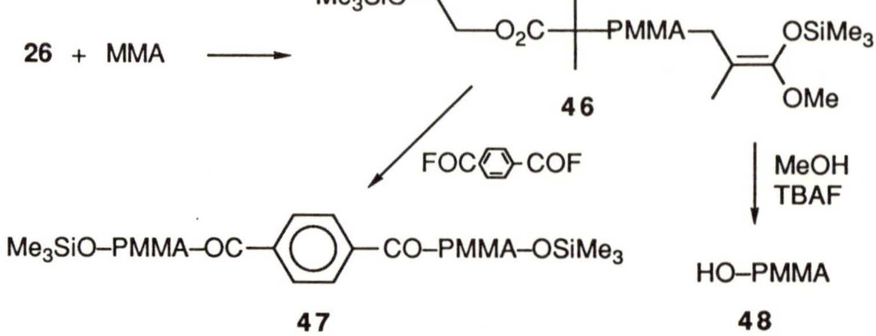

SCHEME 1.10.

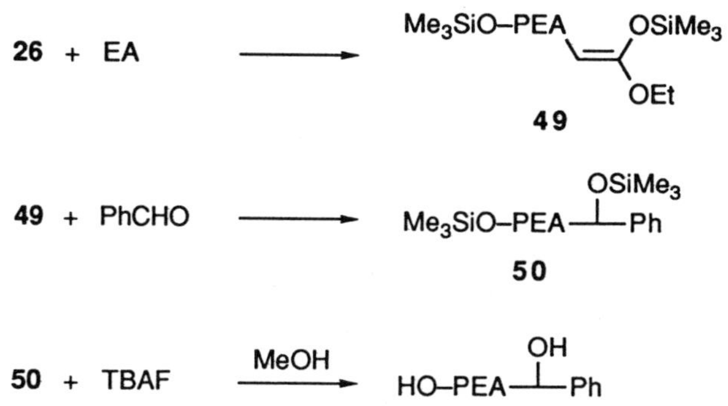

SCHEME 1.11.

cannot be made by GTP.[9] However, random copolymers within any given family of monomers can be made, such as MMA and BMA.

(b) Block-Copolymers. Block polymer synthesis probably gives more useful products than any other modification of polymer chain architecture. Living polymerizations including GTP are especially useful for obtaining block polymers. One merely adds a new monomer when the first one is depleted. In GTP this procedure only works if one uses both monomers from the same family, i.e., all methacrylate or all acrylate. AB block polymers made with MMA/BMA A blocks and ring-opened GMA B blocks are excellent pigment dispersants.[44] The polar GMA derived block adsorbs on the pigment surface and the nonpolar block provides steric stabilization.

One can prepare an AB block polymer from two dissimilar monomers if the block from the least reactive monomer is formed first, for example, PMMA/PEA.[7,25]

A way to make block polymers from totally different types of monomers is to first form a block from one monomer and convert its living end to an initiator for the next monomer. Examples of this strategy using GTP are the synthesis of an ABA block polymer with a polycaprolactone B block and PMMA A blocks by first converting poly(caprolactone) to a

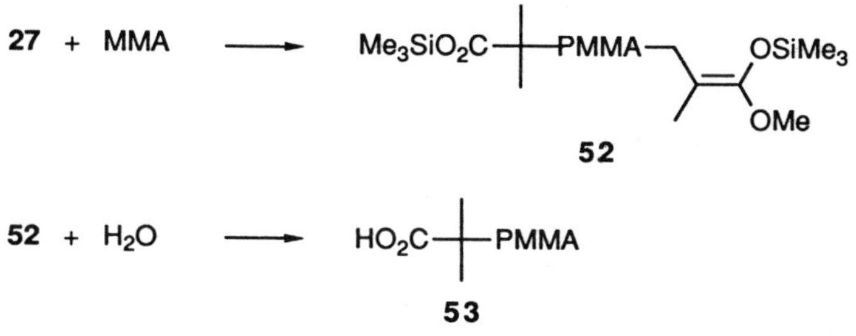

SCHEME 1.12.

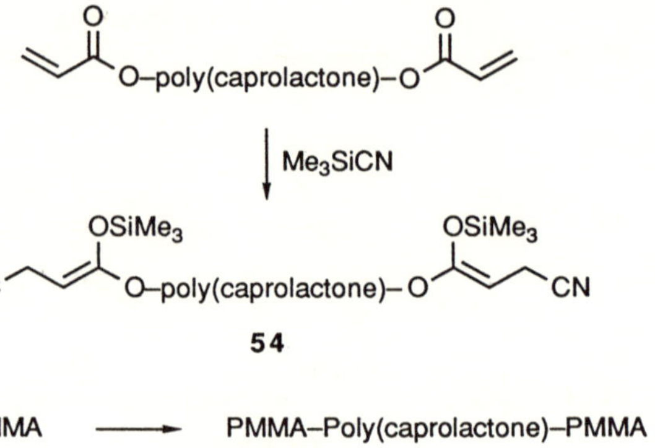

SCHEME 1.13.

GTP/initiator **54** by addition of Me_3SiCN to diacrylate ended poly(caprolactone). **54** was then used to initiate MMA[9] (Scheme 1.13).

A block copolymer of poly(dimethylsiloxane) and PMMA was formed by converting a methacrylate ended polysiloxane **55** to silyl ketene acetal ended polymer **56** with R_sSiH and using **56** to initiate MMA[45] (Scheme 1.14).

A combination of free radical initiators with GTP has been used to make a PMMA/polystyrene block polymer.[46] Living PMMA polymer was converted to the α-bromo derivative **57** by treatment with bromine. This in turn was used to generate free radical macroinitiator by treatment with dimanganesedecacarbonyl in the presence of styrene, thereby obtaining PMMA-*b*-polystyrene **58** (Scheme 1.15). PMMA-*b*-PS has also been synthesized by conversion of living PS to macroinitiator **59** and subsequent polymerization of MMA[47] (Scheme 1.16).

A block polymer of poly(propylene sulfide) and PMMA was made by initiation of MMA with **14** [Eq. (1.18)].

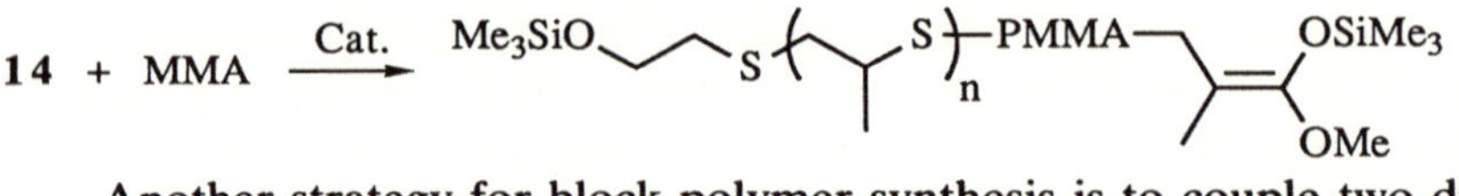

(1.18)

Another strategy for block polymer synthesis is to couple two different polymer chains. This method is not used very often since, usually they are

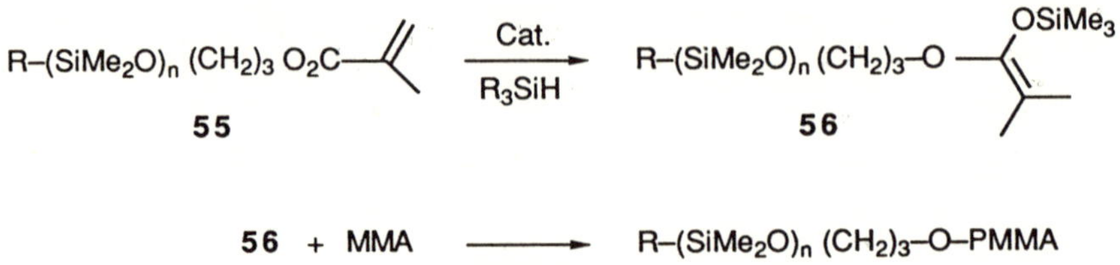

SCHEME 1.14.

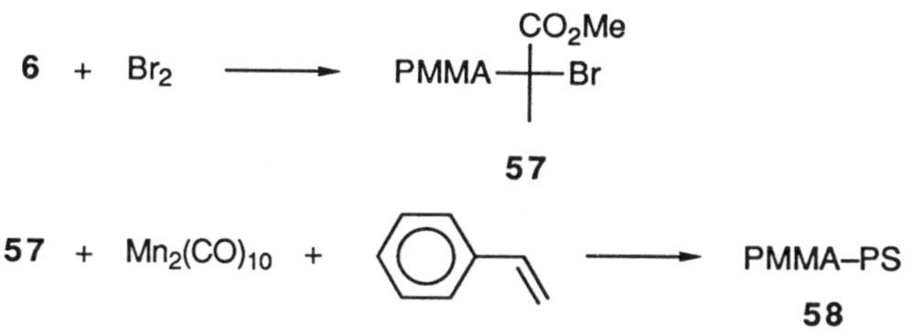

SCHEME 1.15.

incompatible and coupling yields are thus low. Poly(methyl methacrylate)/*b*-Poly(vinyl ether) (**61**) was made this way by coupling living PMMA **6** with the aldehyde group on **9**[(11)] (Scheme 1.17).

(c) Graft Polymers. Graft polymers, also called comb polymers, are made by polymerization or copolymerization of macromonomers. The macromonomers are obtained by living polymerization or by conversion of a functionalized polymer end to an end containing a polymerizable vinyl group. The following examples of graft polymer synthesis will illustrate these principles. Macromonomer **45** was polymerized by radical initiation using AIBN to get polystyrene comb polymer with PMMA teeth[28] **62**. Polymerization of a methacrylate ended polystyrene macromonomer by GTP gave the reverse comb with a polymethacrylate backbone and polystyrene teeth[(48)] **63**.

62 63

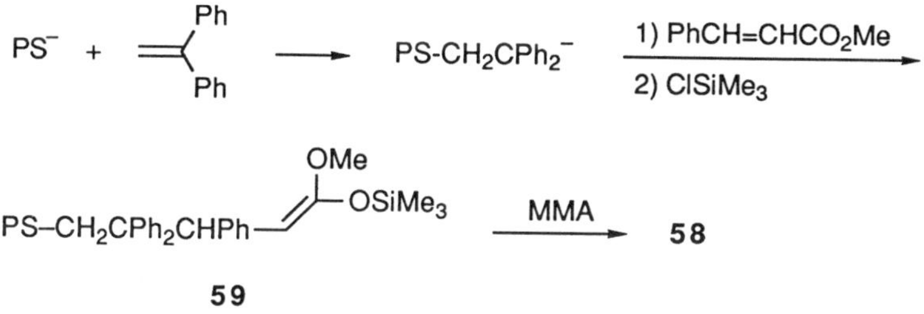

SCHEME 1.16.

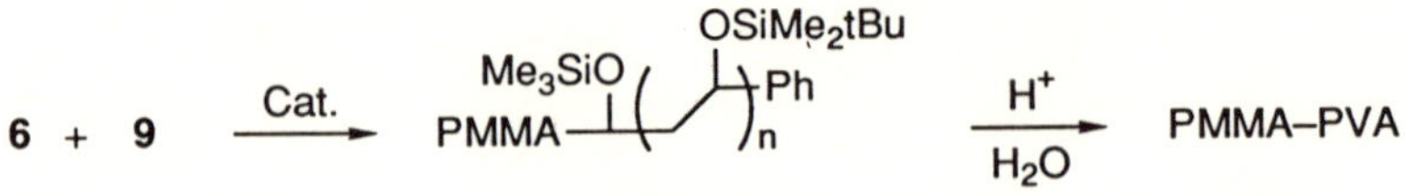

60 **61**

SCHEME 1.17.

Treatment of hydroxy-ended PMMA **48** with methacryl chloride gives PMMA macromonomer **64**. When **64** is polymerized with MMA by anionic initiation at −78 °C, well-defined graft polymers containing nearly monodisperse branches of controlled molecular weight result[49] (Scheme 1.18).

One of the end uses of macromonomers is the synthesis of nonaqueous dispersions, NADs. They are made by copolymerizing the macromonomer with other acrylic or styrene monomers containing small amounts of divinyl crosslinking agents.[44] Microgel particles **65** are formed that stay dispersed due to steric stabilization from the macromonomer chains extending into the solution [Eq. (1.19)]. NADs are used for rheology control and reinforcement of coating systems.[44]

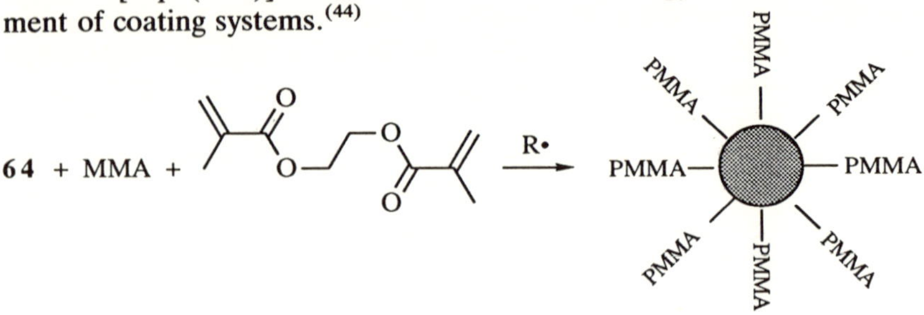

65

(1.19)

Mesogenic polymers and copolymers similar in structure to comb polymers were made by GPC of methacrylates carrying mesogenic group attached through a chain of six methylene units[50] **66** [(Eq. (1.20)].

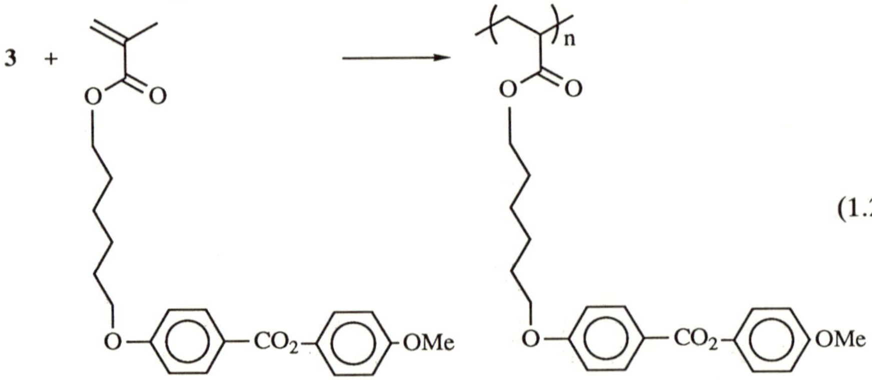

66

(1.20)

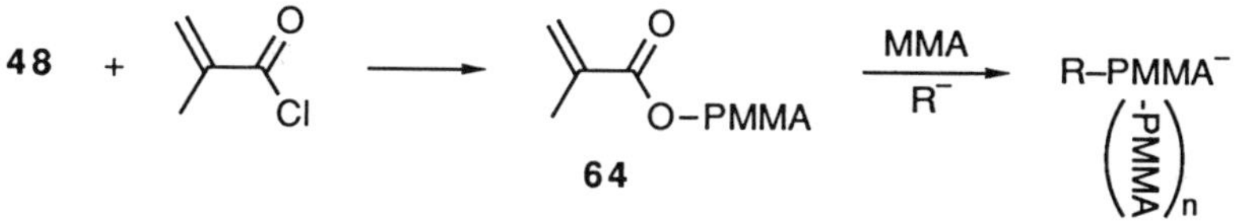

SCHEME 1.18.

1.3.5.5. Star Polymers

The term star polymer has evolved to denote a polymer with branches emanating from a common source or core. The first star polymers were synthesized by coupling living polymers with polyfunctional tie molecules for example, living polystyrene with $SiCl_4$[51] [Eq. (1.21)]. Later it was

$$PS^- + SiCl_4 \longrightarrow Si(PS)_4 \qquad (1.21)$$

discovered that treatment of a living polymer with a divinyl monomer would crosslink the end of the polymer to form a core with several hundred arms emanating into the solution. Examples of this kind of star polymer formation were first reported for polybutadiene[52] [Eq. (1.22)]. GTP is an

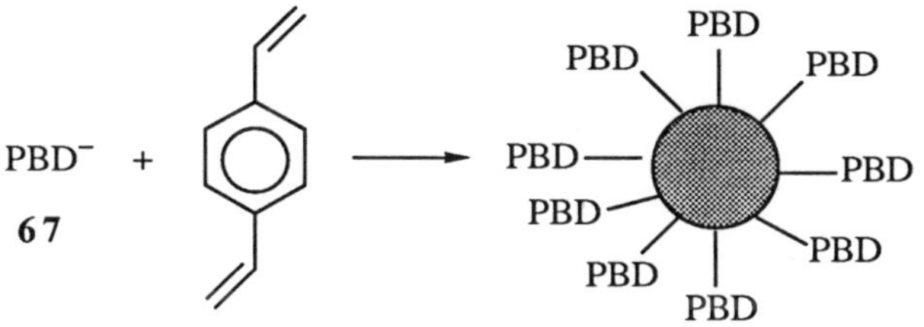

(1.22)

ideal system for synthesis of star polymers. Three- and four-armed poly(ethyl acrylate) star polymers can be synthesized by initiation with tri and tetra functional initiators.[9] The procedure is illustrated for a three-armed star (Scheme 1.19).Trimethanol propane triacetate (**68**) is treated with **3** to form initiator **69**, which is used to initiate ethyl acrylate to give the three-armed star **70**.

Multiarmed star polymers with tightly crosslinked cores result from treatment of a living methacrylate from GTP with a difunctional MA, such as ethylene glycol dimethacrylate **71**[44,53] [Eq. (1.23)]. These stars, **72**, are

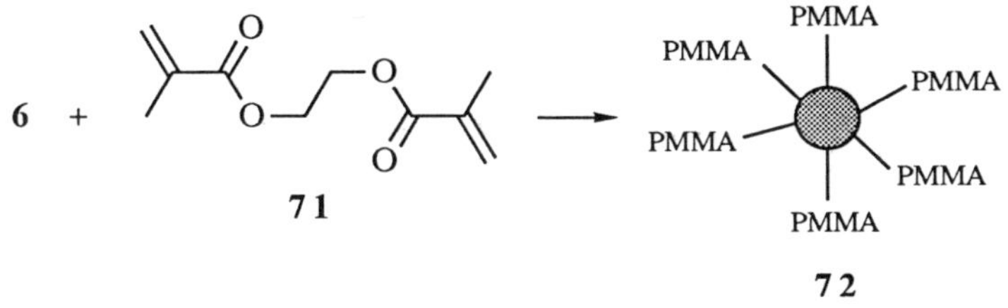

(1.23)

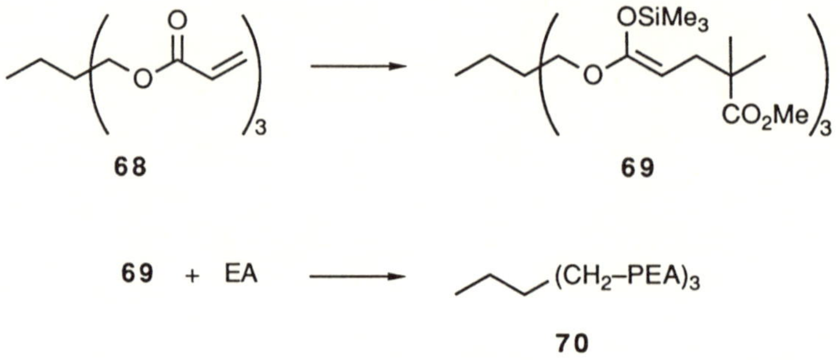

SCHEME 1.19.

similar in structure to NAD **65** but have lower molecular weights. The arms, which radiate out from the core, may be homopolymers, copolymers, or block polymers and may have functional groups attached anywhere along the chain. A flexible-hard film suitable for coating automobile bumpers and also steel can be made from hydroxy functional star blended with enamel.[(44)]

1.3.5.6. Ladder Polymers and Cyclic Polymerization

When a difunctional acrylic monomer with a connector group that is small is polymerized by GTP, a soluble uncrosslinked polymer is obtained by cyclic polymerization. Thus polymerization of *N*-phenyldimethacrylimide by GTP gives **73**[(54)] [Eq. (1.24)]. The molecular weight of the polymer obtained was rather low. However, the best GTP conditions were not used (see Section 1.5).

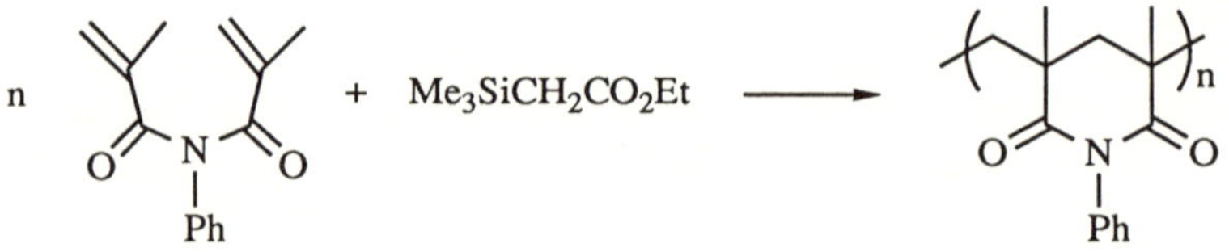

(1.24)

73

If the spacer between the two methacrylate groups is 4 to 7 methylene units **74**, a soluble polymer is still formed if a difunctional initiator **75** and dilute conditions are used [Eq. (1.25)]. A monofunctional initiator gives a crosslinked polymer. Thus a ladder structure **76** has been proposed for these polymers.[55]

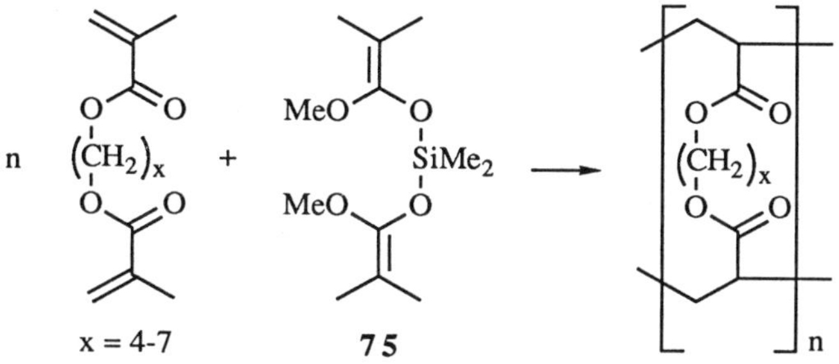

74 **76**

(1.25)

1.4. REACTION MECHANISM

1.4.1. Proposed Reaction Pathway

1.4.1.1. Anionic Catalysts

A mechanism for anionic catalysis of GTP should explain the following: GTP works better (fewer dead ends are produced) when one maintains a low level of catalyst[9] and a high level of monomer.[31,35,40] The mechanism must take into account that the trialkylsilyl group remains with the polymer chain it started with.[56] A dissociative mechanism is favored by the authors as depicted in Scheme 1.20. The catalyst forms a reversible adduct with trimethylsilyl ketene acetal initiator. This adduct dissociates reversibly to

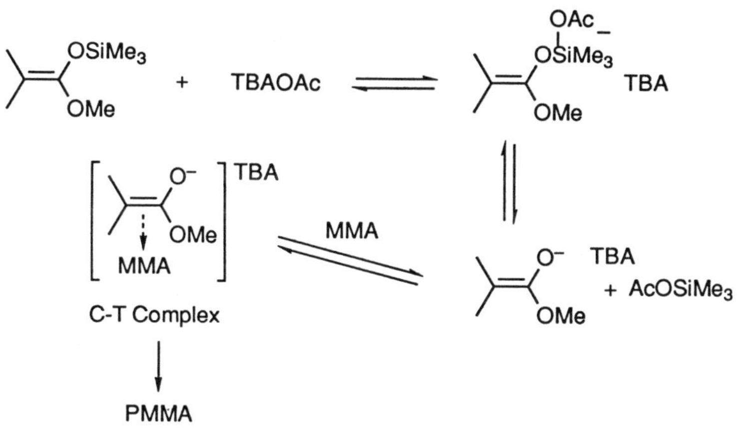

SCHEME 1.20.

ester enolate and trimethylsilyl carboxylate. The ester enolate reacts with monomer via a charge-transfer (C-T) complex to give polymeric enolate. To complete this cycle the enolate is recapped by the trimethylsilyl carboxylate which is still in the solvent cage. Low levels of catalyst which generate low levels of enolate favor the C-T complex formation with monomer which is present in relatively large amounts. A C-T complex has been proposed for living cationic polymerizations to account for a similar stabilizing effect of monomer with active ends.[57] The C-T complex and large gegenion[22] stabilize ester enolate ends until they react with monomer and are recapped by the trimethyl silyl carboxylate. Another explanation for the stabilizing effect of monomer could be that chain termination is a first-order process and chain propagation a second-order process. The ratio of chain termination to chain propagation would therefore increase at low monomer concentrations.[40] Added trimethylsilyl carboxylate enhances livingness of GTP by complexation with catalyst and possibly by capping enolate ends to further reduce their concentration. A crucial experiment would be to see if added silyl carboxylate exchanges with chain ends faster than polymerization is occurring.

1.4.1.2. *Lewis Acid Catalysts*

The mechanism of Lewis acid catalysis of GTP appears to be less complicated than that of anionic catalysis, possibly because it has not been studied in detail. Since relatively large amounts of catalyst are required (up to 10% based on monomer) it is assumed that the catalyst is activating the monomer.[8] large amounts are needed because the polymer chain ester groups compete with monomer for the Lewis acid. The mode of action of HgI_2 as a catalyst for acrylate polymerization appears to be different.[33] Here only trace amounts are needed and the HgI_2 is thought to be interacting with the silyl ketene acetal initiator.

In a combination of anionic polymerization of MMA (with TBA counterion) and GTP, monodisperse polymer was obtained and the molecular weight was controlled by the ratio of silyl ketene acetal to monomer.[58] This is a surprising result, since one would expect exchange of chain ends to slow down drastically as molecular weight increased resulting in broadening of the molecular-weight dispersity or bimodal polymer.

1.5. *SYNTHETIC PROCEDURES AND REACTION CONDITIONS*

For all of the synthetic procedures described below, the polymerization equipment was dried at 150 °C for 24 h and cooled to room temperature under a stream of dry oxygen-free argon. Solvent and reagent transfers were made by syringes. Tetrahydrofuran (THF) was distilled from sodium and benzophenone immediately prior to use. Acetonitrile, methylene chloride, and 1,2-dichloroethane were purified by distillation from P_2O_5. TAS^+ bifluoride was prepared by hydrolysis of TAS^+ $SiMe_3F_2^-$.[(9)] The TBA carboxylates were prepared by neutralization of TBAOH with the carboxylic acid.[(31)] The silyl ketene acetals were prepared by treatment of the ester enolate with Me_3SiCl.[(9)] A preparation from α-haloesters has recently appeared.[(59)]

1.5.1. Methacrylate Polymerization

1.5.1.1. GTP of Methyl Methacrylate with TBAF as Catalyst[(9)]

A 500-mL, four-necked round-bottomed flask, fitted with a magnetic stirring bar, argon inlet, thermocouple well, serum cap, and 125-mL pressure-equalizing dropping funnel equipped with a serum cap, was assembled under argon while hot. The entire apparatus was then heated with a heat gun while being flushed with argon. The argon flush was continued until the apparatus was cool. The assembly was then maintained under a slight positive pressure of argon. THF (225 mL) was transferred to the flask via a cannula. MMA (75 mL, 0.69 mol) was transferred to the dropping funnel via a cannula. Then 0.37 mL (1.16 mmol) of **3** and 58 μL of tetrabutylammonium fluoride trihydrate (0.02 M in THF) were introduced with syringes. MMA was then added dropwise over a period of 90 min. After stirring for 16 h, the reaction mixture was quenched with 10 mL of methanol, stirred for 1 h, and poured very slowly into a fivefold volume of hexane that was being agitated strongly in a blender. After filtration and washing with hexane, the polymer was air-dried at room temperature and then dried for 48 h to constant weight in a vacuum oven at 65 °C, under a slight nitrogen purge. The PMMA weighed 71.2 g (quantitative conversion), which is slightly more than theory due to the presence of traces of solvent (observed only in TGA): M_n(GPC) 62300, M_w 71500, M_w/M_n 1.15, M_n(theory) 60180.

A comparison of conditions for polymerization of methacrylates is given in Table 1.1 and block polymers in Table 1.2.

1.5.1.2. GTP of MMA with TBA ^-OAc/TBA $H(OAc)_2^-$ as Catalyst[(31)]

To a 100-mL round-bottom flask containing a stirred solution of 30 mL of tetrahydrofuran and 0.20 mL (1.0 mmol) of **3** was added 0.02 mmol of tetrabutylammonium $OAc^-/H(OAc)_2^-$ catalyst in acetonitrile (0.3 M). After 5 min, 5.0 mL (46.7 mmol) of methyl methacrylate was added dropwise at a rate of 1.0 mL/min. The reaction mixture exothermed. One hour from

TABLE 1.1. Comparison of Conditions for Preparation Polymethacrylates GTP

No.	Polymer	Initiator	Catalyst	Solvent	$M_n \times 10^{-3}$ (Theory)	$M_n \times 10^{-3}$ (GPC)	M_w/M_n	Ref.
1	PMMA	**3**	$TBAF \cdot H_2O$	THF	60.0	62.3	1.15	9
2	PMMA	**3**	$TASHF_2$	THF	10.1	10.2	1.17	9
3	PMMA	**3**	$ZnBr_2$	$ClCH_2CH_2Cl$	3.4	6.02	1.20	9
4	PMMA	$Me_3SiCH_2CO_2Et$	$TASHF_2$	THF	5.1	28.6	3.61	9
5	PMMA	Me_3SiSMe	Et_4NCN	CH_3CN	3.07	11.8	1.64	9
6	PMMA	**29**	TBACN	THF	50	48.1	1.34	58
7	PMMA	**3**	TBAOAc/AcOHOAc	THF	4.78	4.86	1.32	31, 32
8	PMMA	**3**	$TBAPhCO_2$	THF	4.78	4.50	1.38	31, 32
9	PMMA	**3**	Ph_4PHF_2	None	28.4	43	1.32	34
10	BMA	**29**	$TASHF_2$	THF	6.11	9.91	1.23	29
11	AMA	**29**	$TASHF_2$	THF	5.20	5.20	1.21	29

TABLE 1.2. Selected Copolymers Synthesized by GTP

No.	Polymer (wt%)	Initiator	Catalyst	$M_n \times 10^{-3}$ (Theory)	$M_n \times 10^{-3}$ (GTP)	M_w/M_n	Ref.
1	P(MMA-co-BMA) (35/65)	**26**	$TASHF_2$	20.2	22.1	1.11	9
2	PMMA-*b*-P(MMA-co-BMA-co-GMA) (2.5/47.5/20/30)	**3**	$TASHF_2$	4.09	3.91	1.10	9
3	PLMA-*b*-PMMA (22/78)	**3**	$TASHF_2$	7.12	6.54	1.14	9
4	PMMA-*b*-PBMA-*b*-PAMA (57/32/11)	**3**	$TASHF_2$	4.06	3.80	1.07	7, 9
5	P(MMA-co-HEMA)-*b*-PLMA-*b*-P(MMA-co-HEMA) [(11/17)/44/(11/17)]	**28**	$TASHF_2$	13.8	9.34	1.20	9
6	PMMA-*b*-Polycaprolactone-*b*-PMMA (40/20/40)	Me_3SiCN	Et_4NCN	4.30	4.12	1.78	9
7	PMMA-*b*-PEA (35/65)	**27**	$TASHF_2$	3.46	3.40	1.92	7
8	PBMA-*b*-PMMA-*b*-PBMA (37.5/25/37.5)	**29**	$TASHF_2$	4.38	7.20	1.21	29
9	PMMA-*b*-PBMA-*b*-PMMA (20.5/59/20.5)	**29**	$TASHF_2$	5.04	8.16	1.27	29
10	PMA-*b*-PMMA-*b*-PMA (31.5/37/31.5)	**29**	$TASHF_2$	4.85	7.29	1.312	29
11	PMMA-*b*-PAMA-*b*-PMMA (37/26/37)	**29**	$TASHF_2$	6.35	9.86	1.29	29
12	PDMS-*b*-PMMA	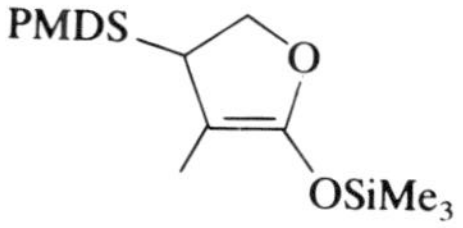	$TBAPhCO_2^-$				45
13	PMMA-*b*-PS-*b*-PMMA (10/80/10)	PMMA-Br $Mn_2(CO)_{10}$			~100		46

the time at which peak temperature was observed, the mixture was quenched with methanol and the solvent removed under reduced pressure. A quantitative yield of polymer was obtained: (GPC) M_n 4860, M_w/M_n 1.32, m_n(theory) 4780.

1.5.1.3. *Bulk Polymerization of MMA by GTP*[34]

3 mL of MMA (28.1 mmol) and 20 μL of **3** (9.9×10^{-2} mmol) were added to a dry reaction tube (internal diameter 8 mm) under N_2 4 × 50 μL aliquots of tetraphenylphosphonium bifluoride in MeCN (0.0208 mmol/mL) were added. Eventually, polymerization occurred as indicated by a slight temperature rise to ~50 °C. Polymerization was allowed to proceed for $\frac{1}{2}$ h when almost complete solidification had occurred. A clear rod PMMA was thus produced [M_n(GPC) 43000, M_w 56780, M_w/M_n 1.32, M_n(theory) 2800].

1.5.1.4. *Aluminum Tetraphenylporphyrin as Initiator*[26]

To a flask containing methyl(5,10,15,20-tetraphenylporphinato)-aluminum [(TPP)AlMe] (0.2 mmol as a 0.025 mol/L solution in CH_2Cl_2) was added 1.08 mL (10 mmol) of methyl methacrylate (10 mmol, 1.08 mL) through a syringe in a dry nitrogen atmosphere. The flask was illuminated at room temperature for 18 h by a 300-W xenon arc lamp through a glass filter to cut out the light of wavelength shorter than 420 nm. Poly(methyl methacrylate) 100% conversion [M_n(GPC) 5500, M_w/M_n 1.14, M_n(theory) 5022] was obtained. When thiolate (TPPAlSMe) was used as initiator, light activation was not necessary for living polymerization to proceed. Under similar conditions, ethyl acrylate polymerized completely in 4 min at room temperature.

1.5.2. *Acrylate Polymerization*

1.5.2.1. *GTP of Methyl Acrylate using HgI_2 Catalysis*[33]

To a stirred mixture of 20 mL toluene, 0.10 g (0.15 mmol) mercuric iodide and 0.40 mL (2.0 mmol) **3** was added 5.0 mL (55 mmol) of methyl acrylate at a rate of 0.5 mL/min. The mixture exothermed on contact with the monomer with the temperature climbing to 47 °C seven minutes after all the monomer had been added. The solvent was removed under reduced pressure, to give 4.9 g (100%) of poly(methyl acrylate): M_n(GPC) 2440, M_w 3360, M_w/M_n 1.38, M_n(theory) 2390. Table 1.3 gives a comparison of conditions for synthesis of polyacrylates by GTP.

1.5.2.2. *Tetraalkylammonium Enolate Initiator for BA*[23]

Butyl acrylate is added to a toluene solution of TBA 1,1-diethoxy-carbonylethide at room temperature. The temperature rises 10–25 °C. After

TABLE 1.3. Comparison of Conditions for Synthesis of Polyacrylates by GTP

No.	Polymer	Initiator	Catalyst	Solvent	$M_n \times 10^{-3}$ (Theory)	$M_n \times 10^{-3}$ (GPC)	M_w/M_n	Ref.
1	PBA	**3**	HgI_2	Hexane	9.34	8.62	1.04	33
2	PEA	**3**	HgI_2	Toluene	9.34	9.91	1.18	33
3	PEA	**3**	HgI_2	CH_3CN	4.5	4.09	1.74	33
4	PBA	**12**	ZnI_2	Toluene	12.5	14.2	1.08	30
5	PEA	**3**	$ZnBr_2$	$ClCH_2CH_2Cl$	10.1	17.0	1.57	8
6	PEA	**26**	ZnI_2	$ClCH_2CH_2Cl$	1.2	1.25	1.10	9
7	PBA	**3**	*i*-Bu_2AlCl	$ClCH_2Cl$	2.66	2.37	1.06	8
8	PEA	**3**	(*i*-$Bu_2Al)_2O$	Toluene	2.1	1.33	1.19	9
9	PBA	$MeSSiMe_3$	$TASHF_2$	CH_3CN	1.0	1.5	1.44	30
10	PBA	**3**	$TASHF_2$	THF	23.8	27.2	2.16	9

TABLE 1.4. Comparison of Conditions for Aldol GTP

No.	Polymer	Initiator	Catalyst	Solvent	$M_n \times 10^{-3}$ (Theory)	$M_n \times 10^{-3}$ (GPC)	M_w/M_n	Ref.
1		PhCHO	$ZnBr_2$	CH_2Cl_2	4.10	3.61	1.21	11
2	"	PhCHO	$ZnBr_2$	CH_2Cl_2	16.0	78.9	1.51	11
3	"	PhCHO	$TASHF_2$	THF	2.65	2.40	1.33	11
4	"	PhCHO	$(i\text{-Bu})_2AlCl$	CH_2Cl_2	3.20	6.64	1.70	11
5	"		$ZnBr_2$	CH_2Cl_2	3.84	4.15	1.24	11
6	"	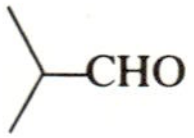	$ZnBr_2$	CH_2Cl_2	3.20	2.90	1.02	11
7	"	Ph_2CHCl	$ZnBr_2$	CH_2Cl_2	7.90	6.50	1.23	11
8	"	$PhCH(OMe)_2$	$ZnBr_2$	CH_2Cl_2	4.30	1.49	2.48	11
9	"	PhCH=CHCHO	$ZnBr_2$	CH_2Cl_2	10.13	7.35	1.50	11
10	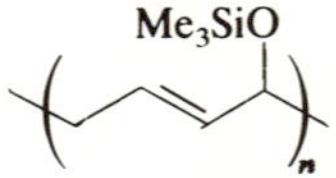	PhCHO	$ZnCl_2$	Benzene	4.20	3.10	1.57	12

the exotherm has subsided the polymer is isolated by distillation of solvent under reduced pressure: M_n(GPC) 20400, M_w 26500, M_w/M_n 1.30, M_n(theory) 20700. (The concentration of reactants and time of reaction were not given.)

1.5.3. Aldol GTP of t-Butyldimethylsilyl Vinyl Ether[11]

A three-necked 50-mL flask, fitted with a stirrer, argon inlet, and a thermocouple, was charged with zinc chloride (anhydrous, 0.035 g, 0.26 mmol), dichloromethane (20 mL), and benzaldehyde (0.25 mL, 2.46 mmol). Then *t*-butyldimethylsilyl vinyl ether (7.24 g, 46.81 mmol) was added via a syringe pump at 0.5 mL/min. The ensuing reaction was accompanied by a temperature rise to 35.4 °C from 26.0 °C. The mixture was stirred for 3 h and evaporated. The residue (7.70 g, 100%) was dissolved in dichloromethane (20 mL) and added to methanol (1 L) to precipitate the polymer. The white precipitate was collected by filtration and dried at 60 °C (0.1 mm Hg) to give 7.50 g (98%) of white powdery polymer: M_n(GPC) 2180, M_w 2800, M_w/M_n 1.28, M_n(theory) 3112. Table 1.4 gives a comparison of conditions for aldol GTP.

1.6. GLOSSARY OF TERMS

AMA	Allyl methacrylate
BA	*n*-Butyl acrylate
BD	Butadiene
BMA	*n*-Butyl methacrylate
EA	Ethyl acrylate
GMA	Glycidyl methacrylate
GPC	Gel permeation chromatography
GTP	Group transfer polymerization
HEMA	2-Hydroxyethyl methacrylate
LMA	Lauryl methacrylate
MA	Methyl acrylate
MAN	Methacrylonitrile
MMA	Methyl methacrylate
PMA	Poly(methyl acrylate)
PBA	Poly(butyl acrylate)
PBD	Polybutadiene
PBMA	Poly(butryl methacrylate)
PDMS	Polydimethylsiloxane
PEA	Poly(ethyl acrylate)
PGMA	Poly(glycidyl methacrylate)
PHEMA	Poly(hydroxyethyl methacrylate)

PMA	Poly(methyl acrylate)
PMAN	Polymethacrylonitrile
PMMA	Poly(methyl methacrylate)
PPS	Poly(propylene sulfide)
PS	Polystyrene
S	Styrene
TAS	Trisdimethylaminosulfonium
TBA	Tetrabutylammonium
THF	Tetrahydrofuran
TPP	Tetraphenylporphinato

REFERENCES

1. K. Narasaka, K. Soai, Y. Aikawa and T. Mukaiyama, *Bull. Chem. Soc. Jpn.* **49**, 779–783 (1976).
2. K. Saigo, M. Osaki and T. Mukaiyama, *Chem. Lett.* 163–164 (1976).
3. M. Kawai, M. Onaka and Y. Izumi, *Bull. Chem. Soc. Jpn.* **61**, 2157–2164 (1988).
4. T. V. Rajan Babu, *J. Org. Chem.* **49**, 2083–2089 (1988).
5. Y. Kito, J. Segawa, J. Haruto, T. Fujii and Y. Tamura, *Tetrahedron Lett.* **21**, 3779–3782 (1980).
6. R. A. Bunce, M. F. Schlecht, W. G. Dauben and C. H. Heathcock, *Tetrahedron Lett.* **24**, 4943–4946 (1983).
7. O. W. Webster, W. R. Hertler, D. Y. Sogah, W. B. Farnham and T. V. Rajan Babu, *J. Am. Chem. Soc.* **105,** 5706–5707 (1983).
8. W. R. Hertler, D. Y. Sogah, O. W. Webster, and B. M. Trost, *Macromolecules* **17**, 1415–1417 (1984).
9. D. Y. Sogah, W. R. Hertler, O. W. Webster, and G. M. Cohen, *Macromolecules* **20**, 1473–1488 (1988).
10. K. Saigo, M. Osaki and T. Mukaiyama, *Chem. Lett.* 989–990 (1975).
11. D. Y. Sogah and O. W. Webster, *Macromolecules* **19**, 1775–1777 (1986).
12. T. Hirabayashi, T. Itoh, and K. Yokata, *Polym. J.* **20**, 1041–1048 (1988).
13. R. P. Quirk and G. P. Bidinger, *Polym. Prepr., Am. Chem. Soc., Div. Polym. Chem.* **29**, 120–121 (1988).
14. R. A. Haggard and S. N. Lewis, *Prog. Org. Coat.* **12**, 1–26 (1984).
15. D. M. Wiles, in *Structure and Mechanism in Vinyl Polymerization,* ed. by T. Tsusuto and K. F. O'Driscoll (Marcel Dekker, Inc., New York, 1960), p. 223.
16. W. E. Goode, F. H. Owens, and W. L. Myers, *J. Polym. Sci.* **47**, 75–89 (1960).
17. G. D. Andrews and L. R. Melby, in *New Monomers and Polymers,* ed. by B. M. Culbertson and C. U. Pittman, Jr. (Plenum Press, New York, 1984), pp. 357–379.
18. R. Fayt, R. Forte, C. Jacobs, R. Jerome, T. Ouhadi, Ph. Teyssie, and S. K. Varshney, *Macromolecules* **20**, 1442–1444 (1987).
19. T. Kitano, T. Fujimoto, and M. Nagasawa, *Polym. J.* **9**, 153–159 (1977).
20. Ph. Teyssie, New Projections in Living Anionic Polymerization of Methacrylic and Acrylic Esters, 9th International Symposium on Cationic Polymerization and Related Ionic Processes, Strasbourg, France (1989).
21. B. C. Anderson, G. D. Andrews, P. Arthur, Jr. H. W. Jacobson, L. R. Melby, A. J. Playtis, and W. H. Sharkey, *Macromolecules* **14**, 1599–1601 (1981).

22. M. T. Reetz, T. Knauf, U. Minet, and C. Bingel, *Angew. Chem., Int. Ed. Engl.* **27**, 1373–1374 (1988).
23. M. T. Reetz, *Angew. Chem., Int. Ed. Engl.* **27**, 994–998 (1988).
24. M. T. Reetz and R. Ostarek, *J. Chem. Soc., Chem. Commun.* 213–215 (1988).
25. S. Inoue, *Polym. Prepr., Am. Chem. Soc., Div. Polym. Chem.* **29**, 42–43 (1988).
26. M. Kuroki, S. Nashimoto, T. Aido, and S. Inoue, *Macromolecules* **21**, 3114–3115 (1988).
27. M. T. Reetz, *Pure Appl. Chem.* **57**, 1781–1788 (1985).
28. R. Asami, Y. Kondo, and M. Takaki, *Polym. Prepr., Am. Chem. Soc., Div. Polym. Chem.* **27**, 186–187 (1986).
29. H. S. Yu, W. J. Choi, K. T. Lim, and S. K. Choi, *Macromolecules* **21**, 2893–2894 (1988).
30. M. T. Reetz, R. Ostarek, K. E. Piejko, D. Arlt, and B. Boemer, *Angew. Chem., Int. Ed. Engl.* **25**, 1108–1109 (1986).
31. I. B. Dicker, W. R. Hertler, G. M. Cohen, W. B. Farnham, E. D. Laganis, and D. Y. Sogah, *Polym. Prepr., Am. Chem. Soc., Div. Polym. Chem.* **28**, 106–107 (1987).
32. I. B. Dicker, W. B. Farnham, W. R. Hertler, E. D. Laganis, D. Y. Sogah, T. W. Del Pessco, and P. H. Fitzgerald, Oxyanion-Catalyzed Polymerization, *U.S. Patent,* 4,588,795 (1986).
33. I. B. Dicker, *Polym. Prepr., Am. Chem. Soc., Div. Polym. Chem.* **29**, 114–115 (1988).
34. A. Pickering and A. J. Thorne, Transfer Polymerization, *U.S. Patent,* 4,791,181 (1988).
35. L. V. Schneider and I. B. Dicker, Enhanced Livingness of Polymerization Using Silylated Oxyanions, *U.S. Patent,* 4,736,003 (1988).
36. A. D. Jenkins, E. Tsartolia, D. R. M. Walton, J. Stejskal, and P. Kratochvil, *Polym. Bull.* **20**, 97–100 (1988).
37. T. Wei and G. E. Wnek, *Polym. Prepr., Am. Chem. Soc., Div. Polym. Chem.* **28**, 252–253 (1987).
38. C. Pugh and V. Percec, *Polym. Bull.* **14**, 109–116 (1985).
39. W. R. Hertler, T. V. Rajan Babu, D. W. Ovenall, G. S. Reddy, and D. Y. Sogah, *J. Am. Chem. Soc.* **110,** 5841–5853 (1988).
40. W. J. Brittian and I. B. Dicker, *Macromolecules* **22**, 1054–1057 (1989).
41. W. R. Hertler, *Macromolecules* **20**, 2976–2982 (1987).
42. D. Braun, M. Herner, U. Johnson, and W. Kern, *Makromol. Chem.* **51**, 15–38 (1962).
43. G. M. Cohen, *Polym. Prepr., Am. Chem. Soc., Div. Polym. Chem.* **29**, 46–47 (1988).
44. J. A. Simms and H. J. Spinnelli, *J. Coat. Technol.* **59**, 125–131 (1987).
45. A. M. Helstern, J. M. DeSimone, and J. E. McGrath, *Polym. Prepr., Am. Chem. Soc., Div. Polym. Chem.* **29**, 148–149 (1988).
46. G. C. Eastman and J. Grigor, *J. Makromol., Rapid Commun.* **7,** 375–379 (1986).
47. Y. Kondo, K. Kojima, T. Shiraishi, M. Takaki, and R. Asami, *Kobunshi Ronbunshu* **46**, 367–373 (1989).
48. R. Asami, M. Takaki, and Y. Moriyama, *Polym. Bull.* **16**, 125–130 (1986).
49. J. M. DeSimone, A. M. Hellstern, E. J. Siochi, T. C. Ward, and J. E. McGrath, *Polym. Prepr., Am. Chem. Soc., Div. Polym. Chem.* **30**, 137–140 (1989).
50. W. Kreuder, O. W. Webster, and H. Ringsdorf, *Macromol. Chem., Rapid Commun.* **7,** 5–13 (1986).
51. B. J. Bauer and L. J. Fetters, *Rubber Chem. Technol.* **51**, 406–436 (1978).
52. M. K. Martin, *Diss. Abstr. Int. B 1982,* **46**, 1867 (1980).
53. J. A. Simms, in *International Symposium on Multiphase Macromolecular Systems,* ed. by W. Culbertson (Plenum Press, New York, 1988).
54. J. J. Kozakiewicz, N. S. Kurose, D. R. Draney, S. Y. Huang, and J. Falzone, *Polym. Prepr., Am. Chem. Soc., Div. Polym. Chem.* **28**, 347–348 (1987).
55. D. Y. Sogah, *Polym. Prepr., Am. Chem. Soc., Div. Polym. Chem.* **29**, 3–5 (1988).
56. D. Y. Sogah and W. B. Farnham, in *Organosilicon and Bioorganosilicon Chemistry : Structures, Bonding, Reactivity and Synthetic Applications,* ed. by H. Sakurai (Wiley, New York, 1985), Chap. 20.

57. P. H. Plesch, Mechanisms of Cationoid Polymerizations of Alkenes, 9th International Symposium on Cationic Polymerizations and Related Ionic Processes, Strasbourg, France (1989).
58. R. P. Quirk and G. P. Bidinger, *Polym. Bull.* **22** 63–67 (1989).
59. W. J. Schulz Jr and J. L. Speier, *Synthesis,* 163–166 (1989).

2

Anionic Ring-Opening Polymerization

Shohei Inoue and Takuzo Aida

2.1. INTRODUCTION

Heterocyclic compounds which can undergo ring-opening polymerization by anionic (nucleophilic) mechanisms are more limited than those polymerized by cationic (electrophilic) mechanisms.[1] Table 2.1 displays the representative types of monomers for anionic ring-opening polymerization. Some of the monomers can be polymerized by both anionic and cationic mechanisms.

The advantages of ring-opening polymerization by anionic mechanisms over cationic mechanisms are (1) the formation of polymers with high molecular weight, (2) the possibility of controlling the molecular weight, (3) the applicability to syntheses of block copolymers and end-reactive polymers, and (4) the formation of stereoregular polymers in some cases. These are usually more difficult to accomplish by cationic polymerization.

Examples of practical applications of anionic ring-opening polymerization include the production of Nylon-6, polysiloxanes, high-molecular-weight poly(ethylene oxide) and poly(propylene oxide), and poly(lactide). The polymerization of α-amino acid *N*-carboxylic anhydride (NCA) losing carbon dioxide to form polypeptide has played an essential role in the syntheses of models of proteins.

Shohei Inoue and Takuzo Aida • Department of Synthetic Chemistry, Faculty of Engineering, University of Tokyo, Hongo, Bunkyo-ku, Tokyo 113, Japan.

New Methods for Polymer Synthesis, edited by W. J. Mijs. Plenum Press, New York, 1992.

TABLE 2.1. **Representative Monomers for Anionic Ring Opening Polymerization**

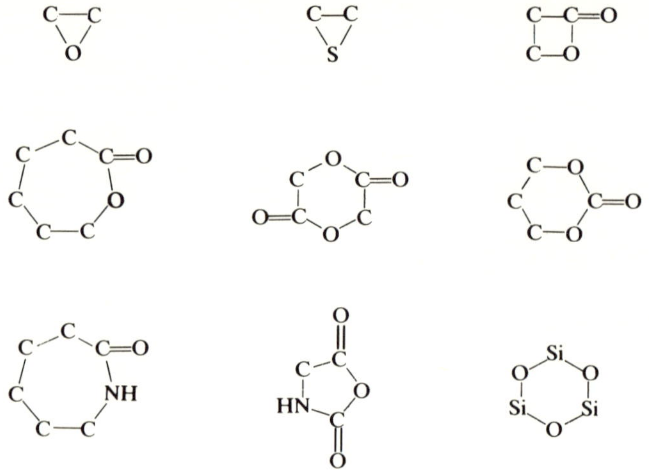

Ring opening of heterocyclic compounds can take place by the attack of a nucleophilic initiator at the carbon atom of the ring adjacent to the hetero atom. This process depends upon the reactivities of the ring and the nucleophile, and results in cleavage of the carbon–hetero atom bond. If the anion of the hetero atom resulting from the ring opening has enough nucleophilicity to again attack the cyclic compound, a repetition of this reaction, i.e., propagation, leads to the formation of a polymer with the hetero atom in the main chain. Typical initiators for the anionic polymerization of epoxides, episulfides, and lactones are summarized in Tables 2.2 and 2.3.[2]

The anion as the initiating and propagating species obviously carries

TABLE 2.2. Initiators for Anionic Polymerization of Epoxides and Episulfides[a]

Derivatives of	Examples
Alkali metal	Na, K, $NaNH_2$, NaOR, KOH, NaOH, NaNaph
Alkaline earth metal	MgR_2, CaO, SrO
Zinc and aluminum	ZnO, ZnR_2/H_2O, $ZnR_2/R'OH$
	$[Zn(OMe)_2\cdot(EtZnOMe)_6]$
	AlR_3, AlR_3/H_2O, AlR_3/H_2O/acetylacetone
	$Al(OR)_3/ZnCl_2$, $(RO)_2AlOZnOAl(OR)_2$
	Al-porphyrin
Other metals	$FeCl_3$, $Fe(OR)_3$, Ph_3SnBr/Ph_3P
Amines	Me_3N

[a] Taken from Ref. 2.

TABLE 2.3. Initiators for Anionic Polymerization of Lactones[a]

Derivatives of	Examples
Alkali metal	Na, NaNaph, Li_2 (or K_2) benzophenone, LiR
	LiH, NaH, K_2CO_3, KOH, NaOR, LiOR
	Li (or Na or K) acetate[b]
Alkaline earth metal	MgR_2, RMgX, $Mg(OR)_2$
Zinc and aluminum	ZnR_2, Et_2Zn/H_2O
	AlR_3, R_2AlX, R_2AlOR', $Al(OR)_3$, $AlEt_3/H_2O$
	Al-porphyrin, $R_2AlOZnOAlR_2$
Other metals	$CdEt_2/H_2O$, Ti phosphate
Amines	R_3N[b]

[a] Taken from Ref. 2.
[b] Active only for β-lactone.

a countercation. The nature of the bond between the anion and the countercation varies widely, from highly ionic to almost covalent. In most cases, the anion is that of oxygen, sulfur, or nitrogen, and the countercation is a metallic species. Since some of the metallic species exhibit Lewis acid character, the countercation can be coordinated by the monomer as a Lewis base prior to the attack by the anion, resulting in the activation of the monomer. Such a coordination sometimes leads to the stereoselection of the monomer in the polymerization.

Some of the anionic ring-opening polymerizations are accompanied by side reactions inherent to the structures of the monomer and the polymer. The presence of hetero atoms in the polymer chain can lead to the attack of a nucleophile at the carbon–hetero atom bond in the polymer chain. In particular, when the reactivities of the carbon–hetero atom bonds in the monomer and the polymer do not differ very much from each other, the reaction occurring on the polymer can be a side reaction which is difficult to avoid, being inherent to the structures of the reactant and the product.

If the attack of the growing species takes place at the carbon–hetero atom bond in the same polymer chain, depolymerization to regenerate the monomer or the formation of a macrocycle will result, depending upon the site of the bond to be attacked. A similar reaction taking place intermolecularly will lead to the redistribution of the polymer chains formed. Such an equilibrium between monomer, polymer, and macrocycle is observed in the polymerization of some lactones and cyclic siloxanes. The extent of these competitive reactions is dependent not only on thermodynamic factors such as ring strain of the monomer, but also on the reactivity of the growing end, which is much affected by the nature of the anion–countercation bond.

Even in cases where side reactions appear difficult to avoid due to similar reactivities of the carbon–hetero atom bonds in the monomer and polymer, the difference in these reactivities can be altered by the structure of the growing end, e.g., the nature of the metallic species in the countercation. In this regard, it is essential to select an appropriate initiator or catalyst, and thus a growing species, in order to control the polymerization reaction. In fact, recent examples described in the following sections take advantage of novel initiator or catalyst systems which are favorable in the above respects.

2.2. EPOXIDES

Polymerization of epoxides takes place via both nucleophilic and electrophilic mechanisms. The polymerization with nucleophilic initiators proceeds via metal alkoxides as the growing species [Eq. (2.1)]. The

(2.1)

nucleophilic initiators are classified into anionic and coordinate anionic initiators, the former mostly comprising alkali metals and the latter aluminum, zinc, and some transition metals. The polymerization of epoxyethane with anionic initiators such as potassium *tert*-butoxide is a representative example of living polymerization, which affords poly(oxyethylene) of a desired molecular weight with a narrow molecular-weight distribution.(3) In contrast, the anionic polymerization of substituted epoxides is generally accompanied by some side reactions, resulting in the formation of low-molecular-weight polyethers. On the other hand, the polymerization of substituted epoxides with coordinate anionic initiators tends to afford high-molecular-weight polyethers, but the molecular-weight distribution is broad. The coordinate anionic polymerization of unsymmetrically substituted epoxides has attracted some attention in view of the possibility to form stereoregular polyethers.

The present section is focused on recent advances in anionic polymerizations and copolymerizations of epoxides from the viewpoint of the control of molecular weight, stereoregularity, and comonomer sequence.(4,5)

2.2.1. Syntheses of Polyethers with Controlled Molecular Weight

2.2.1.1. With Anionic Initiators. Scope and Limitations

Alkali metal alkoxides and the salts of naphthalene, anthracene, fluorene, and carbazole are the representative anionic initiators which bring about the living polymerization of epoxyethane, a nonsubstituted epoxide, under suitable conditions.(6) However, use of these initiators for the polymerization of substituted epoxides results in the formation of polyethers of low molecular weight with broad molecular-weight distribution due to the side reactions associated with the substituents of the monomers. For example, in the case of the anionic polymerization of 1,2-epoxypropane with alkali-metal-based initiators, proton abstraction from the methyl group of the monomer to the growing species occurs concomitantly [Eq. (2.2)],

(2.2)

and the formation of low-molecular-weight polyether with an allylic terminal group results.(3,7) Thus, the applicability of the alkali-metal initiators to the living polymerization of epoxides is rather limited.

2.2.1.2. *With Metalloporphyrin Initiators. Living and Immortal Polymerizations and Application to Block-Copolymer Syntheses*

Polyethers with uniform, controlled molecular weight have recently become available from substituted epoxides since the discovery of metalloporphyrin initiators.(8)

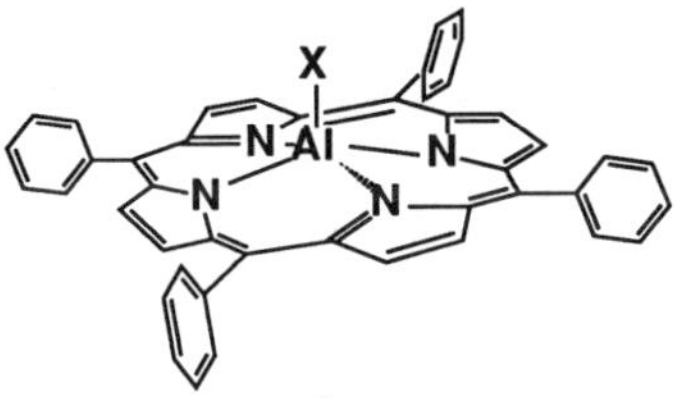

1

Aluminum tetraphenylporphyrin (**1**) is a representative initiator, which effectively brings about living polymerization of various epoxides such as epoxyethane, 1,2-epoxypropane, 1,2-epoxybutane, and 1-chloro-2,3-epoxypropane, affording the corresponding polyethers with a molecular-weight distribution (M_w/M_n) ranging from 1.05 to 1.2.(9,10) The number of polymer molecules formed is equal to that of the molecules of the aluminum porphyrin initiator, owing to the 100% efficiency of the initiation. Accordingly, the number-average molecular weight can be controlled by the mole ratio of the epoxide and the initiator. Sequential polymerizations of different epoxides with aluminum porphyrin initiator give the corresponding AB- and ABC-type block-copolyethers of controlled block lengths. Aluminum porphyrins also bring about living polymerizations of other monomers such as lactones with various ring sizes,(11) lactide,(12) vinyl monomers such as methacrylic-,(13) acrylic esters and methacrylonitrile, and living alternating copolymerizations of epoxides with cyclic anhydrides(14) or carbon dioxide.(15) By virtue of such a wide applicability of the aluminum porphyrin

TABLE 2.4. Polyether-Based Block-Copolymers of Uniform Block Lengths Prepared with Metalloporphyrin Initiators

Polyether–Polyester (Lactone)

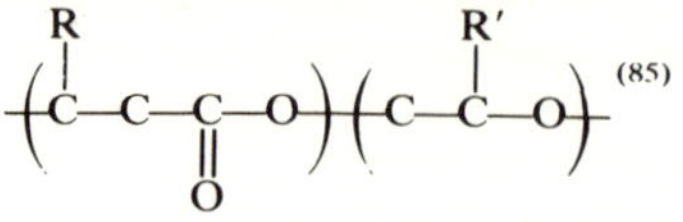

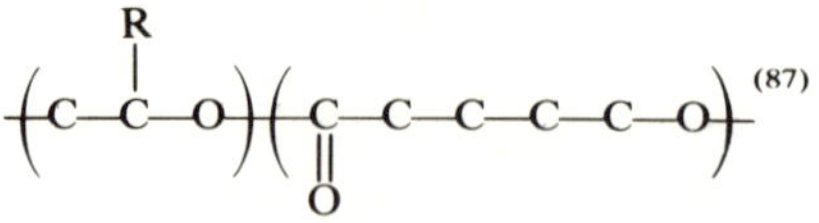

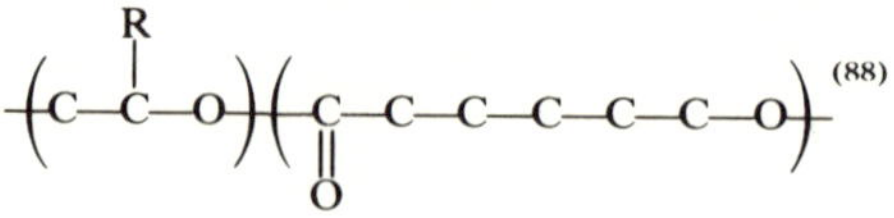

Polyether–Polyester (Lactide)

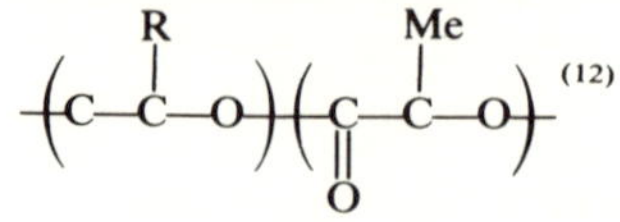

Polyether–Polyester (Anhydride–Epoxide)

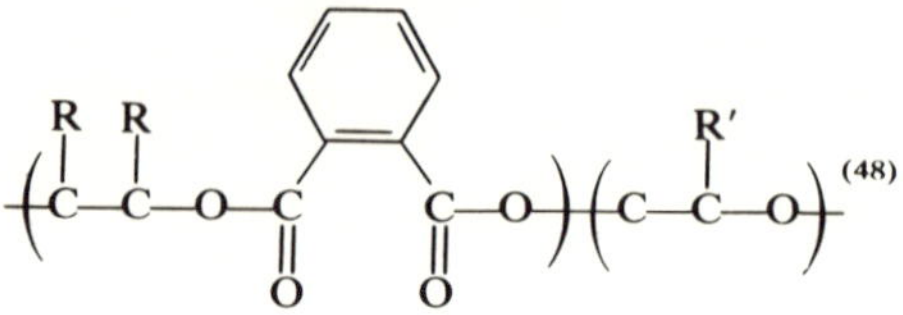

Polyether–Polythioether

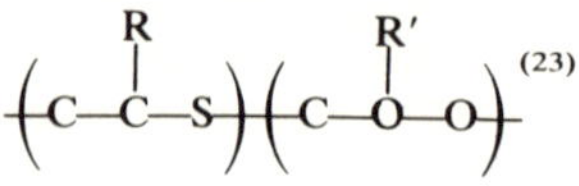

Polyether–Polycarbonate (CO_2–Epoxide)

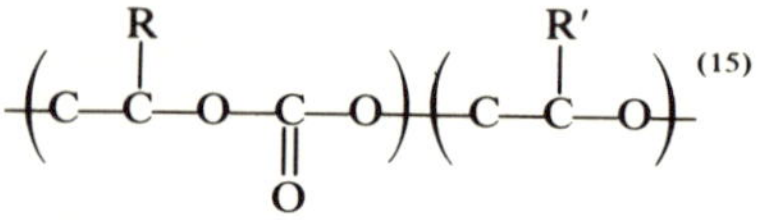

Polyether–Poly(methacrylate)

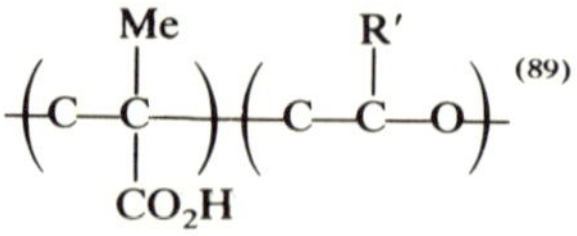

initiators, various novel tailored block copolymers consisting of structurally different segments can be easily synthesized (Table 2.4).

The polymerization of epoxide with aluminum porphyrin has been demonstrated to be initiated by the attack of the Al—X bond of the initiator onto the monomer, and proceeds via a (porphinato)aluminum alkoxide (**2**) as the growing species [Eq. (2.3)].[16] Accordingly, the rate of polymerization is affected by the structure of the porphyrin moiety of the initiator. Very powerful initiators developed so far are aluminum tetraphenylporphyrins carrying bulky, polar groups such as methoxy and chlorine groups at every *ortho*-position of the peripheral phenyl rings.[17]

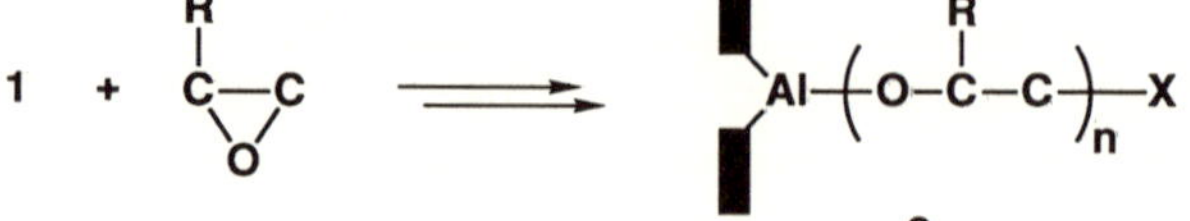

2

(2.3)

In the course of the above studies, a novel concept, *immortal polymerization,* has been discovered, which enables the synthesis of monodisperse polyethers from epoxides with the number of the polymer molecules exceeding that of the initiator molecules.[18] The basic concept of immortal polymerization of epoxides has been established on the basis of the reversible exchange phenomena of the molecules of (porphinato)aluminum alkoxide with alcohol.[19] In the polymerization of epoxide with aluminum porphyrin in the presence of alcohol, the chain growth of the polymer molecules is accompanied by the exchange reaction between the molecules of the growing species, (porphinato)aluminum alkoxide, and alcohol [Eq. (2.4)]. The alkoxide formed in the exchange reaction also reacts with epoxide to initiate the polymerization [Eq. (2.5)], and the newly formed growing polymer exchanges with the polymer alcohol [Eq. (2.6)]. Repetition of the reactions [Eqs. (2.4)–(2.6)] results in the

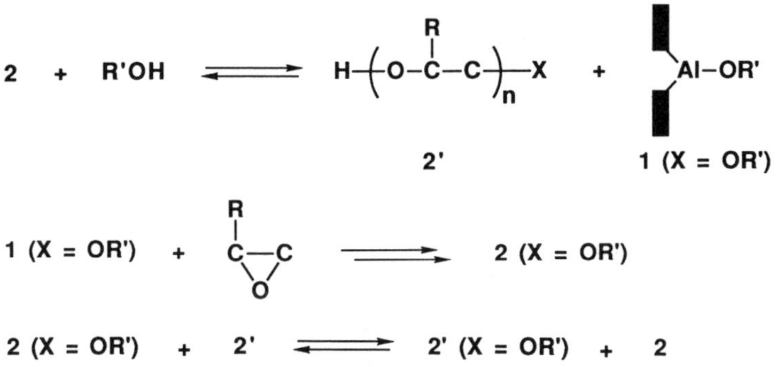

(2.4)

(2.5)

(2.6)

formation of polymer molecules from all the molecules of alcohol as well as of the initiator. Since the exchange reaction takes place much more rapidly than the reaction of the growing species with epoxide (chain growth), a polymer of uniform molecular weight is formed with the number of polymer molecules equal to the sum of the molecules of aluminum porphyrin initiator and alcohol. Quite interesting is the fact that this polymerization is not killed even by acids such as carboxylic acids or hydrogen chloride, which are the terminating agents for the polymerization of epoxides with alkali metal initiators. For example, in the case of the polymerization in the presence of hydrogen chloride, the growing species, (porphinato)aluminum alkoxide (**2**), is converted into (porphinato)aluminum chloride (**1**, X = Cl) by a rapid reaction with hydrogen chloride. This reaction is irreversible, and the polymer with a hydroxyl chain end cannot revive to the growing species by the backward reaction. However, (porphinato)aluminum chloride, in contrast with alkali metal chlorides, is so nucleophilic that it can reinitiate the polymerization of epoxide, and the reactions similar to Eqs. (2.4)–(2.6) ensue. Thus, the polymerization of epoxides with aluminum porphyrin may be called immortal polymerization.

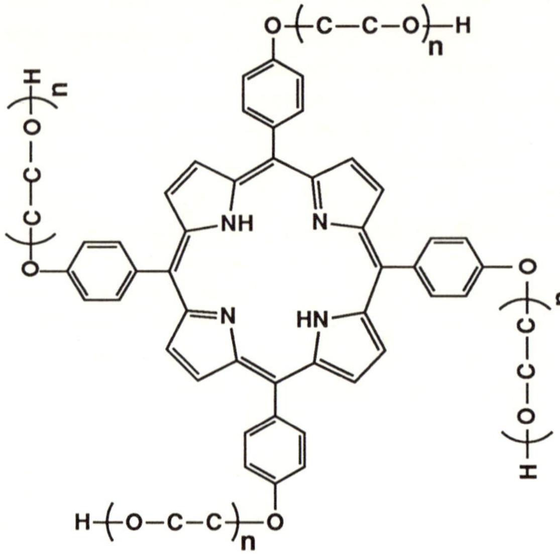

3

In the immortal polymerization of epoxides, a group originating from the protic compound added is introduced into the polymer terminal. By taking advantage of this, a variety of telechelic polyethers and block copolyethers are accessible by using appropriate protic compounds.[19,20] Use of unsaturated protic compounds such as acrylic acid, methacrylic acid, and 2-hydroxyethyl methacrylate gives polyethers carrying polymerizable end groups. Use of multifunctional protic compounds for the immortal polymerization of epoxides enables one to synthesize star-shaped polyethers. In this relation, the immortal polymerization of epoxyethane with aluminum porphyrin initiator in the presence of tetrakis(*p*-hydroxyphenyl)porphine as protic chain transfer agent affords a novel amphiphilic porphyrin carrying four poly(oxyethylene) side chains of the same chain length (**3**), which spontaneously forms cofacial aggregates in aqueous media depending on the length of the water-soluble polyether side chains.[21]

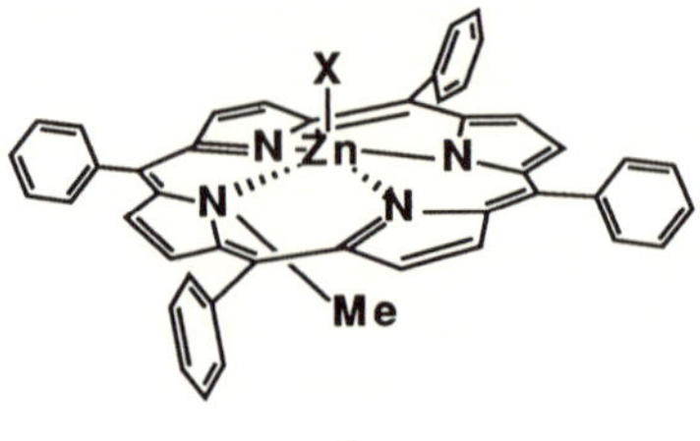

4

In addition to aluminum porphyrins, manganese(III) porphyrins[22] and zinc N-substituted porphyrins (**4**)[23] can be used as initiators for the living and immortal polymerizations of epoxides. Interesting to note is the polymerization of epoxides with a zinc thiolate complex of N-substituted porphyrin (**4**, X = SC_3H_7), which is a novel, visible light-induced living polymerization [Eq. (2.7)].

(2.7)

4 (X = SR)

Example 1. Immortal Polymerization of 1,2-*Epoxypropane with the Aluminum Porphyrin* (**1**, *X* = *Cl*)–*Methanol System*[18]

To a round-bottom flask containing **1** (X = Cl) (1 mmol) under nitrogen, a mixture of 1,2-epoxypropane (0.5 mol, 35 mL) and methanol (24 mmol) is added by means of a hypodermic syringe, and the mixture is stirred magnetically at room temperature. After 48 h, the polymerization mixture is evaporated to dryness under reduced pressure at room temperature to yield poly(1,2-epoxypropane) quantitatively as a viscous liquid with M_n and M_w/M_n, respectively, equal to 1200 and 1.08. The number of polymer molecules relative to that of **1** (X = Cl) is 24.2, which is close to the ratio $([\mathbf{1}]_0 + [\text{methanol}]_0)/[\mathbf{1}]_0$ of 25.

2.2.2. Syntheses of Stereoregular Polyethers

Syntheses of stereoregular polyethers from substituted epoxides have been a matter of fundamental significance in relation to coordination polymerization of α-olefins with Ziegler catalysts.

2.2.2.1. Historical Developments and Scope

Polymerization of mono-substituted epoxides such as 1,2-epoxypropane with nucleophilic initiators (anionic and coordinate anionic initiators) generally proceeds via ring opening of the monomer at the oxygen–methylene bond, where the configuration of the asymmetric carbon atom of the monomer remains intact. Thus, preferential incorporation of one of the antipodes of a monomer into the growing terminal affords the stereoregular polyether. Polymerization of 1,2-epoxypropane, for example, with anionic initiators such as sodium and potassium alkoxides generally affords the atactic polymer,[24] while coordinate anionic initiators afford a polymeric mixture containing a crystalline fraction with high stereoregularity. The first example of a stereoselective polymerization of substituted epoxides has been reported in 1955 for 1,2-epoxpyropane initiated with the $FeCl_3$/1,2-epoxypropane system.[25,26] Later, aluminum isopropoxide/$ZnCl_2$,[27,28] organozinc/water[29] or alcohol,[30] and organoaluminum/water/acetylacetone[31] have also been found effective to give crystalline poly(1,2-epoxypropane) rich in isotactic sequences. The tacticities of polyethers such as poly(1,2-epoxypropane)[32] and poly(1,2-epoxystyrene)[33] can be determined directly by ^{1}H and ^{13}C NMR spectroscopies. The tacticity of poly(1-chloro-2,3-epoxypropane) can be similarly determined via the transformation of the polymer into poly(1,2-epoxypropane).[34]

2.2.2.2. Mechanism of Stereoselective Polymerization

For the formation of stereoregular polyethers in the coordinate anionic process, the "enantiomorphic catalyst site control mechanism" has been proposed.[35] This mechanism is based on the assumption that the initiator involves an equal number of active sites that react preferentially with either *R*- or *S*-antipodes. Thus, the polymerization via the enantiomorphic catalyst site control mechanism will give an equimolar mixture of (*R*)-rich and (*S*)-rich polymers. Statistical treatment of the enantiomorphic catalyst site control mechanism provides the relationship $H = 2S$, where H and S represent the mole fractions of heterotactic and syndiotactic triad sequences, respectively.[36] The polymerization of 1,2-epoxypropane initiated with diethylzinc coupled with a chiral alcohol such as d-(+)-borneol is the first successful example of an asymmetric selective polymerization of epoxide, which affords an optically active polyether ($[\alpha]_D$ −5.9° in benzene, +7.4° in $CHCl_3$).[37] This observation has been understood in terms of the unequal generation of the enantiomorphic active sites in the presence of the chiral alcohol. Very recently, the enantioselective polymerization of 1,2-epoxypropane, using as initiator the reaction product between diethylaluminum chloride and a chiral schiff base (**5**), has been reported to proceed with an enantiomer selectivity ratio of 1.3.[38] In relation to the enantiomorphic catalyst site control mechanism, a well-defined crystalline zinc complex $[Zn(OCH_3)_2 \cdot (C_2H_5ZnOCH_3)_6]$ **6** has been isolated from the reaction mixture of diethylzinc and methanol, which has been shown by X-ray

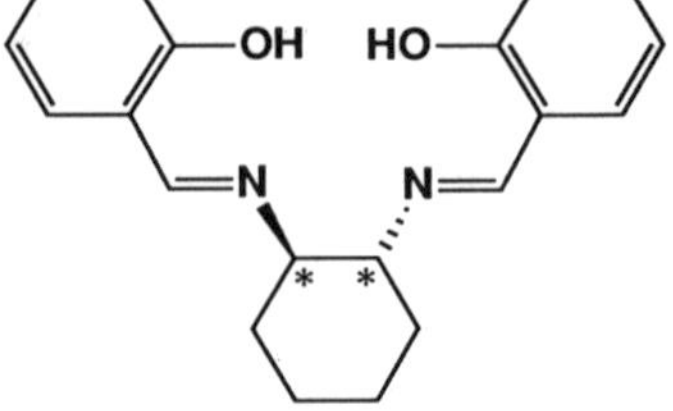

5

analysis to consist of two enantiomorphic distorted cubes sharing a zinc atom at the corner.[39] The polymerization of 1,2-epoxypropane initiated with **6** affords a polymer with fractions of isotactic, syndiotactic, and heterotactic triads being comparable to those obtained with the diethylzinc–water system, where the tacticity obeys the statistics of the "enantiomorphic catalyst site control mechanism" ($H = 2S$).[40,41] Thus, the complex **6** is a model for the active sites proposed for the "enantiomorphic catalyst site control mechanism."

Another mechanism for stereoselective polymerization is the "growing chain control mechanism," where the configuration of the incoming monomer is differentiated by that of the last n units (n-order, $n = 1, 2, \ldots$). Therefore, accidental incorporation of the monomer with the opposite configuration into the growing chain end will result in the formation of a stereoblock copolymer. Statistical treatment of the growing chain control mechanism provides the relationships $I = i^2$ and $S = s^2$ (i and s are the mole fractions of isotactic and syndiotactic diads, respectively).[42] The polymerizations of 1,2-epoxypropane with aluminum porphyrin initiators[43,44] and 1,2-epoxy-3,3-dimethylbutane (*tert*-butylethylene oxide) initiated with potassium *tert*-butoxide[45] have been claimed to proceed via the growing chain control mechanism.

2.2.3. Syntheses of Alternating Copolymers

Anionic copolymerizations of epoxides and those with other homopolymerizable monomers give copolymers of complicated structures as a consequence of competitive homopolymerizations. But the combination of epoxides with some nonhomopolymerizable unsaturated compounds such as cyclic anhydrides, carbon dioxide, and sulfur dioxide results in the formation of alternating copolymers.

2.2.3.1. Copolymerization with Cyclic Anhydrides

Alternating copolymerizations of epoxides with cyclic anhydrides to give polyesters proceed in the presence of catalysts such as alkali metal hydroxides, carboxylates, halides and alkyls, organo-magnesium, -zinc, and -aluminum compounds, and their systems coupled with protic compounds (water and alcohols), zinc and aluminum acetylacetonates, tertiary amines, phosphines, and their quaternary salts, organotin halides and carboxylates, and titanium alkoxides [Eq. (2.8)].[46] Copolymerization of epoxides and cyclic anhydrides with these initiators requires rather high temperatures.

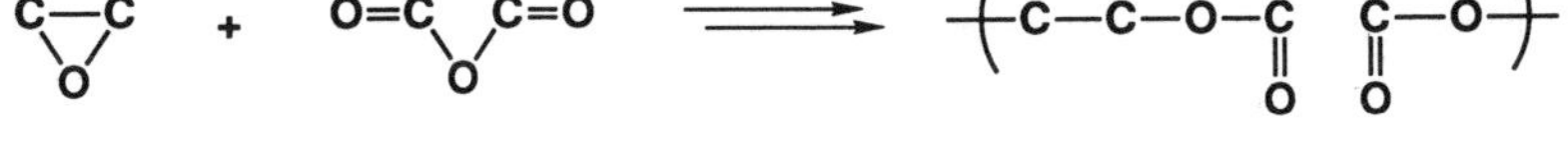

(2.8)

Organic bases such as tertiary amines and phosphines are convenient initiators, which afford the growing copolymer with a zwitterionic structure.[47] In this case, the copolymerization is possibly accompanied by a chain coupling reaction (polycondensation), regenerating the initiator molecule, which may allow the formation of a new copolymer chain (chain transfer reaction). Thus, the copolymer with broad molecular-weight distribution is formed.

Rather recently, the system of aluminum porphyrin coupled with quaternary ammonium and phosphonium salts has been found to be an excellent initiator for the alternating copolymerization of phthalic anhydride with a variety of epoxides such as epoxyethane, 1,2-epoxypropane, 1,2-epoxybutane, 1,2-epoxystyrene, 1,2-epoxy-3-phenyloxypropane (phenyl glycidyl ether), *cis*- and *trans*-2,3-epoxybutanes, and 1,2-epoxycyclohexane at room temperature.[14,48] The polyesters thus formed are characterized by a narrow molecular-weight distribution, as indicated by the ratio of the weight- and number-average molecular weights ranging from 1.08 to 1.16. NMR studies have demonstrated the formation of an active six-coordinate aluminum porphyrin species (**7**) upon mixing aluminum porphyrin (**1**) with onium salt [Eq. (2.9)], and the copolymerization proceeds with living character on both sides, respectively, of an aluminum porphyrin plane.

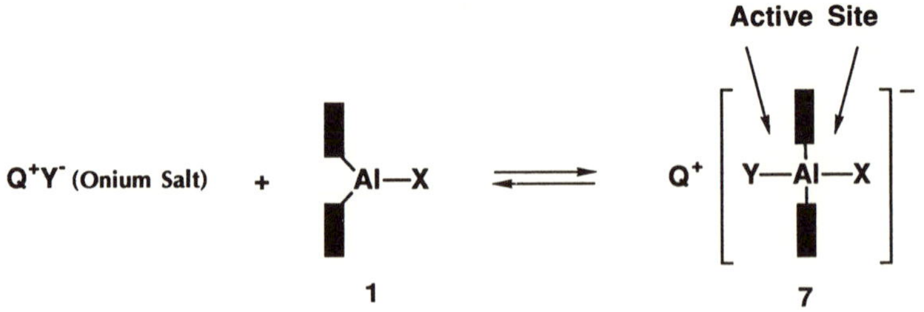

(2.9)

Sequential copolymerizations of phthalic anhydride with different epoxides, or copolymerization of phthalic anhydride with an epoxide followed by the homopolymerization of epoxides or lactones with the aluminum porphyrin/onium salt system, give the corresponding block copolymers of uniform block lengths.[48]

Systems of diethylzinc reacted with alcohols are also effective initiators for the alternating copolymerization of epoxides and cyclic anhydrides at a relatively high temperature, such as 80 °C. Use of the system of dialkylzinc coupled with an optically active alcohol such as d-(+)-borneol as initiator for the copolymerization of racemic 1,2-epoxypropane and 3-phenyl-Δ^4-tetrahydrophthalic anhydride (**8**) affords a polyester with optical activity, which results from the asymmetric selection of the optical antipodes of the anhydride by the chiral initiator.[49]

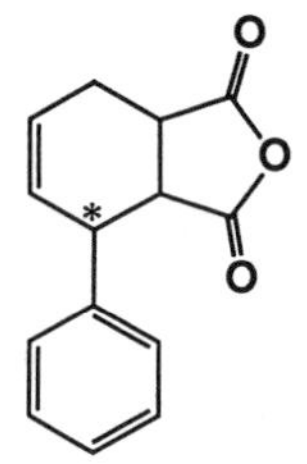

8

2.2.3.2. *Copolymerization with Carbon Dioxide*

Copolymerization of epoxides with carbon dioxide takes place in the presence of organozinc-based initiators, which allows the formation of alternating copolymers, aliphatic polycarbonates (Table 2.5), under moderate pressure of carbon dioxide at ambient temperature [Eq. (2.10)].[50,51]

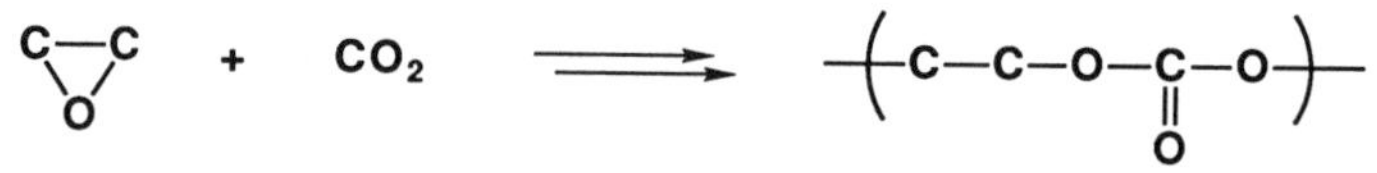

(2.10)

Inorganic zinc compounds such as zinc oxide and zinc hydroxide coupled with carboxylic acids also operate as the effective initiators.[52–54] On the

TABLE 2.5. Epoxides for Copolymerization with Carbon Dioxide[50,51]

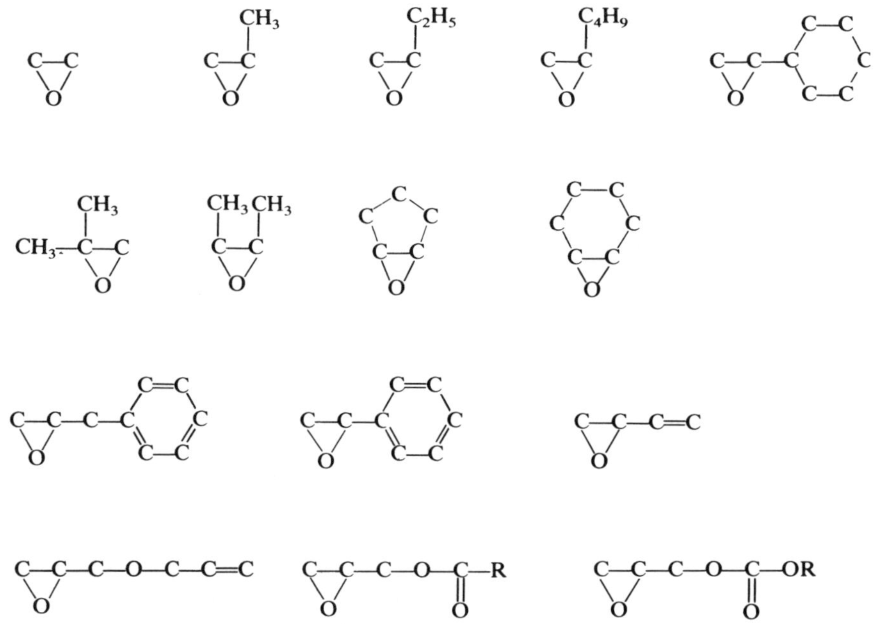

contrary, the systems based on organoaluminum compounds so far examined give only copolymer rich in epoxide units.[55]

Copolymerization of epoxides and carbon dioxide with the organozinc-based initiators are generally accompanied by the formation of cyclic carbonates, which however do not polymerize under these conditions. Therefore, the copolymerization of epoxides and carbon dioxide is considered to proceed by the repeated additions of epoxide and carbon dioxide to the growing chain end of the copolymer.

The molecular weight of the copolymer formed with the system based on organozinc compounds is generally high, but the molecular-weight distribution is very broad. On the contrary, use of the aluminum porphyrin/onium salt system for the copolymerization reaction enables the formation of polycarbonate with uniform, controlled molecular weight.[15] The copolymerization is considered to proceed similarly to Eq. (2.9). The synthesis of polyester–polycarbonate block copolymer is also possible by the sequential copolymerizations of epoxide with phthalic anhydride followed by carbon dioxide.

Aliphatic polycarbonates from epoxides and carbon dioxide have caught some attention due to their interesting properties, such as low O_2-permeability and biodegradability.[56–58]

Example 2. Copolymerization of Carbon Dioxide with 1,2-Epoxycyclohexane using the **1** *(X = Cl)–$C_2H_5(C_6H_5)_3PBr$ System*[15]

To a stainless-steel autoclave containing a CH_2Cl_2 solution (20 mL) of the mixture of **1** (X = Cl) and $C_2H_5(C_6H_5)_3PBr$ (1 mmol/1 mmol), 1,2-epoxycyclohexane (100 mmol) is added by a syringe. The mixture is pressurized by carbon dioxide to 50 kg cm^{-2} and stirred magnetically at room temperature. After 14 days, the autoclave is opened, and the content is evaporated under reduced pressure at room temperature to leave poly(1,2-cyclohexene carbonate) in 100% yield as a powdery solid, the M_n and M_w/M_n being 6800 and 1.06, respectively.

2.2.3.3. Copolymerization with Sulfur Dioxide

Copolymerization of epoxides with sulfur dioxide takes place in the presence of metal halides or amines, where aromatic tertiary amines such as pyridine and quinoline as initiators give the virtually alternating copolymer, poly(alkylene sulfite) [Eq. (2.11)].[59,60] Since the equimolar mixture of pyridine, sulfur dioxide, and 1,2-epoxycyclohexane gives, upon stirring at 20 °C, the zwitterionic salt **9** as a precipitate, the copolymerization is considered to proceed via a stepwise mechanism [Eq. (2.12)].

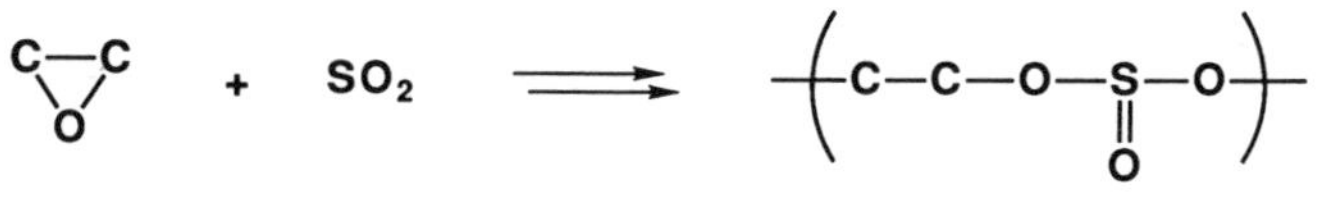

(2.11)

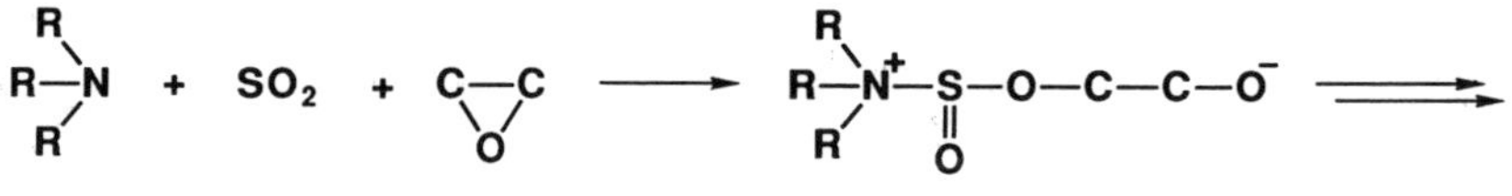

(2.12)

9

2.2.3.4. *Copolymerizations with Other Unsaturated Compounds*

Copolymerizations of epoxides with carbon disulfide,[61] carbon monoxide,[62] α-aminoacid *N*-carboxylic anhydride,[63] benzonitrile,[64] and phenylisocyanate,[65] so far attempted, do not produce alternating copolymers but rather complicated products.

2.3. *EPISULFIDES*

Polymerizations of episulfides with nucleophilic initiators have been studied in relation to the polymerization of epoxides, and to date numerous initiators have been explored.[66] In comparison with the polymerization of epoxides, one of the characteristic features of the polymerization of episulfides is a possible desulfurization reaction of the monomer to give the corresponding olefin as a by-product.[67]

2.3.1. *Synthesis of Polythioethers with Controlled Molecular Weight*

The living polymers of episulfides can be obtained with a wider range of nucleophilic initiators than for epoxides. Alkali metal initiators can bring about the living polymerization of 1,2-epithiopropane (propylene sulfide) owing to the reduced tendency of chain transfer reaction from the monomer [see Eq. (2.2) for the polymerization of 1,2-epoxypropane]. For example, carbazyl sodium and potassium initiate the living polymerization of 1,2-epithiopropane in tetrahydrofuran to afford the polymer with a degree of polymerization (D_p) close to $[\text{Monomer}]_{\text{reacted}}/[\text{Initiator}]_0$ and a ratio M_w/M_n less than 1.1.[68] Sodium naphthalene, a conventional initiator for the living polymerization of styrene, can initiate the polymerization of 1,2-epithiopropane, giving a polymer with two living ends per molecule with a D_p close to $2[\text{Monomer}]_{\text{reacted}}/[\text{Initiator}]_0$.[69] The polymerization of 1,2-epithiopropane with triphenylmethylsodium as initiator also gives bifunctional living polymer, since the actual initiating species is sodium disulfide.[70] Alkali metal initiators coupled with crown ethers or cryptands are effective to bring about the living polymerization of episulfides,[71] where the aggregation of the initiators and/or the growing species is greatly minimized. The polymerizations with sodium and potassium hydroxides and alkoxides are generally slow, due to their limited solubilities in the polymerization media.

Cadmium or zinc thiolates [e.g., Metal $(SCH_2CH{=}CH_2)_2$] also initiate the polymerization of 1,2-epithiopropane in solvents such as benzene, toluene, and tetrahydrofuran at 0–20 °C, affording polymers with M_w/M_n of 1.03–1.53.[72] The polymerization initiated with $CdCO_3$ in the presence of ethanethiol as chain transfer agent has been reported to give a polymer with narrower molecular-weight distribution ($M_w/M_n = 1.5$) than in the absence of ethanethiol ($M_w/M_n = 2.5$–3.5).[73]

Zinc complexes of N-substituted porphyrins (**4**), which are excellent initiators for the living polymerization of epoxides,[23] also give rise effectively to the living polymerization of 1,2-epithiopropane, in which the polythioether with M_w/M_n below 1.1 is formed with D_p in agreement with the ratio $[\text{monomer}]_{\text{reacted}}/[\text{Zn}]_0$.[74] Furthermore, when the polymerization is carried out in the presence of thiol as chain transfer agent, a polymer of uniform molecular weight is obtained. The number of the polymer molecules equals the sum of the molecules of initiator and thiol (*immortal* character). The NMR studies have demonstrated that the polymerization proceeds via a (N-substituted porphinato)zinc thiolate as the growing species. When the mixture of the living polymer of episulfide thus formed and 1,2-epoxypropane is irradiated with visible light, polymerization of 1,2-epoxypropane takes place to give the polythioether–polyether block copolymer of uniform block lengths.

Example 3. Polymerization of 1,2-*Epithiopropane Initiated with* **4** ($X = SCH_2CH_2CH_3$)[74]

To a round-bottom flask containing a CH_2Cl_2 solution (2 mL) of **4** ($X = SCH_2CH_2CH_3$) (0.1 mmol), 1,2-epithiopropane (40 mmol) is added by a syringe, and the mixture is stirred magnetically at 25 °C. After 80 min, the volatile fraction is removed from the reaction mixture under reduced pressure at room temperature to leave poly(1,2-epithiopropane) quantitatively as a viscous liquid, M_n and M_w/M_n being 47,000 and 1.11, respectively.

2.3.2. Synthesis of Stereoregular Polythioethers

For the stereoselective polymerization of substituted episulfides,[75] organometallic compounds [$Zn(C_2H_5)_2$, $Cd(C_2H_5)_2$, and $Al(C_2H_5)_3$] reacted with protic compounds such as water, alcohols (glycols), thiols, and amines are effective initiators. Here the "enantiomorphic catalyst site control mechanism" is operative, as mentioned previously.

The system of $Zn(C_2H_5)_2$ reacted with (*R*)-(−)-3,3-dimethyl-1,2-butanediol [$(CH_3)_3CCH(OH)CH_2OH$] is effective for the enantioselective polymerization of racemic 1,2-epithiopropane. The optical purity of the monomer remained at half-reaction ($O.P._{50\%\ conv.}$) was 30.0% (*S*).(76) The analogous systems $Zn(C_2H_5)_2$/(*R*)-(−)-$(CH_3)_2CHCH(OH)CH_2OH$ and $Zn(C_2H_5)_2$/(*R*)-(−)-$C_6H_5CH(OH)CH_2OH$ exhibit $O.P._{50\%\ conv.}$ of 28 and 20%, respectively, under similar conditions. In sharp contrast, the stereoselectivity of the polymerization almost disappears when (*R*)-(−)-$(CH_3)_3CCH(OH)CH_2OCH_3$ is used in combination with $Zn(C_2H_5)_2$. The systems of $Zn(C_2H_5)_2$/(*R*)-(−)-$(CH_3)_3CCH(OH)CH_2OH$, coupled with coordinate compounds such as dicyclohexyl-18-crown-6, sparteine, (*S*)-(+)-$CH_3CH_2CH(CH_3)CH_2SCH_2CH_3$, and *R*-(+)-limonene, exhibit enhanced stereoselectivities (k_R/k_S ratios of 3.9, 4.0, 4.5, and 4.3, respectively) in the polymerization of 1,2-epithiopropane.(77) In contrast to the above cases, preferential polymerization of the (*S*)-antipode has been observed when $Cd(C_2H_5)_2$ in place of $Zn(C_2H_5)_2$ is used as initiator coupled with (*R*)-(−)-$(CH_3)_3CCH(OH)CH_2OH$.(78)

The system of $Zn(C_2H_5)_2$ combined with an atropisomeric chiral compound such as (*S*)-(−)-1,1′-bi-2-naphthol is a very effective initiator for the enantioselective polymerization of racemic methyl- and ethyl-substituted episulfides. The k_S/k_R ratios of 15–20 have been reported.(79,80) For example, in the case of the polymerization of 1,2-epithiopropane, the optical purity of the monomer remained at half-reaction was 85%, which is the highest value observed up to now. A crystalline polymer with $[\alpha]_D$ of −198° (c = 0.42, $CHCl_3$, 25 °C) is obtained by fractionating the polymerization mixture (polymerization: at −33 °C) with toluene.

2.3.3. Syntheses of Alternating Copolymers

Copolymerization of episulfides and carbon dioxide has been attempted in relation to the copolymerization of epoxides with carbon dioxide (see Section 2.2.3.2), but the alternating copolymer (polythiocarbonate) has never been obtained. For example, the polymerization of 1,2-epithiopropane (50 equiv) and carbon dioxide (30 kg cm^{-2}) with the $Al(C_2H_5)_3$/1,2,3-trihydroxybenzene (1 equiv/2 equiv) system in dioxane at 45 °C for 96 h gives the copolymer with the CO_2 content of 42% in 11.6% yield.(81) The copolymerization with the organozinc-based initiators, which afford the alternating copolymer of carbon dioxide and epoxides, gives virtually the homopolymer of 1,2-epithiopropane under similar conditions.

Copolymerization of episulfides such as epithioethane and carbon disulfide is also possible [Eq. (2.13)]. For example, the copolymerization of

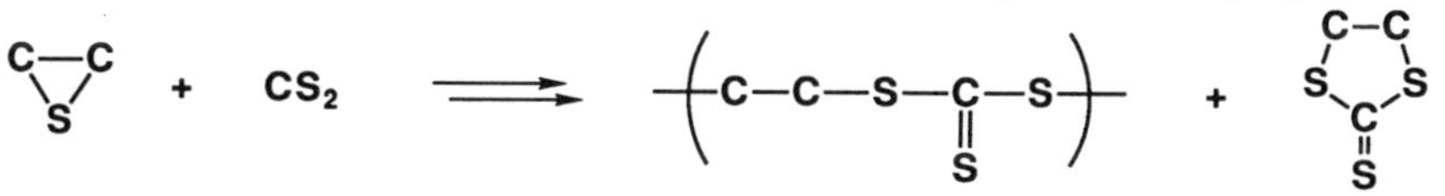

(2.13)

epithioethane and carbon disulfide with triethylamine as a catalyst proceeds at room temperatures to give an alternating copolymer in 63% yield for 4 days.[82] In this case, concomitant formation of ethylene trithiocarbonate has been observed. The copolymerization with $Hg(SC_4H_9)_2$ gives a copolymer with a rather low thiocarbonate content (50–70%).[83] In the copolymerizations initiated with organometallic compounds such as $Al(C_2H_5)_3$, $Zn(C_2H_5)_2$, and $Cd(C_2H_5)_2$, the thiocarbonate content of the produced copolymer falls down considerably to the range of 0–18%.

Copolymerization of episulfides with elemental sulfur with an eight-membered ring takes place in aromatic hydrocarbon solvents below the floor temperature for radical homopolymerization of elemental sulfur. $CdCO_3$ and alkali metal thiophenoxide–crown ether systems are representative catalysts, which allow the formation of copolymer with the number-average molecular weight and sulfur content up to 50,000 and 85%, respectively.[84] The copolymers of episulfides and elemental sulfur are very reluctant to undergo depolymerization, and form transparent films by casting from solution.

2.4. LACTONES AND RELATED MONOMERS

Lactones with four-, six-, and seven-membered rings can be polymerized anionically, but the polymerization of five-membered lactones with anionic initiators has never been successful. Polymerization of the four-membered cyclic ester, β-lactone, usually proceeds by ring cleavage at the alkyl–oxygen bond, differently from lactones with a larger ring, the growing species being a carboxylate [Eq. (2.14)]. Polymerization of lactones with a larger ring proceeds by the normal mode of cleavage of the ester bond, i.e., acyl–oxygen bond scission [Eq. (2.15)]. Cyclic dimers of α-hydroxy acids such as glycolide and lactide, six-membered cyclic carbonates, and morpholinediones are the related monomers that can be also polymerized with nucleophilic initiators.

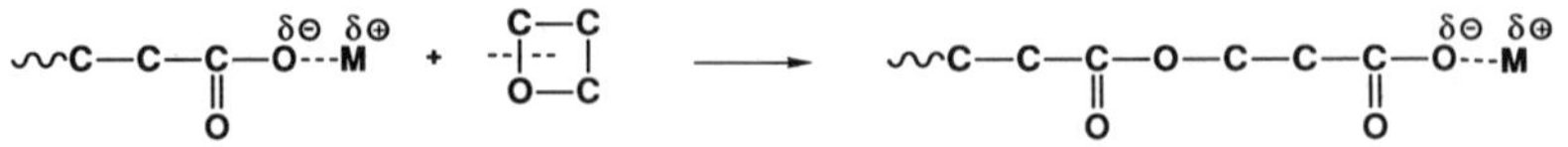
(2.14)

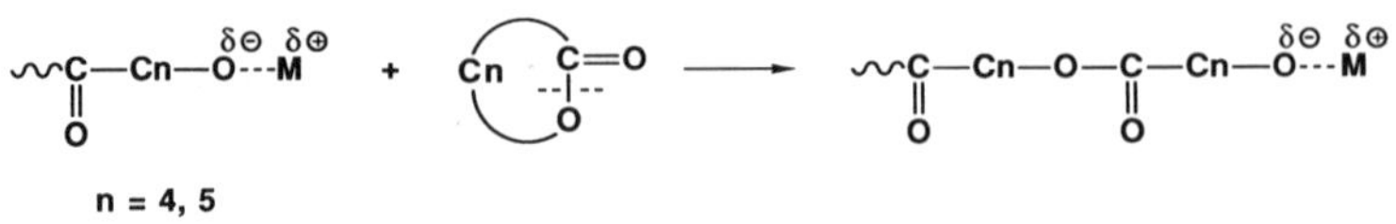
(2.15)

2.4.1. Lactones

2.4.1.1. Syntheses of Polyesters with Uniform Molecular Weight

For the synthesis of polyesters of uniform, controlled molecular weight, aluminum porphyrin is one of the most effective initiators and is capable of initiating living polymerizations of four-,[11,85,86] six-,[87] and seven-membered lactones.[88] The growing species of the polymerization of four-membered lactones with aluminum porphyrin initiator is a (porphinato)aluminum carboxylate,[85] while the polymerizations of higher lactones proceed via (porphinato)aluminum alkoxides as the growing species.[87,89] Furthermore, these polymerizations are of an immortal character in the presence of appropriate protic compounds. Accordingly, polyesters of uniform molecular weight are formed with the number of polymer molecules equal to the sum of the molecules of the aluminum porphyrin and protic compound.[19,88] Tailored block copolymers consisting of polyesters and polyethers can also be synthesized by sequential polymerizations of lactones and epoxides using aluminum porphyrin initiators. Living poly(methyl methacrylate) prepared with alkylaluminum porphyrins brings about the polymerization of a six-membered lactone to afford a poly(methacrylic ester)–polyester block copolymer with uniform block lengths.[39]

Bimetallic μ-oxo alkoxides with the formula $(RO)_2AlOZnOAl(OR)_2$[90] and dialkylaluminum alkoxides[91] are also very effective initiators for the living polymerization of a seven-membered lactone such as ε-caprolactone. Use of $(RO)_2AlOZnOAl\ (OR)_2$ reacted with polystyrene or polybutadiene carrying a hydroxyl end group as initiator for the polymerization of ε-caprolactone affords the corresponding polyvinyl–polyester block copolymers.[92]

Alkali metal solutions containing crown ether or cryptand have been reported to initiate the polymerization of four-membered lactones such as β-propio- and β-butyrolactones, where the lactone enolate species are formed at the initial stage, and polymerization has been proposed to be initiated via C—C bond cleavage of the momoner [Eq. (2.16)].[93,94]

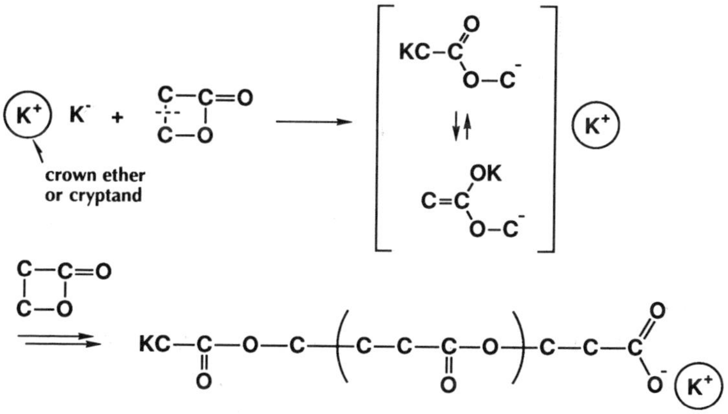

(2.16)

2.4.1.2. Syntheses of Stereoregular Polyesters

Stereoselective polymerization of lactones to give crystalline polyesters has attracted some attention in relation to the structure of naturally occurring poly(α-hydroxy ester).[75]

Enantioselective polymerizations of racemic α,α-disubstituted β-propiolactones such as α-ethyl-α-methyl-β-propiolactone and α-methyl-α-propyl-β-propiolactone have been attempted with the $ZnEt_2/(R)$-(−)-$(CH_3)_3CCH(OH)CH_2OH$ system as initiator at room temperature. The k_S/k_R ratios of 1.02–1.07 and 1.25, respectively, have been observed.[95]

For the polymerization of β-substituted β-propiolactones, the $ZnEt_2/H_2O$ (1/1) system is a more favorable initiator than the $AlEt_3/H_2O$ (1/1) system in terms of the crystallinity of the produced polymers.[96] Poly(β-butyrolactone) prepared with the $ZnEt_2/H_2O$ (1/1) system gives, upon fractionation with chloroform, a crystalline polymer as an insoluble fraction, whose X-ray diffraction pattern is virtually identical to that of the naturally occurring poly(β-hydroxybutyrate). The polymerization of β-butyrolactone with the $AlEt_3/H_2O$/epichlorohydrin (1-chloro-2,3-epoxypropane) system affords a polyester with a very high crystallinity.[97] The polymerization of racemic β-butyrolactone with the $ZnEt_2/(R)$-(−)-$(CH_3)_3CCH(OH)CH_2OH$ system is one of the most successful examples of enantioselective polymerization of β-substituted lactones. The unreacted monomer recovered at 84% conversion is rich in the (S)-enantiomer with 46% ee.[98] The methanol-insoluble fraction of the produced polymer, as determined by ^{13}C NMR, contains 72% isotactic diad sequences. Systematic studies on the polymerization of a series of β-substituted β-propiolactones with coordinate anionic initiators such as $AlEt_3/H_2O$/epichlorohydrin at 60 °C indicate that the crystallinity of the produced polymer tends to become low when the monomer bears a bulky substituent such as the *tert*-butyl or trichloromethyl group.[97] Additionally, the mode of ring opening of four-membered lactone has been studied in the case of polymerization of (R)-benzyl malolactonate initiated with triethylamine. It was demonstrated that the polymerization proceeds via alkyl–oxygen bond cleavage with inversion of the configuration of the asymmetric carbon atom to give poly((S)-malic acid benzyl ester).[99]

2.4.2. Cyclic Dimer of α-Hydroxy Acids (Glycolide and Lactide)

Glycolide and lactide, cyclic dimers of α-hydroxy acids (**10**), can be polymerized with anionic initiators to give polyesters [Eq. (2.17)], which are

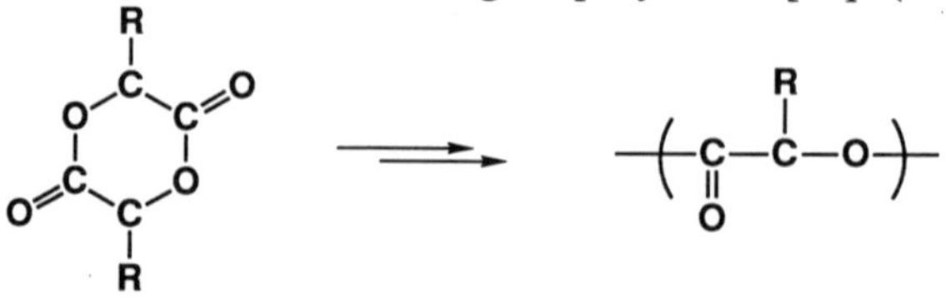

10

(2.17)

of practical importance for biomedical applications owing to their inherent biodegradability. Representative anionic and coordinate anionic initiators for the polymerization of these monomers include quaternary ammonium or phosphonium salts, aluminum isopropoxide, and dibutyltin dimethoxide, the latter two giving high-molecular-weight polyesters.[100] For the synthesis of poly(lactide) (**10**, R = CH_3) with narrow molecular-weight distribution, aluminum porphyrin initiators are effective.[12] A typical example is the polymerization of D-lactide with (tetraphenylporphinato)aluminum alkoxide with the mole ratio of 100 in CH_2Cl_2 at 100 °C, which proceeds to 94% conversion in 96 h to give a polymer with M_n and M_w/M_n of 16,400 and 1.12, respectively. In this case, M_n of the produced polymer increases linearly with conversion, retaining the M_n/M_w ratio of 1.1, and the number of polymer molecules relative to the number of initiator molecules remains constant, close to 1.0. In the presence of a protic chain transfer agent such as methanol, the polymerization of lactide with aluminum porphyrin initiator is of an immortal character. Thus, poly(lactide) with uniform molecular weight is formed with the number of polymer molecules more than that of the aluminum porphyrin initiator.[101] It should be of further interest to note that the sequential immortal polymerizations of δ-valerolactone and lactide afford the corresponding polyester–poly(lactide) block copolymer of uniform block length. Bimetallic μ-oxo alkoxides $[(RO)_2AlOZnOAL(OR)_2]$ have also been reported to be effective as initiators for the synthesis of poly(lactone)–poly(lactide) block copolymers.[102]

2.4.3. *Cyclic Carbonates (1,3-Dioxan-2-Ones)*

Polymerization of ethylene carbonate, a five-membered cyclic carbonate, is accompanied by considerable decarboxylation reaction, resulting in the formation of a copolymer of ethylene oxide and ethylene carbonate.[103] Metallic initiators such as butyllithium, alkali metal carbonates, dialkyltin alkoxides, and zirconium alkoxides, and organic bases such as aromatic amines and phosphines have been used. The highest mole fraction of ethylene carbonate units (45–49%) have been attained by using dialkyltin alkoxides and aromatic phosphines as initiators.

In sharp contrast, six-membered cyclic carbonates (**11**) undergo anionic ring-opening polymerization without decarboxylation under appropriate conditions, giving poly(trimethylene carbonate)s in excellent yields [Eq. (2.18)]. An example is shown by the polymerization of 5,5-dimethyl-1,3-dioxan-2-one (**11**; R^1, R^2 = CH_3, 0.74 M) with *sec*-butyllithium (2.5 ×

10^{-3} M) as initiator in toluene at −10 °C in 5 min, which affords the

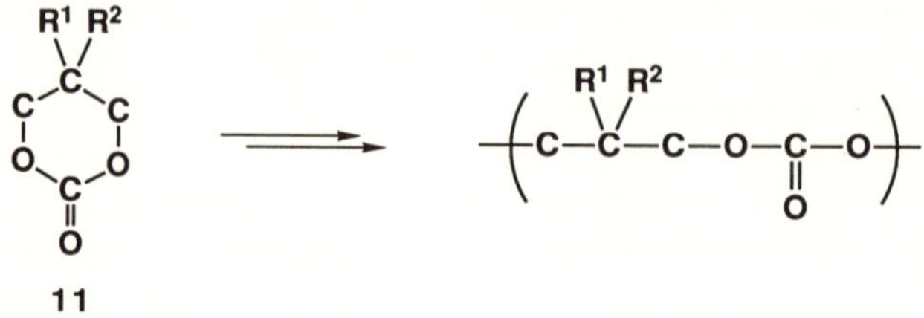

(2.18)

corresponding poly(trimethylene carbonate) with M_n and M_w/M_n of 36,400 and 1.2, respectively, in 91.9% yield.[104] Some polyvinyl–polycarbonate block copolymers can also be synthesized by using as initiators living vinyl polymers such as polystyrene and polybutadiene. Cyclic oligocarbonates with various ring sizes are concomitantly formed in 15–20% total yield when tetrahydrofuran is used as solvent.

2.4.4. Morpholinediones

Morpholinediones (**12**) can be polymerized by coordinate anionic initiators such as $Zn(C_2H_5)_2$, ZnO,[105] and Sn(2-ethylhexanoate)$_2$,[106] affording polydepsipeptides consisting of alternating sequences of hydroxy acid and amino acid units [Eq. (2.19)]. Polydepsipeptides have attracted

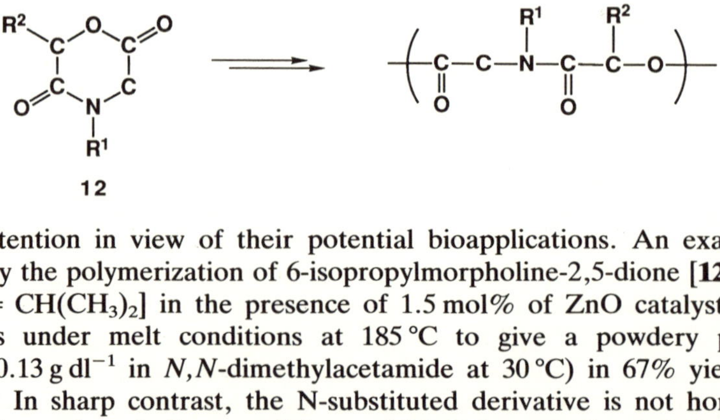

(2.19)

some attention in view of their potential bioapplications. An example is shown by the polymerization of 6-isopropylmorpholine-2,5-dione [**12**, R^1 = H, R^2 = $CH(CH_3)_2$] in the presence of 1.5 mol% of ZnO catalyst, which proceeds under melt conditions at 185 °C to give a powdery polymer (η_{inh} = 0.13 g dl^{-1} in *N,N*-dimethylacetamide at 30 °C) in 67% yield after 40 h.[105] In sharp contrast, the N-substituted derivative is not homopolymerizable under similar conditions. 6-Methylmorpholine-2,5-dione (**12**, R^1 = H, R^2 = CH_3) is copolymerizable with lactide in the presence of Sn(2-ethylhexanoate)$_2$ ([**12**]$_0$/[lactide]$_0$ = 1, [comonomers]$_0$/[Sn]$_0$ = 2500) at a temperature such as 135 °C, affording the copolymer (M_n = 16,000, M_w/M_n = 1.75) in 70% yield with an increased content of hydroxy acid units (76%).[107]

2.5. LACTAMS

In the anionic polymerization of lactam, cyclic amide, the amide hydrogen with some acidic character plays an important role. In the presence of a strong base, the amide hydrogen of the monomer is abstracted to form the anion of the monomer [Eq. (2.20)], which is able to attack the carbonyl group of the monomer. Particularly in the presence of an N-acylated lactam, the anion of the monomer reacts with the more reactive carbonyl group to generate a linear amide anion and an N-acyllactam group [Eq. (2.21)]. The linear amide anion abstracts amide hydrogen from the monomer to form the monomer anion [Eq. (2.22)] which reacts with the N-acyllactam group, resulting in the propagation of a polymer chain. Thus the polymerization may be regarded as the attack of "activated monomer" to the growing end of the polymer chain.

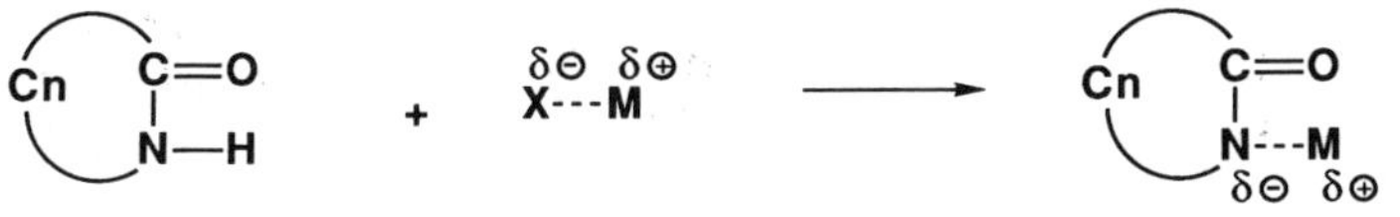

(2.20)

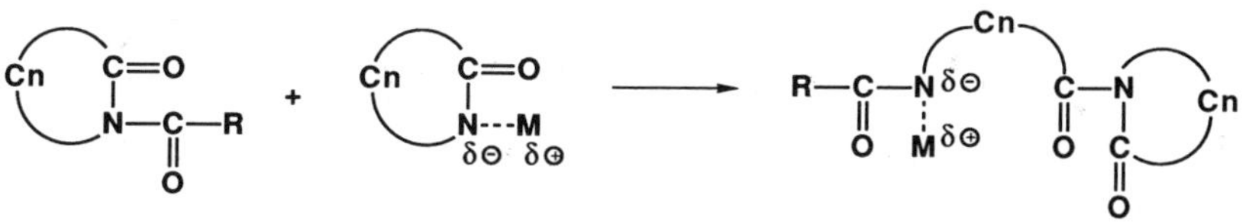

(2.21)

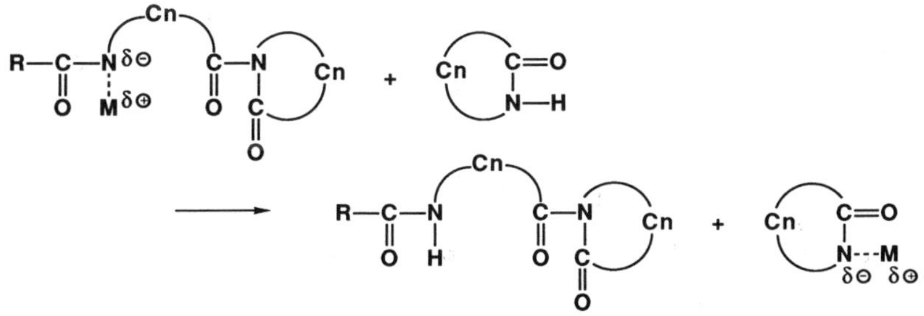

(2.22)

Anionic ring-opening polymerization of lactams makes use of a sodium salt of lactam for the formation of "activated monomer," and an N-acyllactam to be attacked by the activated monomer at the initial stage of the polymerization. Experimental details have been described in the literature for the syntheses of polypentanamide (5-nylon)[108,109] and polybutanamide (4-Nylon).[110] There are various side reactions including those due to the amide anion on the polymer chain, which have been studied and described in detail.[111]

More recent examples include the polymerization of bicyclic lactams. In the anionic polymerization of 2-azabicyclo[2,2,1]heptan-3-one (**13**), a polyamide (**14**) containing cyclopentane rings in the main chain is obtained [Eq. (2.23)].[112] The anionic polymerization of a bicyclic oxalactam 8-oxa-6-

azabicyclo[3,2,1]octan-7-one (**15**) proceeds readily at room temperature to give a polymer (**16**) of high molecular weight[113] [Eq. (2.24)].

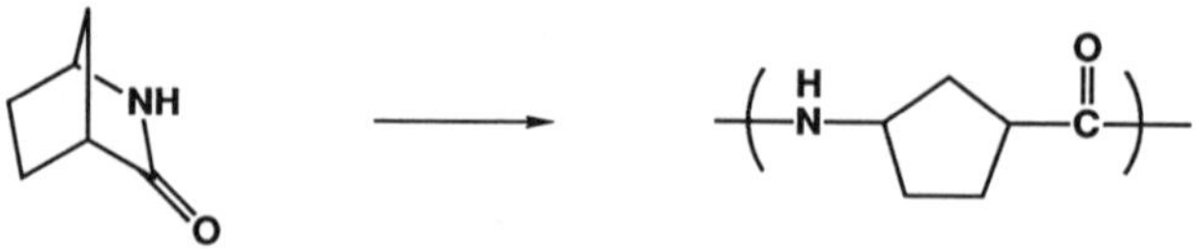

13 **14** (2.23)

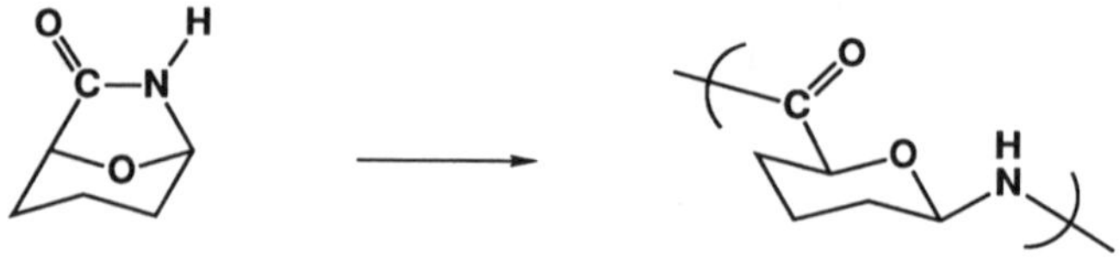

15 **16** (2.24)

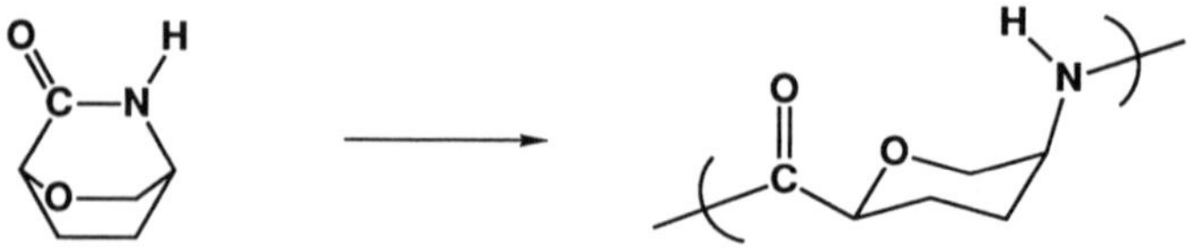

17 **18** (2.25)

The reactivity of another bicyclic oxalactam, 2-oxa-5-azabicyclo[2,2,2]octan-6-one (**17**), in anionic polymerization [Eq. (2.25)] is much lower than that of **15**.[114] Besides the expected *cis* structure of the bonds in the tetrahydropyran ring, the polymer contains 5 ~ 30% of the *trans* structure, indicating the isomerization of the *cis* unit once formed into thermally more stable *trans* unit under the polymerization conditions.

2-Oxa-6-azabicyclo[2,2,2]octan-5-one (**19**), which is isomeric to **17**, does not give polyamide under similar conditions,[115] probably due to the isomerization of the lactam anion to more stable alkoxide anion [Eq. (2.26)].

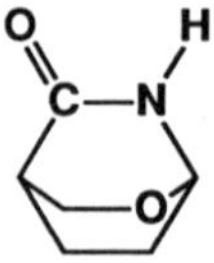

19

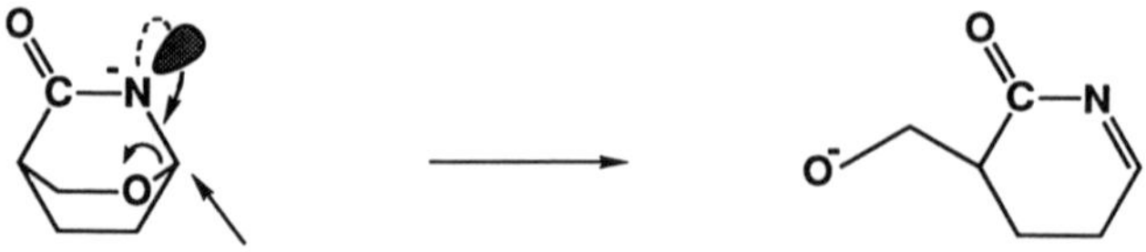

20 **21** (2.26)

2.6. N-CARBOXYANHYDRIDES OF α-AMINO ACIDS (NCA)

Polymerization of α-amino acid N-carboxylic anhydride (NCA) can be initiated by primary amine, tertiary amine, sodium methoxide, and some organometallic compounds [Eq. (2.27)].[116] Much work has been carried out as to the mechanism of the polymerization including the stereochemical aspects. Polymerization of NCA proceeds either by the direct addition of a nucleophile to the carboxylic anhydride group or by the "activated monomer" mechanism similar to lactam polymerizations [Eqs. (2.28) and (2.29)]. NCA monomer with a highly reactive anhydride carbonyl group can be attacked by the activated monomer.

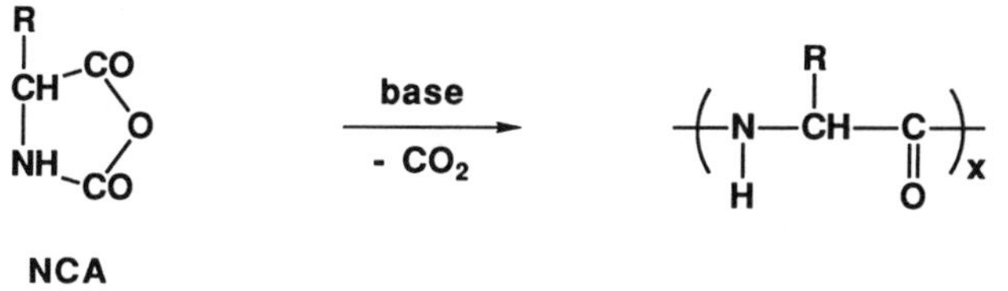

(2.27)

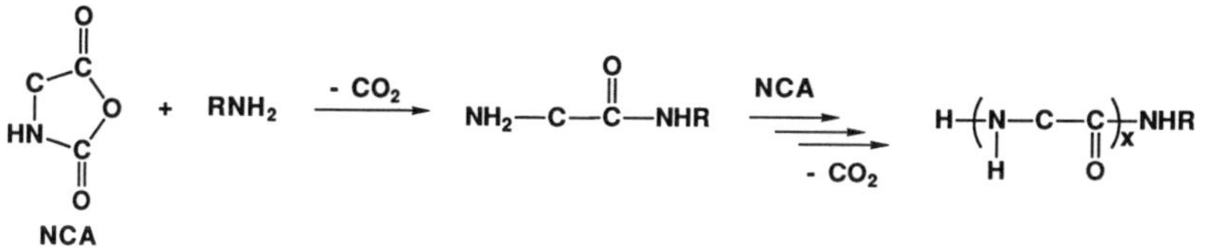

(2.28)

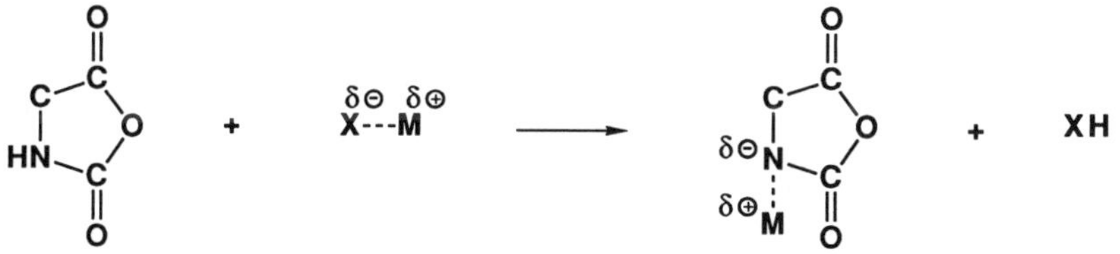

(2.29)

2.6.1. A New Method of Synthesis of NCA

The usual procedure for the synthesis of NCA makes use of phosgene, which is a highly toxic gas, and special precautions should be taken. A more convenient reagent that can be used in place of phosgene is trichloromethyl chloroformate (TCF: $Cl_3C—O—CO—Cl$), which is a liquid, the composition corresponding to the "dimer" of phosgene.[117] A mixture of an amino acid, a 40% excess of TCF, and activated carbon in tetrahydrofuran when heated gives NCA.

More recently, bis(trichloromethyl) carbonate or "triphosgene" has been reported as a good reagent for the synthesis of NCA,[118] even better than TCF [Eq. (2.30)]. The reaction is almost quantitative, and does not require excess of the reagent, which can be handled more easily because of the lower volatility.

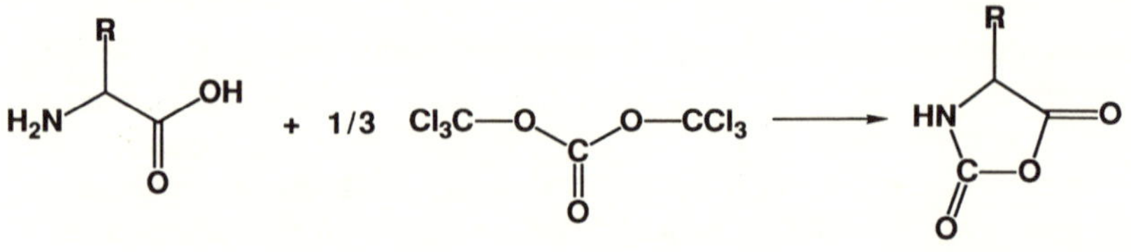

(2.30)

For the synthesis of NCA, a mixture of an amino acid and triphosgene (1.04–1.26 equivalent) in tetrahydrofuran is heated at 40–50 °C until the dissolution of the amino acid, followed by the addition of the mixture to hexane in order to separate and collect NCA. Examples include the NCAs of γ-stearyl-L-glutamate, DL-2-aminostearic acid, γ-benzyl-L-glutamate, *O*-benzyl-L-tyrosine, L-phenylalanine, L-leucine, L-alanine, and DL-valine, the yield in most cases being higher than 80%.

2.6.2. NCA for the Synthesis of Block and Graft Copolymers Containing Polypeptide Chains

NCA has been used for the synthesis of a variety of block and graft copolymers containing polypeptide chains, in connection with various bioapplications of these materials. There are numerous examples,[116] but here the work of the author's group will be described as an example.

We have synthesized novel polyvinyl–polypeptide graft copolymers which are expected to form a microphase separated structure in the film or the membrane, in order to mimic the structure and function of biomembranes composed of lipids and proteins.

The synthetic strategy consists of the incorporation of pendant primary amino groups in a vinyl polymer (resulting in structure **23**), followed by the polymerization of NCA initiated by the amino group to form polypeptide branches in a graft copolymer of structure **24**.[119] Butyl methacrylate was selected as the monomer for the backbone, and benzyl L-glutamate or aspartate NCA to form the branches.

Example 4. Synthesis of Polyvinyl–Polypeptide Graft Copolymer **24**

A tetrahydrofuran solution of benzyl-L-glutamate NCA or benzyl-L-aspartate NCA (0.3–2.0 g) was mixed with a dichloromethane or THF

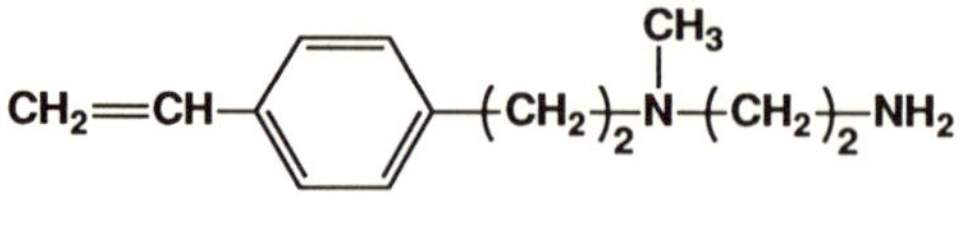

22

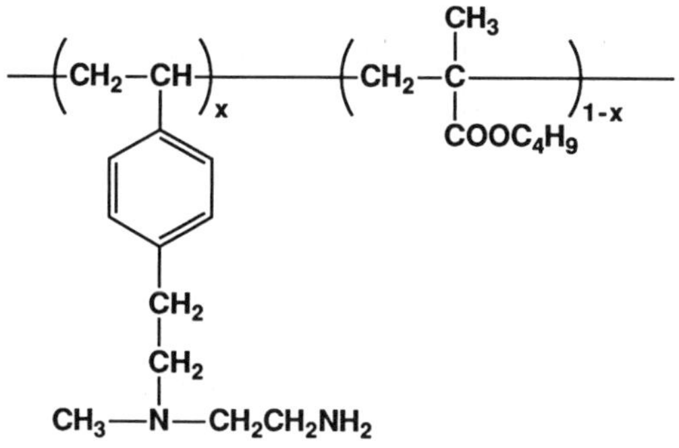

23

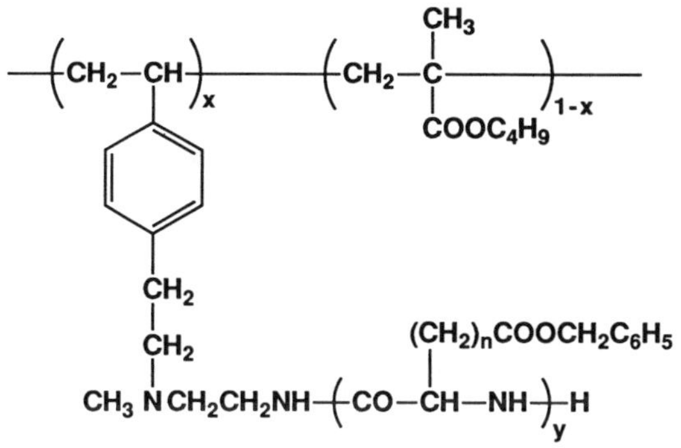

24 (n = 1, 2)

solution (20 mL) of butyl methacrylate–**22** copolymer (**23**) (0.5–1.0 g) at room temperature, and the mixture was stirred for 1 day. The reaction mixture was homogeneous throughout the reaction. Then the reaction mixture was poured into a large excess of cold (2–3 °C) hexane (750–1000 mL), and the precipitated polymer was filtered off and dried in vacuo. The average degree of polymerization of the poly(amino acid) chain grafted onto **24** was determined by ^{1}H NMR; since the content of **22** in **23** is known, the average number of amino acid residues in a side chain of the graft copolymer (**24**) is calculated by using the area ratio of the signal of —$COOCH_2$– of the butyl group in the butyl methacrylate–**22** copolymer backbone to that of benzyl CH_2 in the poly(β-benzyl L-aspartate) (for **24**, $n = 1$) or poly(γ-benzyl L-glutamate) branch (for **24,** $n = 2$).

The membrane from the graft copolymer solution was found to have a microphase separated structure as revealed by electron microscopic studies. When the membrane was subjected to debenzylation of the polypeptide branches under appropriate conditions, the polypeptide microdomains consisting of poly-L-glutamic or aspartic acid became hydrophilic to form a transmembrane pathway for hydrophilic substances. The permeability has been found to be regulated with the conformational change of the polypeptide chains caused by external stimuli such as pH, divalent ions, urea, and quaternary ammonium salts.[120]

2.7. CYCLIC SILOXANES

Representative initiators for the anionic polymerization of cyclic siloxanes are alkali metal compounds such as alkyllithium and potassium hydroxide. The polymerization is usually accompanied by equilibrium involving monomer and polymer. Living polymerization has so far been possible only in the polymerization of hexamethylcyclotrisiloxane initiated with organolithium or lithium silanolate systems.

In connection with recent interests in the synthesis of block and graft copolymers containing polysiloxane chains,[121] the preparation of end-functional polysiloxane with well-defined molecular weight has been the subject of a number of studies. Among these is the recent development in the synthesis of bifunctional polysiloxanes.

An example makes use of a dilithium dimethylsilanediolate (**25**), which has a similar structure as the growing species.[122]

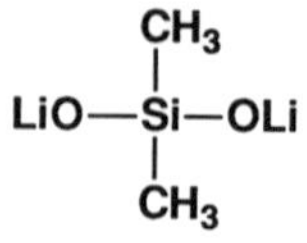

25

Example 5. Polymerization of Hexamethylcyclotrisiloxane

Polymerizations were carried out in anhydrous tetrahydrofuran medium. Dimethylsilanediol was introduced, but remained insoluble. Its metalation was performed by the addition of a stoichiometric amount of butyllithium. The temperature should remain below −40 °C to avoid side reactions of BuLi with tetrahydrofuran. A few drops of styrene serve as colored indicator. Once metalation was completed, the medium became homogeneous and remained colorless. The concentration of active species ranged from 3 to 6 mmol/L. The hexamethylcyclotrisiloxane solution in benzene was then added within a few minutes, and the polymerization was conducted at room temperature. It was allowed to proceed for 3 to 6 h, depending upon the molecular weight desired. Silanolate sites were likely to react—though slowly—with siloxane bonds of the chain, yielding a redistribution of molecular weights. To prevent this, the living ends were deactivated before total conversion was reached.

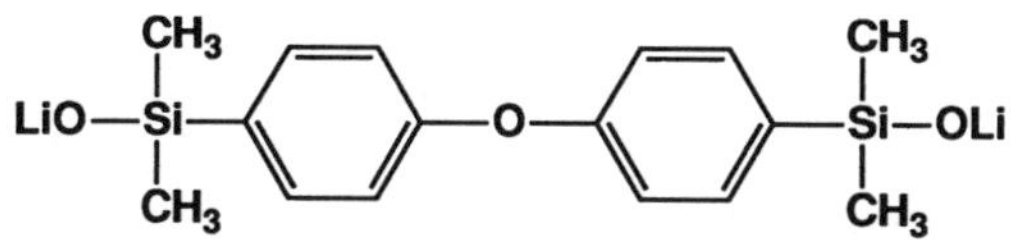

26

Under appropriate conditions polymers with M_n 2500–6500 and M_w/M_n 1.07–1.2 could be obtained. The bifunctional living polymers thus formed were converted to α,ω-diunsaturated polydimethylsiloxane macromonomers by end-capping reactions.

Another example utilizes an initiator system consisting of dilithium salt of bis(*p*-hydroxydimethylsilyl)phenyl ether (**26**) and *N*-methylpyrrolidine in tetrahydrofuran.[123]

Bifunctional macromonomers can be derived from end-capping of the bifunctional living polydimethylsiloxane. Polymers with M_n 3400–8300 are obtained with narrow molecular weight distribution.

2.8. DISILABICYCLO COMPOUNDS

A novel interesting reaction to form polysilane has been reported recently by anionic ring-opening polymerization of a disilabicyclo compound, 1-phenyl-7,8-disilabicyclo[2,2,2]octa-2,5-diene (**27**) (Eq. (2.31)).[124] The monomer can be regarded as a masked disilene (Si═Si) which is usually unstable.

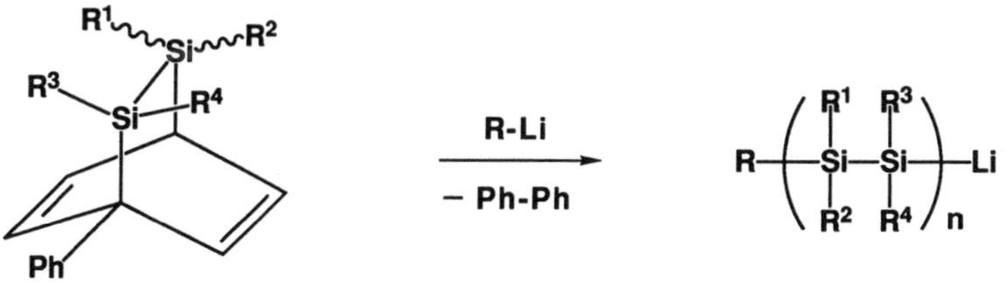

(2.31)

27

e.g., $R^1 = R^2 = R^3 = CH_3$, $R^4 = n\text{-}C_4H_9$

Example 6. Polymerization of Disilabicyclo Compound **27**

The polymerization of **27** was carried out in breakable seal apparatus under high vacuum conditions ($<10^{-5}$ mm Hg). To a tetrahydrofuran solution of the monomer was added a catalytic amount of alkyllithium such as butyllithium, and the mixture was kept in the dark at −110 °C, then brought up to room temperature, over a period of several hours. By the addition of the initiator the transparent mixture turned red, and the color disappeared within 30 minutes. Finally the reaction was quenched by the addition of ethanol, and the polymer was separated by reprecipitation followed by freeze-drying. Another product was biphenyl. The molecular weight of the polymer was 3.4–14.0 × 10^4, yield being 52–79%.

Polysilanes obtained by this method are characterized by relatively narrow molecular-weight distribution when compared with those prepared by the classical method of condensation of dialkyldichlorosilane with sodium. M_w/M_n of polysilanes obtained here ranged from 1.3 to 1.9.

The polymerization is considered to proceed by the attack of silyl anion to Si of the cyclic compound. Even after the consumption of all monomers, the polymer anion still remains active, and the addition of methyl methacrylate leads to the formation of a block copolymer.

REFERENCES

1. *Ring-Opening Polymerization,* ed. by K. C. Frisch and S. L. Reegen (Marcel Dekker, New York, 1969); *Ring-Opening Polymerization,* ed. by K. J. Ivin and T. Saegusa (Elsevier Applied Science Publishers, London, 1984); *Comprehensive Polymer Science,* ed. by G. C. Eastmond, A. Ledwith, S. Russo, and P. Sigwalt (Pergamon Press, Oxford, 1989), Vol. 3, pp. 457–569.
2. T. Tsuruta and Y. Kawakami, in *Comprehensive Polymer Science,* ed. by G. C. Eastmond, A. Ledwith, S. Russo, and P. Sigwalt (Pergamon Press, Oxford, 1989), Vol. 3, pp. 460, 463.
3. C. C. Price and D. D. Carmelite, *J. Am. Chem. Soc.* **88**, 4039–4044 (1966).
4. Y. Ishii and S. Sakai, in *Ring-Opening Polymerization,* ed. by K. C. Frisch and S. L. Reegen (Marcel Dekker, New York, 1969), pp. 12–109.
5. S. Inoue and T. Aida, in *Ring-Opening Polymerization,* ed. by K. J. Ivin and T. Saegusa (Elsevier, London and New York, 1984), Vol. 1, pp. 185–298.
6. S. Boileau, in *Comprehensive Polymer Science,* ed. by G. C. Eastmond, A. Ledwith, S. Russo, and P. Sigwalt (Pergamon Press, Oxford, 1989), Vol. 3, pp. 467–487.
7. D. M. Simons and J. J. Vervanc, *J. Polym. Sci.* **44,** 303–311 (1960).
8. T. Aida, R. Mizuta, Y. Yoshida, and S. Inoue, *Makromol. Chem.* **182**, 1073–1079 (1981).
9. T. Aida and S. Inoue, *Makromol. Chem., Rapid Commun.* **1**, 677–680 (1980).
10. T. Aida and S. Inoue, *Macromolecules* **14**, 1162–1166 (1981).
11. T. Yasuda, T. Aida, and S. Inoue, *Makromol. Chem., Rapid Commun.* **3,** 585–588 (1982).
12. L. R. Trofimoff, T. Aida, and S. Inoue, *Chem. Lett.* 991–994 (1987).
13. M. Kuroki, T. Aida, and S. Inoue, *J. Am. Chem. Soc.* **109,** 4737–4738 (1987).
14. T. Aida and S. Inoue, *J. Am. Chem. Soc.* **107**, 1358–1364 (1985).
15. T. Aida, M. Ishikawa, and S. Inoue, *Macromolecules* **19**, 8–13 (1986).
16. T. Aida and S. Inoue, *Macromolecules* **14**, 1166–1169 (1981).
17. H. Sugimoto, T. Aida, and S. Inoue, *Macromolecules* **23**, 2869–2875 (1990).
18. S. Asano, T. Aida, and S. Inoue, *J. Chem. Soc., Chem. Commun.* 1148–1149 (1985).
19. T. Aida, Y. Maekawa, S. Asano, and S. Inoue, *Macromolecules* **21**, 1195–1202 (1988).
20. S. Inoue and T. Aida, *Makromol. Chem., Macromol. Symp.* **6**, 217–224 (1986).
21. T. Aida, A. Takemura, M. Fuse, and S. Inoue, *J. Chem. Soc., Chem. Commun.* 391–393 (1988).
22. M. Kuroki, T. Aida, and S. Inoue, *Makromol. Chem.* **189**, 1305–1313 (1988).
23. Y. Watanabe, T. Aida, and S. Inoue, *Macromolecules* **23**, 2613–2617 (1990).

24. T. Uryu, H. Shimazu, and K. Matsuzaki, *J. Polym. Sci., Polym. Lett. Ed.* **11,** 275–278 (1973).
25. C. C. Price and M. Osgan, *J. Am. Chem. Soc.* **78**, 690–691 (1956).
26. R. O. Colclough, G. Gee, W. C. E. Higginson, J. B. Jackson, and M. Litt, *J. Polym. Sci.* **34**, 171–179 (1959).
27. M. Osgan and C. C. Price, *J. Polym. Sci.* **34**, 153–160 (1959).
28. Z. Jedliński, A. Dworak, and M. Bero, *Makromol. Chem.* **180**, 949–952 (1979).
29. J. Furukawa, T. Tsurata, R. Sakata, and T. Saegusa, *Makromol. Chem.* **32**, 90–94 (1959).
30. M. Ishimori and T. Tsuruta, *Makromol. Chem.* **64**, 190–206 (1963).
31. E. J. Vandenberg, *J. Polym. Sci.* **47**, 486–489 (1960).
32. N. Oguni, K. Lee, and H. Tani, *Macromolecules* **5**, 819–820 (1972).
33. J. Kasperczyk and Z. J. Jedliński, *Makromol. Chem.* **187**, 2215–2221 (1986).
34. A. Dworak, *Makromol. Chem., Rapid Commun.* **6**, 665–669 (1985).
35. T. Tsurura, *Pure Appl. Chem.* **53**, 1745–1751 (1981).
36. R. A. Shelden, T. Fueno, T. Tsunetsugu, and J. Furukawa, *J. Polym. Sci., Part B,* **3**, 23–26 (1965).
37. S. Inoue, T. Tsuruta, and J. Furukawa, *Makromol. Chem.* **53**, 215–218 (1962).
38. V. Vincens, A. L. Borgne, and N. Spassky, *Makromol. Chem., Rapid Commun.* **10**, 623–628 (1989).
39. M. Ishimori, T. Hagiwara, T. Tsuruta, Y. Kai, N. Yasuoka, and N. Kasai, *Bull. Chem. Soc. Jpn.* **49**, 1165–1166 (1976).
40. M. Ishimori, T. Hagiwara, and T. Tsuruta, *Makromol. Chem.* **179**, 2337–2342 (1978).
41. T. Hagiwara, M. Ishimori, and T. Tsuruta, *Makromol. Chem.* **182**, 501–511 (1981).
42. F. A. Bovey, *J. Polym. Sci.* **44**, 173–182 (1960).
43. N. Takeda and S. Inoue, *Makromol. Chem.* **179**, 1377–1381 (1978).
44. A. L. Borgne, N. Spassky, C. L. Jun, and A. Montaz, *Makromol. Chem.* **189**, 637–650 (1988).
45. A. Sato, T. Hirano, and T. Tsuruta, *Makromol. Chem.* **176**, 1187–1190 (1975).
46. J. Lustoň and F. Vašš, *Adv. Polym. Sci.* **56**, 91–133 (1984).
47. J. Lustoň and Z. Manásek, *J. Macromol. Sci.* **A21**, 983–994 (1978).
48. T. Aida, K. Sanuki, and S. Inoue, *Macromolecules* **18**, 1049–1055 (1985).
49. T. Tsuruta, K. Matsuura, and S. Inoue, *Makromol. Chem.* **81**, 258–260 (1965).
50. S. Inoue, in *Organic and Bio-organic Chemistry of Carbon Dioxide,* ed. by S. Inoue and N. Yamazaki (Kodansha, Tokyo, 1981), pp. 167–176.
51. S. Inoue and T. Aida, in *Ring-Opening Polymerization,* ed. by K. J. Ivin and T. Saegusa (Elsevier, London, 1984), Vol. 1, pp. 283–287.
52. S. Inoue, T. Takada, and H. Tatsu, *Makromol. Chem., Rapid Commun.* **1**, 775–777 (1980).
53. K. Soga, E. Imai, and I. Hattori, *Polym. J.* **13**, 407–410 (1981).
54. Y. Hino, Y. Yoshida, and S. Inoue, *Polym. J.* **16**, 159–163 (1984).
55. H. Koinuma and H. Hirai, *Makromol. Chem.* **178**, 1283–1294 (1977).
56. M. Takanashi, Y. Nomura, Y. Yoshida, and S. Inoue, *Makromol. Chem.* **183**, 2085–2092 (1982).
57. T. Kawaguchi, M. Nakano, K. Juni, S. Inoue, and Y. Yoshida, *Chem. Pharm. Bull.* **30**, 1517–1520 (1982).
58. T. Kojima, M. Nakano, K. Juni, S. Inoue, and Y. Yoshida, *Chem. Pharm. Bull.* **32**, 2795–2802 (1984).
59. K. Soga, I. Hattori, J. Kinoshita, and S. Ikeda, *J. Polym. Sci., Polym. Chem. Ed.* **15**, 745–753 (1977).
60. K. Soga, K. Kiyohara, I. Hattori, and S. Ikeda, *Makromol. Chem.* **181**, 2151–2159 (1980).
61. N. Adachi, Y. Kida, and K. Shikata, *J. Polym. Sci., Polym. Chem. Ed.* **15**, 937–944 (1977).

62. J. Furukawa, Y. Iseda, T. Saegusa, and H. Fujii, *Makromol. Chem.* **89**, 263–268 (1965).
63. T. Tsuruta, K. Matsuura, and S. Inoue, *Makromol. Chem.* **81**, 258–260 (1965).
64. K. Lee, M. Yanagida, Y. Kanesaka, and Y. Minoura, *Polymer* **18**, 308–310 (1977).
65. J. Furukawa, S. Yamashita, M. Maruhashi, and K. Harada, *Makromol. Chem.* **85**, 80–90 (1965).
66. P. Sigwalt and N. Spassky, in *Ring-Opening Polymerization*, ed. by K. J. Ivin and T. Saegusa (Elsevier, London and New York, 1984), Vol. 2, pp. 603–714.
67. F. G. Bordwell, H. M. Anderson, and B. M. Pitt, *J. Am. Chem. Soc.* **76**, 1082–1085 (1954).
68. S. Boileau and P. Sigwalt, *Bull. Soc. Chim. France* 1418–1421 (1968).
69. S. Boileau, G. Champetier, and P. Sigwalt, *Makromol. Chem.* **69**, 180–192 (1963).
70. J. C. Favier, S. Boileau, and P. Sigwalt, *Europ. Polym. J.* **4**, 3–12 (1968).
71. S. Boileau, *ACS, Symp. Ser.* **166**, 283–305 (1981).
72. P. Guerin, S. Boileau, and P. Sigwalt, *Eur. Polym. J.* **10**, 13–18 (1974).
73. W. Cooper, *Br. Polym. J.* **3**, 28–35 (1971).
74. T. Aida, K. Kawaguchi, and S. Inoue, *Macromolecules* **23**, 3887–3892 (1990).
75. T. Tsuruta and Y. Kawakami, in *Comprehensive Polymer Science,* ed. by G. C. Eastmond, A. Ledwith, S. Russo, and P. Sigwalt (Pergamon Press, Oxford, 1989), Vol. 3, pp. 489–500.
76. M. Sepulchre, N. Spassky, and P. Sigwalt, *Isr. J. Chem.* **15**, 33–38 (1976/77).
77. N. Spassky, A. Leborgne, and M. Sepulchre, *Pure Appl. Chem.* **53**, 1735–1744 (1981).
78. N. Spassky, A. Leborgne, A. Momtaz, and M. Sepulchre, *J. Polym. Sci., Polym. Chem. Ed.* **18**, 3089–3099 (1980).
79. M. Sepulchre and N. Spassky, *Makromol. Chem., Rapid Commun.* **2**, 261–266 (1981).
80. M. Sepulchre, *Makromol. Chem.* **188**, 1583–1596 (1987).
81. W. Kuran, A. Rokicki, and W. Wielgopolan, *Makromol. Chem.* **179**, 2545–2548 (1978).
82. G. A. Razuvaev, V. S. Etlis, and L. N. Gribov, *Zh. Obshch. Khim.* **33**, 1366 (1963).
83. K. Soga, H. Imamura, M. Sato, and S. Ikeda, *J. Polym. Sci., Polym. Chem. Ed.* **14**, 677–684 (1976).
84. A. Duda and S. Penczek, *Makromol. Chem.* **181**, 995–1001 (1980).
85. T. Yasuda, T. Aida, and S. Inoue, *Macromolecules* **16**, 1792–1796 (1983).
86. T. Yasuda, T. Aida, and S. Inoue, *Macromolecules* **17**, 2217–2222 (1984).
87. K. Shimasaki, T. Aida, and S. Inoue, *Macromolecules* **20**, 3076–3080 (1987).
88. M. Endo, T. Aida, and S. Inoue, *Macromolecules* **20**, 2982–2988 (1987).
89. M. Kuroki, S. Nashimoto, T. Aida, and S. Inoue, *Macromolecules* **21**, 3114–3115 (1988).
90. Ph. Teyssié, J. P. Bioul, A. Hamitou, J. Heuschen, L. Hocks, R. Jérôme, and T. Ouhadi, *ACS Symp. Ser.* **59**, 165–177 (1977).
91. A. Duda, *Prepr. 9th International Symp. on Cationic Polym. and Related Ionic Processes,* Strasbourg, France (1989), p. 79.
92. J. Heuschen, R. Jérôme, and Ph. Teyssié, *Macromolecules* **14**, 242–246 (1981).
93. Z. Jedliński, P. Kurcok, and M. Kowalczuk, *Macromolecules* **18**, 2679–2683 (1985).
94. Z. Jedliński, A. Misiolek, and P. Kurcok, *J. Org. Chem.* **54**, 1500–1501 (1989).
95. A. Leborgne, D. Grenier, R. E. Prud'homme, and N. Spassky, *Eur. Polym. J.* **17**, 1103–1106 (1981).
96. D. E. Agostini, J. B. Lando, and J. R. Shelton, *J. Polym. Sci., Polym. Chem. Ed.* **9**, 2775–2787 (1971).
97. K. Teranishi, M. Iida, T. Araki, S. Yamashita, and H. Tani, *Macromolecules* **7**, 421–427 (1974).
98. A. Leborgne, N. Spassky, and P. Sigwalt, *Prepr. 27th International Symp. on Macromolecules,* Strasbourg, France (1981), Vol. 1, p. 152.
99. P. Guerin, J. Francillette, C. Braud, and M. Vert, *Makromol. Chem., Macromol. Symp.* **6**, 305–314 (1986).

100. E. E. Kohn, J. G. van Ommen, and J. Feijen, *Eur. Polym. J.* **19**, 1081–1088 (1983) and references cited therein.
101. K. Shimasaki, T. Aida, and S. Inoue, *Polym. Prepr. Jpn.* **37**, E 106 (1988).
102. X. D. Feng, C. X. Song, and W. Y. Chen, *J. Polym. Sci., Polym. Lett. Ed.* **21**, 593–600 (1983).
103. L. Vogdanis and W. Heitz, *Makromol. Chem., Rapid Commun.* **7**, 543–547 (1986).
104. H. Keul, R. Bächer, and H. Höcker, *Makromol. Chem.* **187**, 2579–2589 (1986).
105. N. Yonezawa, F. Toda, and M. Hasegawa, *Makromol. Chem., Rapid Commun.* **6**, 607–611 (1985).
106. J. Helder, F. E. Kohn, S. Sato, J. W. Vanden Berg, and J. Feijen, *Makromol. Chem., Rapid Commun.* **6**, 9–14 (1985).
107. J. Helder, J. Feijen, S. J. Lee, and S. W. Kim, *Makromol. Chem., Rapid Commun.* **7**, 193–198 (1986).
108. H. Sekiguchi and B. Coutin, *Macromol. Synth.* **6**, 57–61 (1977).
109. T. Konomi and H. Tani, *Macromol. Synth.* **5**, 73–76 (1974).
110. R. W. Wynn, S. A. Glickman, and M. E. Chiddix, *Macromol. Synth.* **3**, 106–108 (1969).
111. J. Sebenda, in *Comprehensive Polymer Science,* ed. by G. C. Eastmond, A. Ledwith, S. Russo, and P. Sigwalt (Pergamon Press, Oxford, 1989), Vol. 3, pp. 511–530.
112. H.-N. Cho, K.-Y. Choi, and S.-K. Choi, *J. Polym. Sci., Polym. Chem. Ed.* **23**, 623–634 (1985).
113. K. Hashimoto and H. Sumitomo, *Adv. Polym. Sci.* **64**, 63–91 (1985).
114. M. Okada, H. Sumitomo, H. Mori, and H. K. Hall, Jr., *Polym. Prepr. Jpn.* **36**, 280 (1987).
115. M. Okada, H. Sumitomo, T. Sassa, and H. K. Hall, Jr., *Polym. Prepr. Jpn.* **36**, 1535–1537 (1987).
116. H. R. Kricheldorf, in *Comprehensive Polymer Science,* ed. by G. C. Eastmond, A. Ledwith, S. Russo, and P. Sigwalt (Pergamon Press, Oxford, 1989), Vol. 3, pp. 531–551.
117. R. Katakai and Y. Iizuka, *J. Org. Chem.* **50**, 715–716 (1985).
118. W. H. Daly and D. Poché, *Tetrahedron Lett.* **29**, 5859–5862 (1988).
119. S. Higuchi, T. Mozawa, M. Maeda, and S. Inoue, *Macromolecules* **19**, 2263–2267 (1986).
120. D. Chung, S. Higuchi, M. Maeda, and S. Inoue, *J. Am. Chem. Soc.* **108**, 5823–5826 (1986).
121. I. Yilgör and J. E. McGrath, *Adv. Polym. Sci.* **86**, 1–86 (1988).
122. Y. Gnanou and P. Rempp, *Makromol. Chem.* **189**, 1997–2005 (1988).
123. H. Kazama, Y. Tezuka, and K. Imai, *Polym. Bull.* **21**, 31–37 (1989).
124. K. Sakamoto, K. Obata, H. Hirata, M. Nakajima, and H. Sakurai, *J. Am. Chem. Soc.* **111**, 7641–7643 (1989).

3

Cationic Ring-Opening Polymerization

Eric J. Goethals and Ronny R. De Clercq

3.1. INTRODUCTION

Ring-opening polymerization of (hetero)cyclic monomers by a cationic mechanism provides the possibility to produce a very large number of polymers with a broad spectrum of physicochemical properties. As shown in Table 3.1, cationic ring-opening polymerization (CROP) can be applied to most types of heterocycles on condition that the monomers contain enough ring strain. Thus it is possible to produce polyethers, polyacetals, polysulfides, polyamines, polyesters, polyamides, polysiloxanes, polyphosphates, and polyphosphazenes. In each of these groups the properties of the polymers are dictated by the nature of the heteroatoms, but they can be additionally influenced by the presence of one or more substituents on the ring or by the use of bicyclic structures.

In modern synthetic polymer chemistry, not only is a given backbone—providing specific properties—of importance, but also the control of molecular weight, dispersity, and often end-group functionality, so that the polymer can be reacted further to yield more sophisticated structures such as block or graft polymers, or polymer networks. In this respect, cationic ring-opening polymerizations have been successfully used notwithstanding some intrinsic drawbacks of the cationic polymerization mechanism, as will

---

Eric J. Goethals and Ronny R. De Clercq • Laboratory of Organic Chemistry, Polymer Division, University of Ghent, Krijgslaan 281 (S4), B-9000 Ghent, Belgium.

New Methods for Polymer Synthesis, edited by W. J. Mijs. Plenum Press, New York, 1992.

TABLE 3.1. Monomers that Polymerize Cationically

Ethers	
Acetals	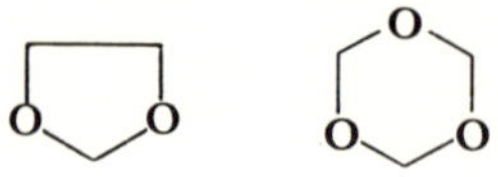
Lactones	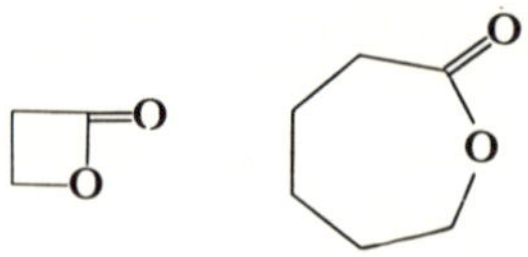
Sulfides	
Amines	
Iminoethers	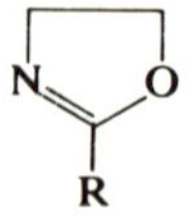
Siloxanes	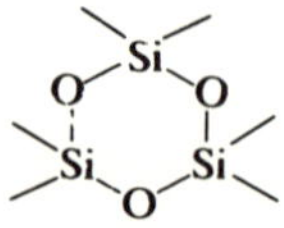
Phosphazenes	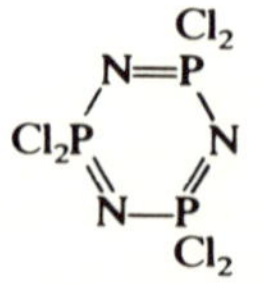

be discussed below. The examples of tailor-made polymers to be given in this chapter are indeed the result of research efforts primarily directed toward gaining detailed knowledge of the mechanism of these polymerizations, from which conditions were deduced eventually leading to the control of molecular weight, dispersity, and end groups.

3.2. GENERAL MECHANISM OF CATIONIC RING-OPENING POLYMERIZATIONS

With the restriction that it is only partly possible to depict "a" mechanism for such a variety of monomers, the following reactions are quite general and can be accepted as valid for most cases.

3.2.1. Initiation

The formation of the active species in CROPs consists of an alkylation of the monomer with the formation of a cyclic onium ion. This alkylation can be a one-step reaction if the initiator is an alkylating agent strong enough to alkylate the heteroatom of the monomer.

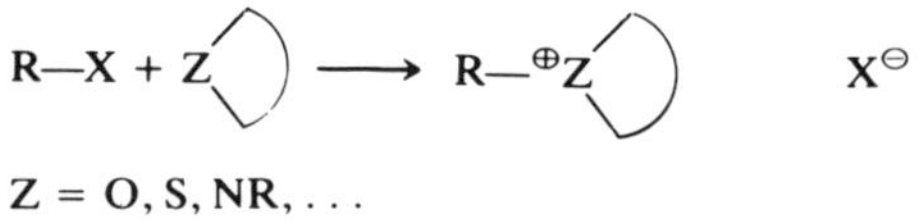

Z = O, S, NR, . . .

With other initiators the alkylation occurs in two steps, for example, with a proton acid: protonation and a first ring-opening by monomer.

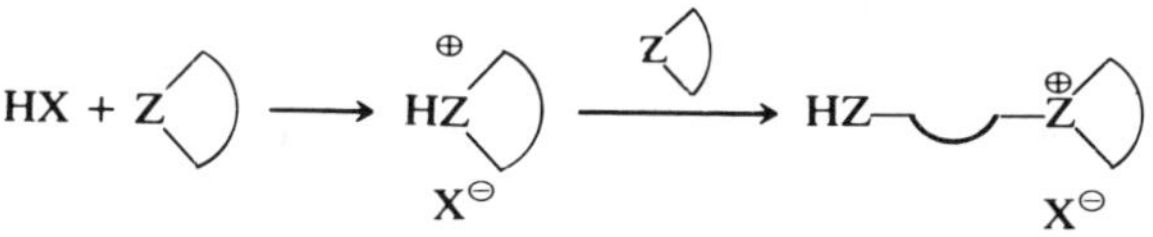

Typical examples of direct initiators are the trialkyloxonium ions ($R_3O^{\oplus}MX_n^{\ominus}$) and esters of strong acids such as FSO_3CH_3 and $CF_3SO_3CH_3$. The ability to initiate will depend on the electrophilicity of the initiator relative to the nucleophilicity of the monomer. Thus, alkyl tosylates will initiate the polymerization of oxazolines and amines but not of ethers or acetals.

Proton acids with nonnucleophilic counterions can be used for the polymerization of most heterocycles. Typical examples are perchloric acid and trifluoromethane sulfonic acid. Since most polymerizations have to be carried out in dry conditions the former is, due to its explosion risks when dry, not often used.

An interesting class of initiators are the oxocarbenium ions. These are generally prepared *in situ* or just prior to polymerization by mixing equimolar amounts of an acylchloride and a silver salt with a non-nucleophilic counterion, for instance

$$R{-}\overset{\overset{\displaystyle O}{\|}}{C}{-}Cl + AgSbF_6 \longrightarrow RC\overset{\oplus}{\equiv}O \quad SbF_6^{\ominus} + AgCl\swarrow$$

These initiators are very active, i.e., they generally give a fast and quantitative initiation, and they allow one to introduce functional end groups by choosing the R.

A short note should be added here on the use of Lewis acids as initiators. It is well established now that these compounds can initiate either through "cocatalysis" with a proton donor such as water, or by direct addition to the heteroatom of the monomer. In the first case the real initiation reaction is by protonation. In the second case, which probably occurs only with highly strained monomers such as epoxides, the first-formed zwitter ionic structure can give rise to exchange reactions producing ill-defined counterions and uncontrollable concentrations of active species.

3.2.2. Propagation

The propagation reactions in CROPs are generally of the S_N2 type whereby the heteroatom of the monomer is the nucleophile and the cyclic onium ion the electrophile:

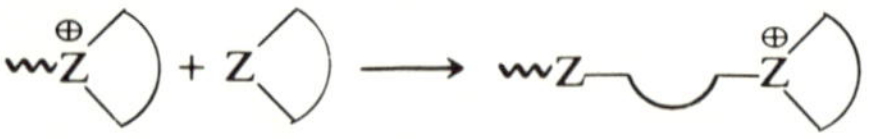

In a few cases, the reaction is believed to occur (at least partially) in two steps: ring opening forming a carbenium ion followed by combination of the carbenium ion with the monomer. Such an S_N1 type of reaction has been proposed for polymerization of monomers which can give rise to a stabilized carbenium ion such as α-gem-disubstituted monomers or acetals:

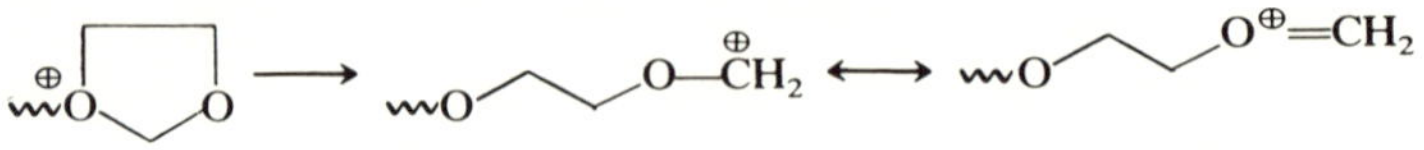

Some systems are known in which the propagation reaction is accompanied by an isomerization. A typical example is that of the oxazolines where the imino-ether structure of the monomer is isomerized to an amide function in the polymer:

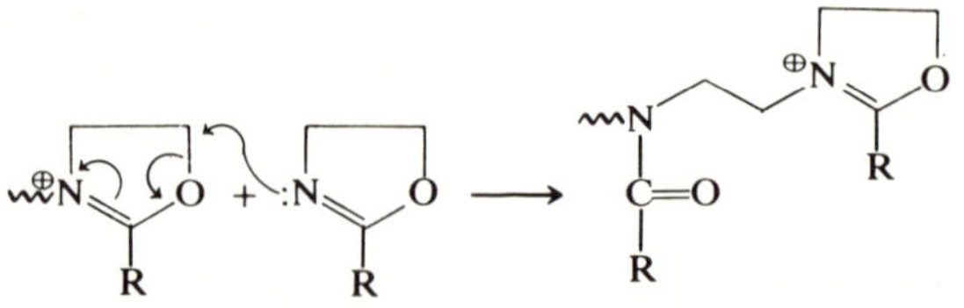

Another interesting example is that of some spiro compounds such as spiro *ortho*-esters leading to ether–ester (carbonate) structures:[1]

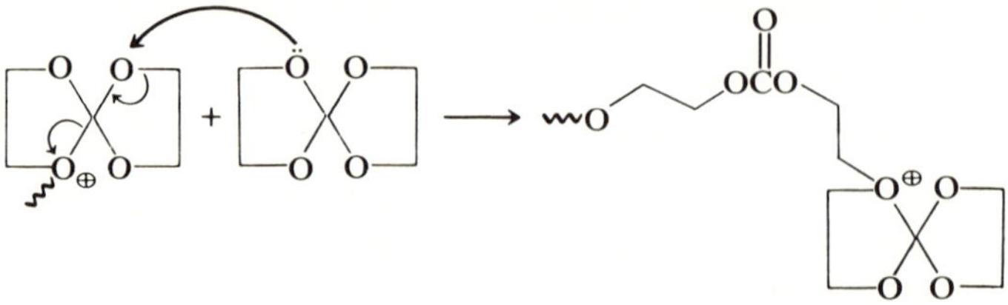

3.2.3. Termination

The active species in CROP can undergo a number of reactions which lead to inactive species. A first reaction is that with counterions. Counterions with a nucleophilicity higher than that of the monomer will lead to nonionic, less reactive or even completely unreactive end groups. In the

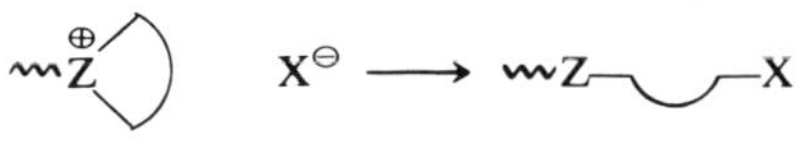

latter case the reaction is a real termination. Most initiator systems therefore use complex counterions without nucleophilicity, such as $BF_4^{\ominus}$, $SbF_6^{\ominus}$, $SbCl_6^{\ominus}$, $AsF_6^{\ominus}$, or $PF_6^{\ominus}$. However, even with these counterions termination can occur by transfer of a halogen ion to the active species. This has been shown to occur, for example, between $SbCl_6^{\ominus}$ and thiiranium ions[2] and between $BF_4^{\ominus}$ and oxonium ions:[3]

$$R_3O^{\oplus} + BF_4^{\ominus} \rightarrow RF + R_2O + BF_3$$

Such reactions do not occur with triflate anions ($CF_3SO_3^{\ominus}$) which explains the growing success of triflate-based initiator systems.

Due to the fact that the polymer chains formed by CROP contain the same heteroatomic functions as the starting monomer (except for the isomerization polymerizations, see above) there will always be a competition between monomer and polymer to react with the active species (Scheme 3.1).

Reaction with a polymer leads to an onium ion that has no ring strain and therefore shows less reactivity than the active species. In many cases these species are still able to reinitiate and are then designated as "dormant" species. In other cases, however, these species are so unreactive that the reaction can be considered to be a real termination reaction. In the former case, the polymerization slows down because of the (sometimes important) decrease in concentration of the real active species, but the polymerization will continue to complete conversion (or to the thermodynamically allowed conversion). In the second case, the real termination leads to limited conversions the values of which depend on the values of k_p,

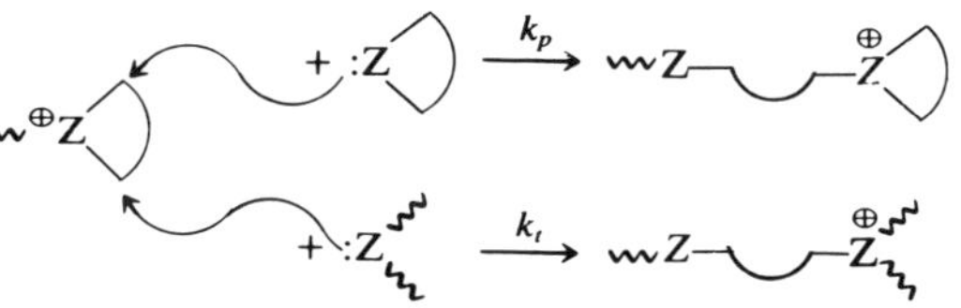

SCHEME 3.1.

k_t, and initiator concentration. Polymerizations showing such "spontaneous termination" have been called "suicidal" polymerizations, because the active species of the polymerization is killed by its own polymer.

Polymerizations which show a temporary, i.e., reversible, termination could, at first sight, be considered as living systems. Indeed, each dormant species being woken up at regular intervals, and all species being subjected to the same equilibrium, one would expect that, except for the rate, the polymerization has the characteristics of a living system. However, as will be described in the next paragraph, this continuous termination–reinitiation cycle leads to a broadening of the molecular weight distribution (MWD), to uncontrolled end-group functionalities, and to the formation of substantial amounts of cyclic oligomers.

Polymerizations which do not show spontaneous termination may lead to living polymerizations and are interesting from a synthetic point of view because, given a well-defined initiator, they provide the possibility to prepare polymers with predictable molar masses and narrow MWD.

Next to the "spontaneous" termination, which is due to the similar nucleophilicities of monomer and polymer, one must also consider the intentional termination that consists in adding a reagent which is more nucleophilic than the monomer to the polymerization mixture, whereby the reaction is stopped and an end group is introduced. This "end capping" is of particular interest for the synthesis of telechelic polymers and macromonomers, i.e., polymers with well-defined functional end groups.

3.2.4. Chain Transfer and Cyclic Oligomer Formation

In ROPs where the occurrence of successive termination and reinitiation reactions is important, i.e., where the rates of these reactions are comparable with the rate of propagation, two important phenomena are observed: broadening of the MWD and cyclic oligomer formation.

Broadening of the MWD is the consequence of the random attack of active species with heteroatoms of polymer chains followed by the random reinitiation (Scheme 3.2). This reaction scheme also shows that the

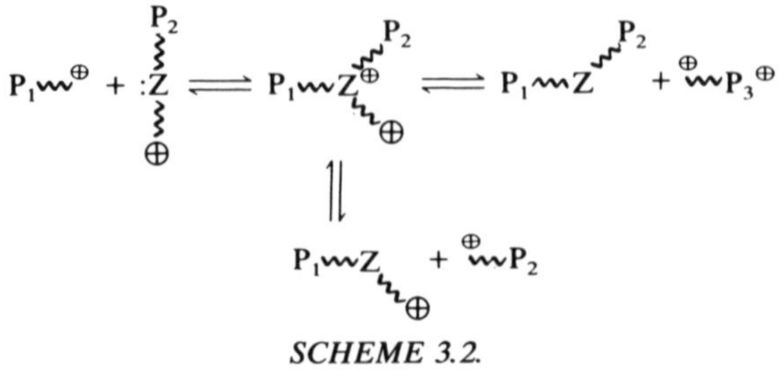

SCHEME 3.2.

originally monofunctionally growing polymer is partially transformed into "dead" and bifunctionally growing polymer.

The formation of cyclic oligomers is due to the same kind of reaction, but now taking place intramolecularly:

Depending on the respective rate constants, these cycles may form during or after the polymerization. In the latter case polymer can be isolated by deactivation of the active species when the polymerization is nearly finished and the degradation has hardly begun. That cyclic oligomer formation is a frequently observed phenomenon in cationic ROP is illustrated in Table 3.2.

The occurrence of rapid exchange reactions between onium ions and heteroatoms of polymer chains can be exploited to control the molecular weight or to introduce well-defined end groups on the polymer chains by

TABLE 3.2. Cyclic Oligomers in Cationic ROP[a]

Monomer	Cyclic oligomers
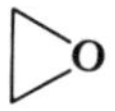	Dimer
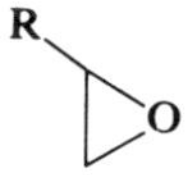	Dimers, tetramers (predominant), pentamers
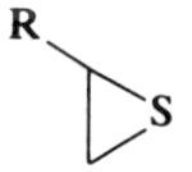	Dimers, tetramers 1,2,5-trithiepanes + alkene
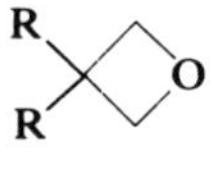	Trimer, tetramer (predominant) and higher cyclics
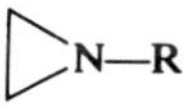	Dimer, tetramer
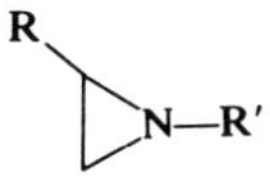	Tetramers, pentamers
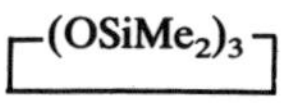	Oligomers containing from 6 to 30 siloxane units

[a] A more complete list can be found in Ref. 4.

carrying out the polymerization in the presence of a low-molecular-weight compound containing the same heteroatom(s) as the monomer. This has been demonstrated experimentally for the polymerization of cyclic siloxanes[5] and acetals.[6]

3.2.5. *The Activated Monomer Mechanism*

Recently, polymerizations occurring via a propagation mechanism involving an activated monomer (AM) instead of an active chain end (ACE) have been elaborated. The main purpose of this technique was to avoid the formation of cyclic oligomers. This polymerization is therefore of special interest for the polymerization of epoxides, where cyclic oligomer formation is often predominant. The principle of the AM mechanism is that the monomer and not the polymer chain end is put in the form of an oxonium ion. The polymerization requires an alcohol as initiator and an acid as activator. Applied to ethylene oxide the mechanism is shown in Scheme 3.3.

If the polymerization occurs by AM mechanism only, at no moment is there any tertiary oxonium ion present and therefore the formation of dioxane or other cyclics (which are necessarily formed via tertiary oxonium intermediates) is inhibited.

The competitive ACE reaction would take place if the protonated monomer is attacked by monomer:

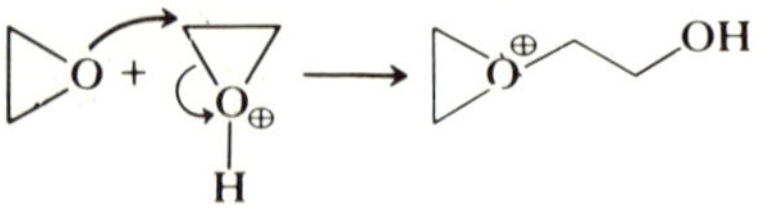

In this way, a tertiary oxonium ion is formed through which cyclic oligomer formation can be initiated. Another possibility to form a tertiary oxonium is

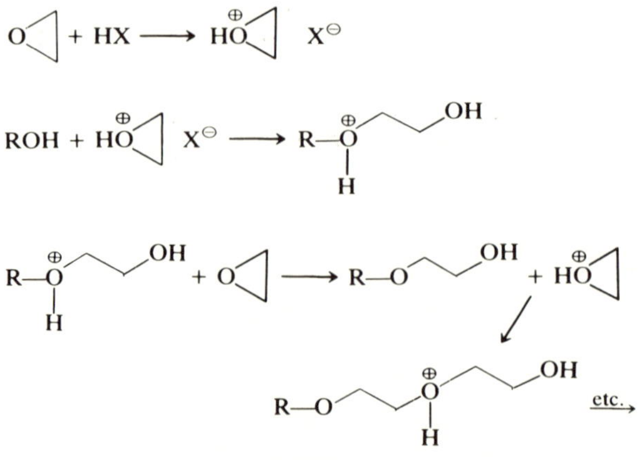

SCHEME 3.3.

that the protonated monomer is attacked by an ether from the polymer chain:

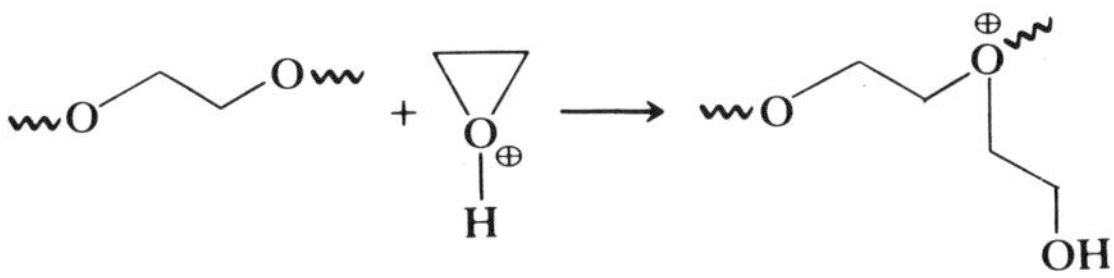

The first reaction can be minimized by keeping the instantaneous monomer concentration low by slow addition of monomer to the polymerizing mixture. The second reaction is limited if monomers with sterically hindering substituents are used and if the ratio [ether]/[OH], i.e., the DP of the polymers, is kept low. Therefore, the AM mechanism is interesting for the production of polymers with relatively low molecular weights derived from substituted monomers. The kinetics of polymerization by the AM mechanism have been described recently.(7)

A further characteristic of the AM mechanism is that by proper choice of the initiator, ROH, functional end groups can be introduced or bi- (or poly-)functional polymers can be formed.

3.2.6. *Equilibrium Polymerizations*

Polymerizations of heterocycles with low ring strain, i.e., mainly five-, six-, and seven-membered rings, have relatively low ceiling temperatures, i.e., the temperatures at which the rate of depolymerization is equal to the rate of propagation. The ceiling temperature (T_c) is determined by the enthalpy (ΔH_p) and entropy (ΔS_p°) of polymerization and by the monomer concentration (m_e):

$$T_c = \frac{\Delta H_p}{\Delta S_p^\circ + R \ln m_e}$$

From this equation it can be deduced that for each temperature there is a corresponding monomer equilibrium concentration, the value of which is given by

$$m_e = \exp\left(\frac{\Delta H_p + T\Delta S_p^\circ}{RT}\right)$$

Typical monomers, which show such a behavior are tetrahydrofurane, dioxolane, and dioxepane. Figure 3.1 shows the values of m_e as a function of temperature for these monomers calculated for the thermodynamic parameters mentioned. It must be noted that these parameters can differ considerably with the solvent due to differences between polymer–solvent

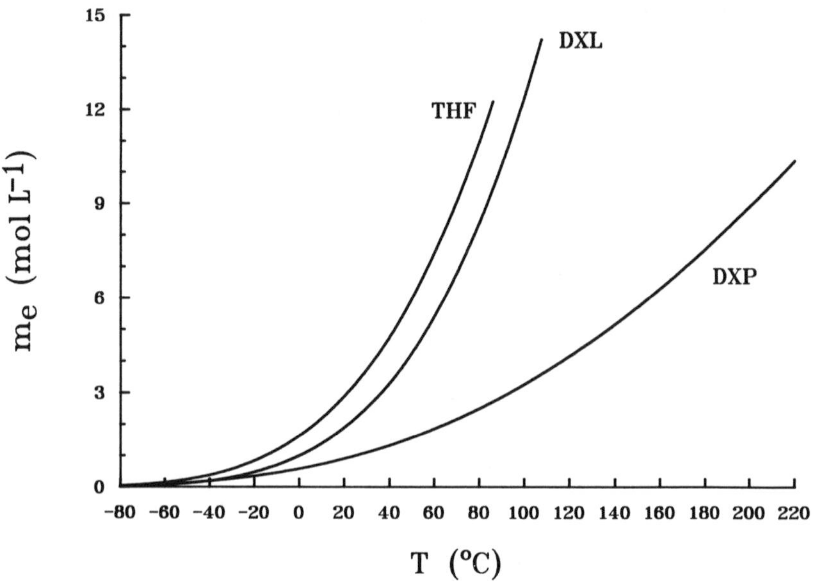

FIG. 3.1. Calculated monomer equilibrium concentrations (m_e) as a function of temperature for tetrahydrofurane (THF), 1,3-dioxolane (DXL), and 1,3-dioxepane (DXP) using the following thermodynamic parameters:

	ΔH_p (kJ mol^{-1})	ΔS_p° (J mol^{-1}K^{-1})
THF	−19.0	−74
DXL	−21.3	−78
DXP	−14.6	−49

interactions and monomer–solvent interactions.[8] The curves of Fig. 3.1 end at the molar concentration of bulk monomer. Above the corresponding ceiling temperatures, polymerization is impossible.

3.3. SCOPE AND LIMITATIONS

Considering the general mechanism described in the previous paragraphs it is clear that CROP cannot be used as a universal method for the clean and controlled synthesis of well-defined polymers. However, from the wealth of research efforts performed in this field a number of cases have emerged where it has indeed been possible to prepare on a preparative scale a number of well-defined polymers. These cases will be presented in more detail in the subsequent sections.

3.3.1. Poly-Tetrahydrofurane

The polymerization of tetrahydrofurane (oxolane) is probably the most studied and also best known cationic ROP. The polymer is produced commercially under the form of the dihydroxy-terminated compound and is sometimes referred to as polytetramethyleneglycol. It is used as soft segment for the production of (linear) polyurethanes and of various poly-block-copolymers (TPEs).

An impressive amount of initiators have been used for the polymerization of THF. Several of them have allowed one to prepare in a controllable way polymers with predictable molecular weights and relatively low dispersities. This is in the first place due to the fact that in the polymerization of THF the chain transfer to polymer is, under usual reaction conditions, unimportant because the polymer units have a lower nucleophilicity than the monomer. Also, cyclic oligomer formation is very limited or nonexistent for the same reason and also due to the fact that dimer formation is thermodynamically unfavorable. Consequently, provided a stable nonnucleophilic counterion, the polymerization can be considered as a typical living one. It must, however, be taken into account that the polymerization of THF has a low ceiling temperature due to the relatively low ring strain of the monomer ($23\,\mathrm{kJ\,mol^{-1}}$). This means that the polymerization cannot proceed until complete monomer consumption. The monomer equilibrium concentration as a function of temperature is given in Fig. 3.1. At 85 °C this equilibrium concentration reaches a value of $12.3\,\mathrm{mol\,L^{-1}}$, which is the molar concentration of pure monomer.

The most useful initiators are the alkylating agents. These include trialkyl oxonium ions and alkylesters of strong acids such as methyl fluorosulfonate ("magic methyl") or trifluoromethanesulfonate (methyl triflate). The polymerization can be carried out in a solvent, but usually it is carried out in bulk, the monomer acting as solvent. Polymers with predictable molecular weight can be obtained by adjusting the initiator concentration and by taking into account the monomer equilibrium concentration at the reaction temperature used. At equilibrium the molecular weight is then given by

$$\mathrm{MW} = \frac{12.3 - m_e}{[\mathrm{In}]} \times 72$$

However, under these reaction conditions the end product generally is a highly viscous mass which is difficult to handle and in which it is difficult to kill the active centers. Therefore, from a practical point of view it is preferable to stop the polymerization by addition of a terminating agent at conversions of 20–30%. The polymer can then easily be isolated either by precipitation in water or by evaporation of the unreacted monomer. To obtain polymer with predictable molecular weight it is necessary to know the conversion as a function of time, since the molecular weight is given by

$$MW = \frac{12.3 - m_t}{[In]} \times 72$$

where [In] is the initiator concentration and m_t the monomer concentration at time t.

Typical time-conversion curves obtained with methyl triflate as initiator at 25 °C are given in Fig. 3.2. The corresponding time–molecular weight curves are shown in Figure 3.3.

The procedure in which the polymerization is stopped at a precise moment with a terminating agent gives the possibility to introduce a functional end group on the polymer chain:

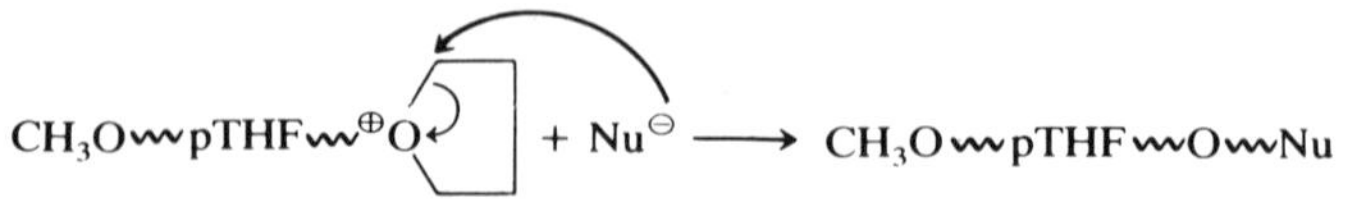

Table 3.3 gives a list of terminating agents (end-capping agents) and the functional groups introduced by them.

Bifunctional living polyTHF can be obtained with bifunctional initiators. Such initiators have been described by Yamashita and Chiba[9] and by Franta *et al.*[10] but the most practical to use is the anhydride of trifluoromethanesulfonic acid, commonly called triflic anhydride, first described by Smith and Hubin.[11] The mechanism of initiation is shown in Scheme 3.4.

According to conductivity measurement during the initiation process,

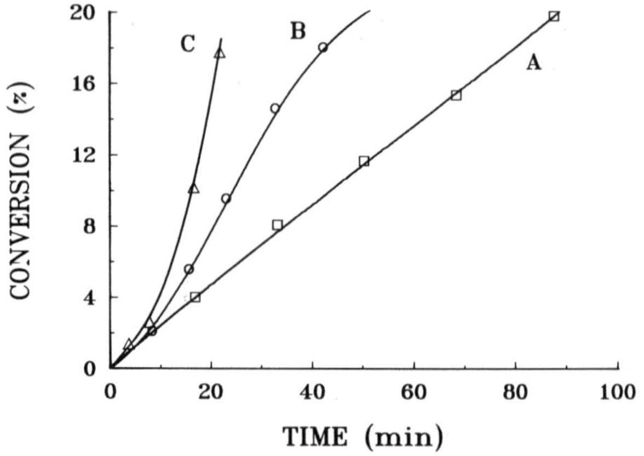

FIG. 3.2. Time-conversion curves for the polymerization of THF with different methyl triflate concentrations in bulk at 25 °C: I_0 equals (A) 0.01233 mol L^{-1}, (B) 0.02466 mol L^{-1}, (C) 0.09932 mol L^{-1}.

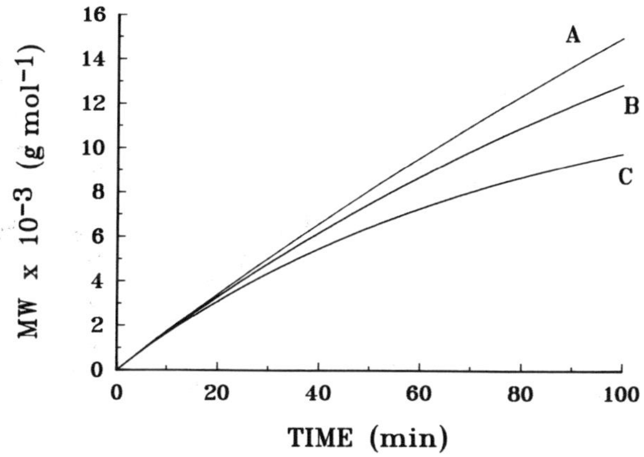

FIG. 3.3. Time–(calculated) molecular weight curves for the same polymerizations as in Figure 3.2.

the formation of ionic species is completed after less than one minute if the reaction is carried out in bulk at 25 °C. Typical time-conversion curves with this initiator are shown in Fig. 3.4 and the corresponding molecular weight–time curves in Fig. 3.5. It is thus possible to prepare well-defined bifunctionally living polyTHFs by choosing the initiator concentration and conversion. These can in turn be converted into telechelic polyTHFs by end

TABLE 3.3. End-Capping Reactions of Living Poly-THF[a]

End capper	End group	Ref.
H_2O	—OH	11
NH_3 (liq.)	$—NH_2$	11
KOCN	—N=C=O	11
H_2S	—SH	11
$HOOC—CH_2—_2COOH$	—COOH	11
LiBr	—Br	11
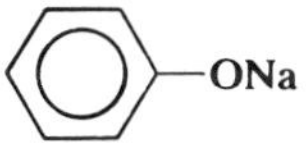	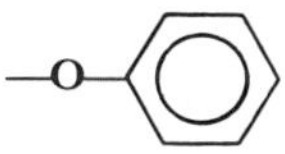	24
R_3N	$—NR_3$	75
	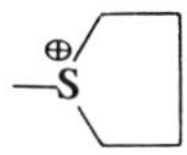	12
$(CH_3)_2C{=}C(OSi(CH_3)_3)(OCH_3)$	$—COOCH_3$ —COOH	25 25

[a] See also Table 3.5 (macromonomers).

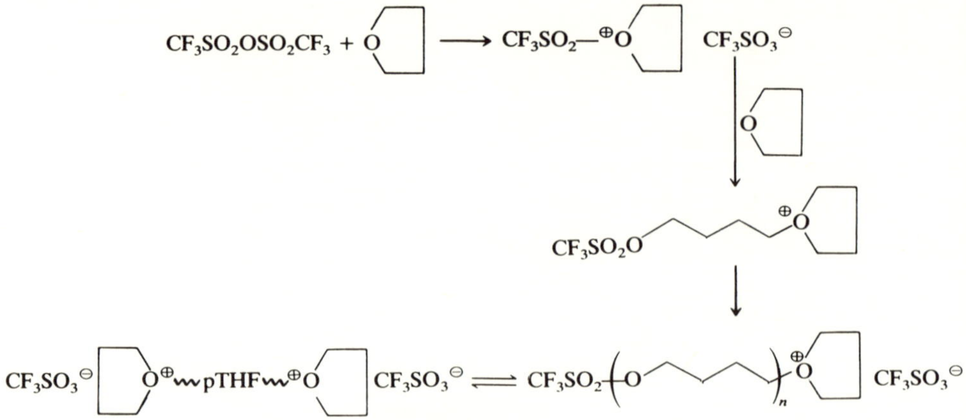

SCHEME 3.4.

capping of the living ends with the appropriate nucleophilic termination agents.[12]

Narrow MWD polyTHFs with MWs ranging from 11^3 to 5×10^5 for use as standards in GPC analysis are commercially available.[13]

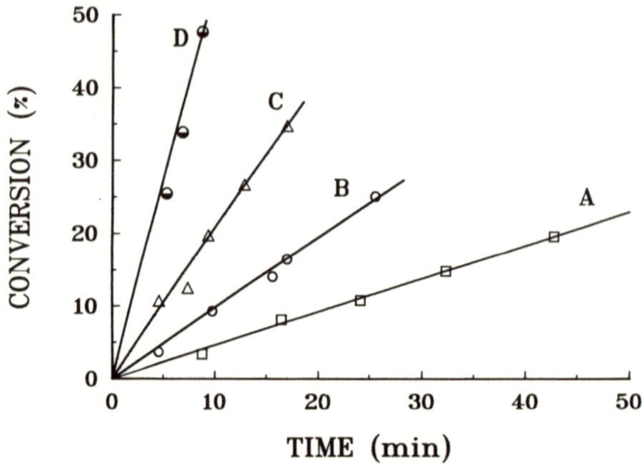

FIG. 3.4. Time-conversion curves for the polymerization of THF with different triflic anhydride concentrations in bulk at 25 °C: I_0 equals (A) 0.01233 mol L^{-1}, (B) 0.02466 mol L^{-1}, (C) 0.04932 mol L^{-1}, (D) 0.1233 mol L^{-1}.

3.3.2. Cyclic Acetals

A number of cyclic acetals are known to polymerize by cationic ROP. Examples are the following:

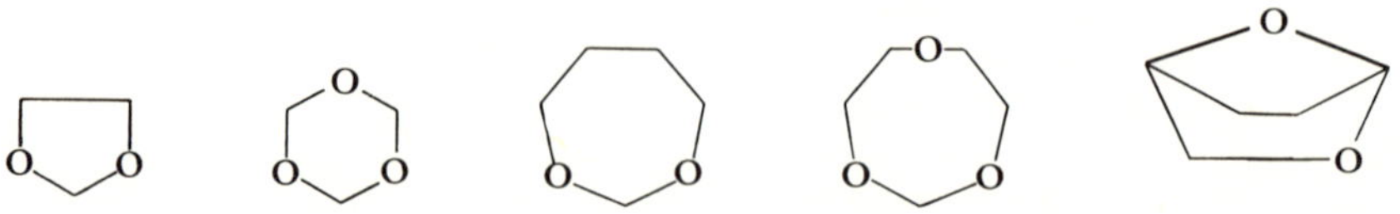

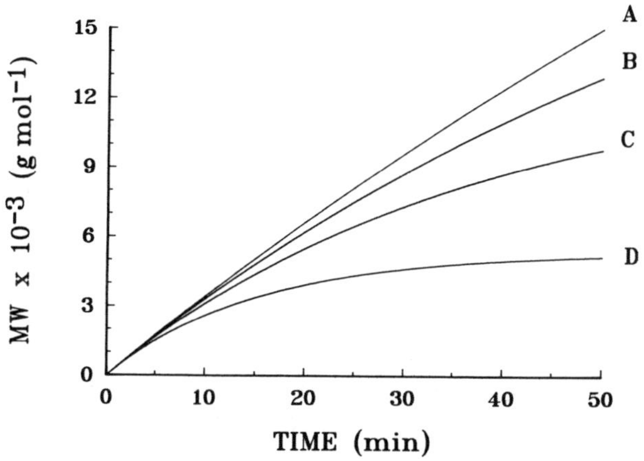

FIG. 3.5. Time–(calculated) molecular weight curves for the same polymerizations as in Fig. 3.4.

All cyclic acetal polymerizations are characterized by the occurrence of transfer to polymer and back-biting reactions forming cyclic oligomers.

3.3.2.1. Trioxane

The most important acetal monomer is trioxane. It is the starting material for polyoxymethylene (POM), an "engineering thermoplastic" which is produced in several countries with an estimated annual production of 160,000 tons per year. POM owes its excellent mechanical properties to its high crystallinity, which provides the material with a heat deflection temperature in the range of 160 °C.

The crystallinity of the polymer plays an important role in the polymerization too, because the heat of crystallization increases the overall change in enthalpy when dissolved or molten monomer is transformed into the crystalline polymer thus allowing a relatively high conversion. The commercial polymer generally contains a small fraction of oxyethylene units, which are introduced by copolymerization either with ethylene oxide or 1,3-dioxolane. These oxyethylene units are distributed randomly in the POM chain and are necessary to stabilize the polymer. In practice, the polymer is stabilized by thermally degrading ("unzipping") the end standing polyacetal segments until the unstable hemiacetal end groups are transformed into stable alcohol end groups:

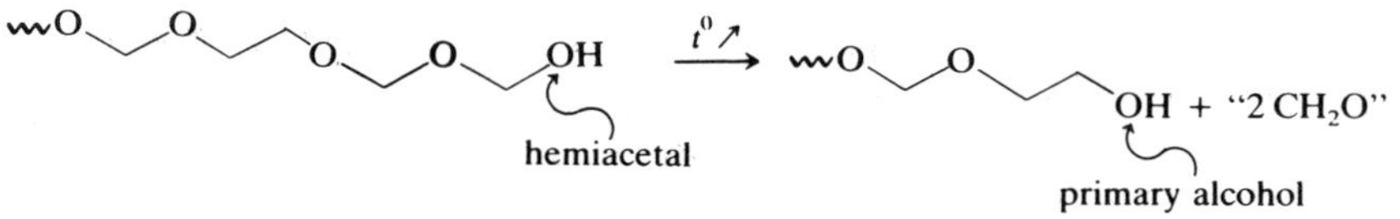

The polymerization of trioxane is very rapid and therefore it is difficult to control the molecular weights by using the appropriate monomer-to-initiator ratio as in the case of THF. End capping is difficult because of the rapid formation of the insoluble crystalline polymer. Addition of a small amount (~1%) of polyethyleneoxide to the reaction mixture prevents aggregation of the precipitated polymer particles and makes isolation of the end product by filtration easier.[14]

The oxyethylene content of the copolymer can be determined by ^{1}H-NMR spectroscopy in deuterated phenol at 120 °C. The spectrum allows one also to distinguish between MMM, EMM, and EME triads (M oxymethylene, E oxyethylene unit). A ^{1}H-NMR spectrum of a copolymer containing 50% of E units (and which is therefore soluble in $CDCl_3$) is shown in Fig. 3.6. For copolymers which are soluble in $CDCl_3$ it is possible to distinguish different pentads and even heptads by ^{13}C NMR in the presence of a europium shift reagent.[15,16]

3.3.2.2. *Dioxolane (DXL)*

The polymerization of DXL is a typical equilibrium polymerization. Several authors have determined the thermodynamic parameters of the polymerization using different experimental methods.[16] From these parameters a bulk ceiling temperature around 100 °C (see Fig. 3.1) can be calculated. Not only monomer but also a series of cyclic oligomers are participating in the equilibrium. The concentration of these oligomers is independent of the initial monomer concentration, so that for the production of linear polymer the initial monomer concentration has to exceed a

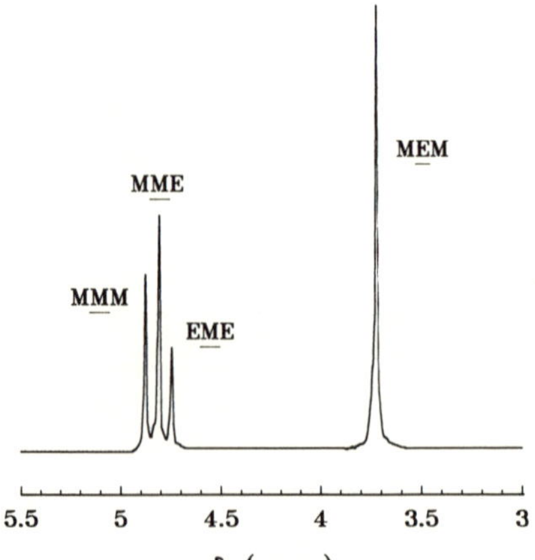

FIG. 3.6. ^{1}H-NMR spectrum (360 MHz) of a statistical 1:1 copolymer polyoxymethylene–polyoxyethylene in $CDCl_3$ at 25 °C.

certain value, called the critical monomer concentration m_c, below which the reaction mixture consists of an equilibrium mixture of monomer and cyclic oligomers without linear polymer. This critical monomer concentration consists of the sum of m_e and the sum of all cyclic oligomers expressed in monomer units m_{ol}. It depends on temperature and on the reaction medium. The values of m_c are thus higher than the values shown in Fig. 3.1, where only m_e was taken into account. Exact values of m_{ol} at various reaction conditions are not well known, but at 25 °C in CH_2Cl_2 it is estimated to be 0.8 mol L^{-1}.[16] The bulk concentration of DXL being 14.3 mol L^{-1} the yield of linear polymer obtained by bulk polymerization at 25 °C can reach a maximum value of $100 \times (14.3 - m_e - m_{ol})/14.3 \approx$ 80%. Provided the initiation is quantitative, the molar mass of the linear polymer at equilibrium is given by $74 \times (14.3 - m_e - m_{ol})/[\text{In}]$. Due to the continuous transacetalization, the MWD lies around 2.

Initiators for the polymerization of DXL must give stable non-nucleophilic anions. The most useful initiators are triflic acid, alkyl triflates, triflic anhydride, oxonium salts (e.g., $Et_3O^{\oplus}BF_4^{\ominus}$), and oxocarbenium salts (e.g., $RCO^{\oplus}SbF_6^{\ominus}$). The polymerizations are carried out at ambient or low temperatures (−40 °C) in solvents such as methylene chloride and preferably at relatively high monomer concentrations. With these initiators, the molar mass can be controlled according to the aforementioned equation. The polymerizations are terminated by adding a nucleophile. Typical capping agents are tertiary amines or phosphines leading to quaternary ammonium or phosphonium end groups, respectively. The nature of the end group plays an important role in the thermal stability of the end product. If the end group is an hemiacetal OH group, the polymers slowly decompose even at room temperature. Polymers provided with quaternary ammonium end groups are stable and decompose only above 200 °C as measured by TGA.[17]

Telechelic polyDXLs have been produced by polymerization with a bifunctional initiator followed by end capping. For this purpose, triflic anhydride or terephtaloyl ditriflate, obtained from terephtaloylchloride and silver triflate, can be used. The latter system, although more complicated to use than triflic anhydride, has the advantage of introducing one terephtaloyl group in the polymer chain which can be used as an internal standard for molecular-weight and end-group determinations (see Scheme 3.5). This initiating system is rapid and quantitative. The molecular weights obtained are in agreement with the calculated values based on monomer/initiator concentration. End capping with a tertiary amine leads to polymers with two quaternary α-aminoether functions (Scheme 3.6). These functions are moderately reactive electrophiles which allow one to perform coupling reactions which carboxylic anions. In this way α,ω-bisacrylates[17] and multiblock-copolymers (using carboxy-terminated polymers) have been prepared.[18]

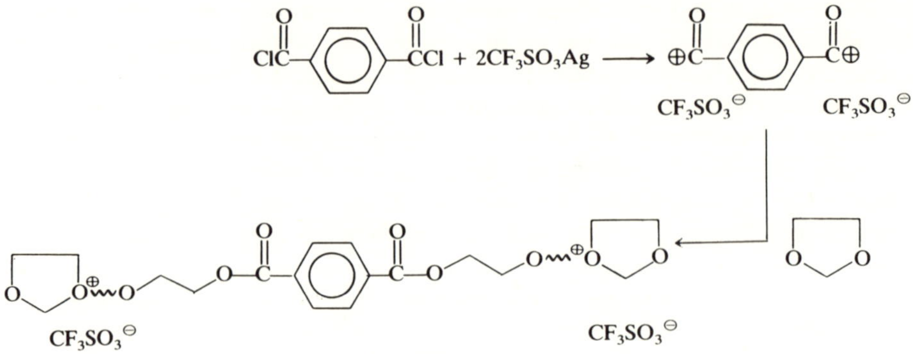

SCHEME 3.5.

Recently, the synthesis of α,ω-hydroxy-terminated polyDXL by proton-initiated polymerization in the presence of ethylene-glycol has been described.(19) Molecular weights between 1000 and 10,000 could be obtained by variation of the mole ratio of monomer to glycol. It was postulated that the polymerization took place by a mixture of the activated monomer mechanism and activated chain end mechanism in combination with chain transfer to the alcohol and reshuffling of the polymer molecules by the *trans*-acetalization reaction. Reaction of these polymers with stoichiometric amounts of a poly-isocyanate lead to the corresponding polyDXL containing polyurethane networks.

3.3.3. Polyamines

3.3.3.1. Polyethylenimine

Most work on the polymerization of cyclic amines has been devoted to the three-membered rings, the aziridines. The parent compound, commonly called ethylenimine, is polymerized commercially in different countries. The polymer polyethylenimine (PEI) has found various applications.(20)

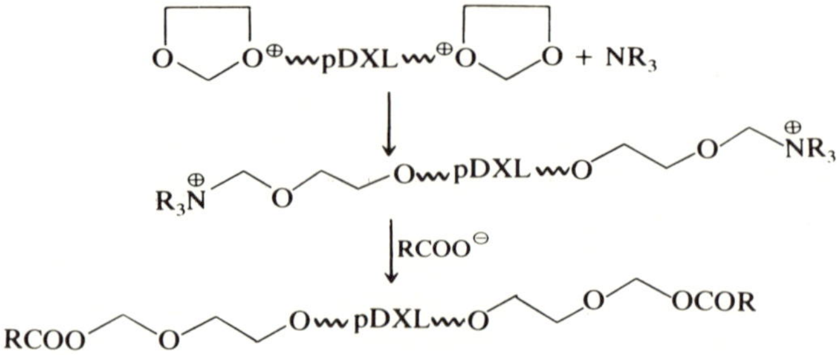

SCHEME 3.6.

The PEI obtained by direct cationic polymerization of EI has a highly branched structure due to the presence of approximately 25% of tertiary amino functions (and an equivalent amount of primary amines):

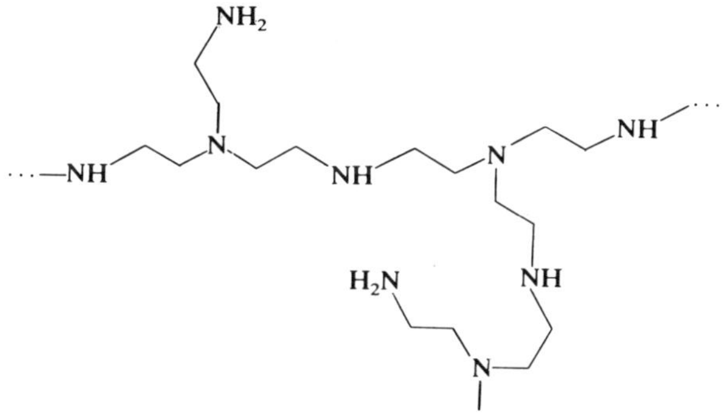

The polymerization is initiated by addition of a proton acid to produce the corresponding aziridinium ion [**1**; Eq. (3.1)]. If an alkylating agent is used as initiator, the alkylated monomer will transfer a proton to a monomer thus producing the same initiating species [Eq. (3.2)]. The first propagation step is the ring opening of the aziridinium ion **1** by nucleophilic attack of monomer [Eq. (3.3)]. The dimer thus formed (**2**) can react in two ways: a

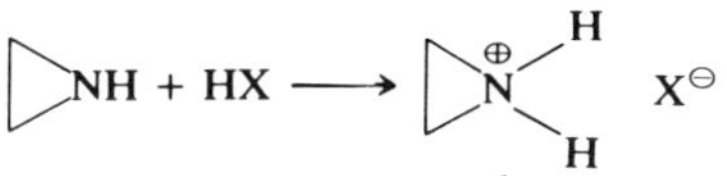

(3.1)

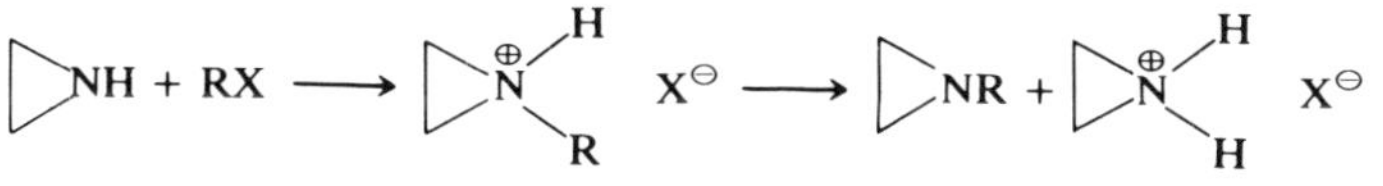

(3.2)

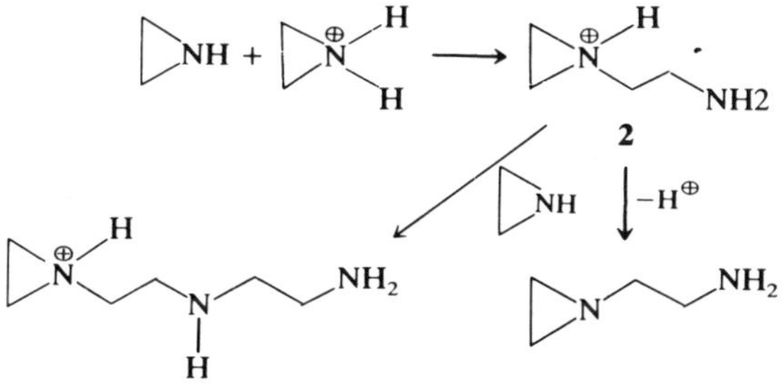

(3.3)

new ring opening of the aziridinium ion by monomer leads to trimer **3**; proton transfer to another amino function present in the reaction mixture leads to the uncharged dimer *N*-(2-aminoethyl)-aziridine (**4**). Compound **4** contains two new nucleophilic groups, a primary amine and an *N*-alkylaziridine, which were not present in the monomer. Chain growth can occur by reaction of these nucleophilic groups with protonated monomer, formed by the initiation or by the proton transfer from, for example, **2** or **3**.

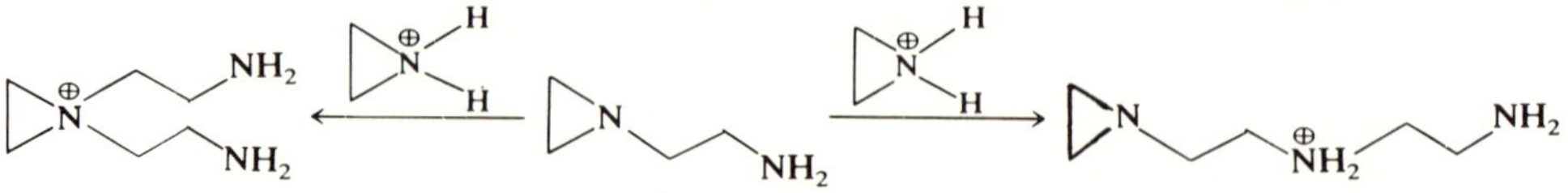

4

The polymerization can thus be considered to occur by a combination of active chain end and activated monomer mechanisms. The tertiary amines in the polymer are formed by attack of an active species by a secondary amine of a polymer chain followed by proton transfer (Scheme 3.7).

Linear polyethylenimine (LPEI) containing secondary amino groups only has been prepared in three different ways. One method is the isomerization–polymerization of a 1,3-oxazoline to the corresponding poly(*N*-acyl iminoethylene) followed by hydrolysis of the amide:[21]

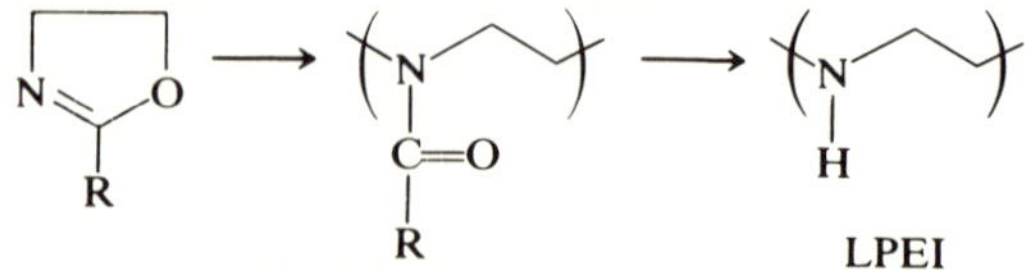

LPEI

A second method starts from EI and proceeds in two stages.[22] The first stage is a limited conversion prepolymerization of EI to a mixture of dimer and trimer predominantly, followed by a proton-catalyzed polycondensation of these oligomers in the presence of a specific amount of water. The process is described as a head-to-tail coupling of these oligomers. The role of water is to form a crystalline hydrate with the linear polyamine as soon as it is formed, thus preventing the reaction of the polymeric amino functions with the aziridinium ions in solution (which is the mechanism leading to branching). The primary amino groups are located at the periphery of the crystalline structures and are therefore still able to react (Scheme 3.8). Finally, LPEI has been prepared by polymerization of *N*-(α-tetrahydropyranyl)aziridine followed by acid hydrolysis of the poly(*N*-tetrahydropyranyl iminoethylene) and neutralization with sodium hydroxide[23] (Scheme 3.9).

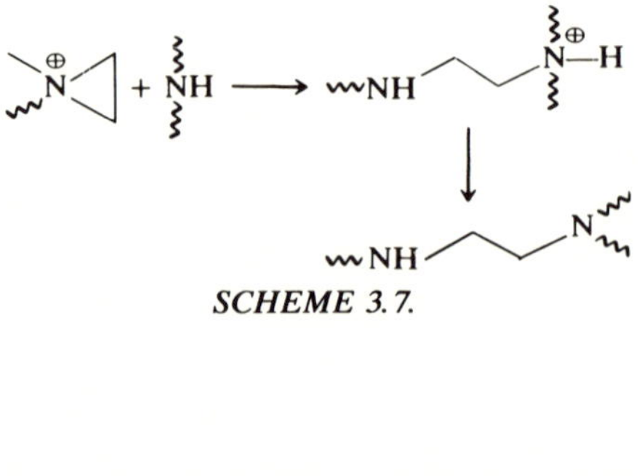

SCHEME 3.7.

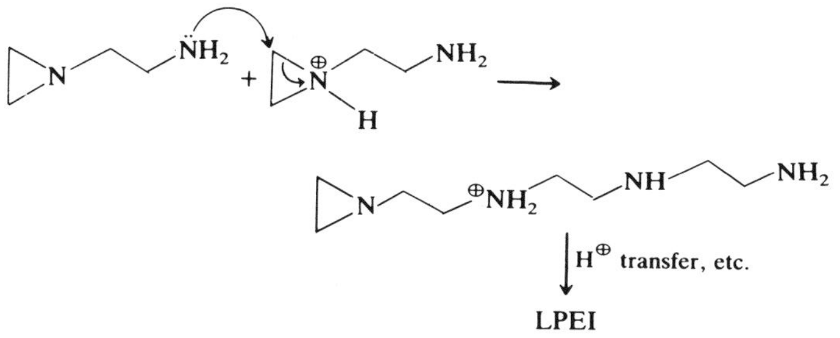

SCHEME 3.8.

The last two methods have the inconvenience that they use as starting material ethylenimine, which is no longer commercially available due to its high toxicity and mutagenicity. Therefore, the most convenient method for the synthesis of LPEI is the oxazoline route. Methyloxazoline and ethyloxazoline are commercially available and the polymerization as well as subsequent hydrolysis are "easy" reactions. However, the molecular weights of the LPEI obtained via these monomers are limited to values of approximately 5000. For high-molecular-weight LPEI, phenyloxazoline can be used. This monomer is not commercial but can be prepared (in moderate yields) from benzoylchloride and β-aminoethanol. With this monomer, molecular weights of LPEI of up to 100,000 have been claimed.(26)

The hydrolysis of the poly(N-acyl iminoethylene) can be carried out by either aqueous base or acid. Two examples of time-conversion curves for the hydrolysis are shown in Fig. 3.7.

LPEI is a crystalline material with a melting point of 40–60 °C. The linearity of the compound is evidenced by its ^{13}C-NMR spectrum, which shows one single signal compared with eight signals for the branched polymer. It forms two different crystalline hydrates containing either one molecule of water or half a molecule of water for each amino function. This is in contrast with branched PEI which is amorphous, highly hygroscopic, and water soluble. Due to its crystallinity, LPEI is insoluble in water below 50 °C.

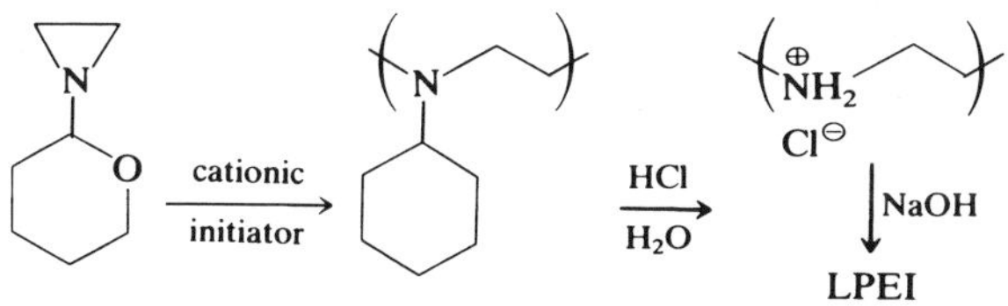

SCHEME 3.9.

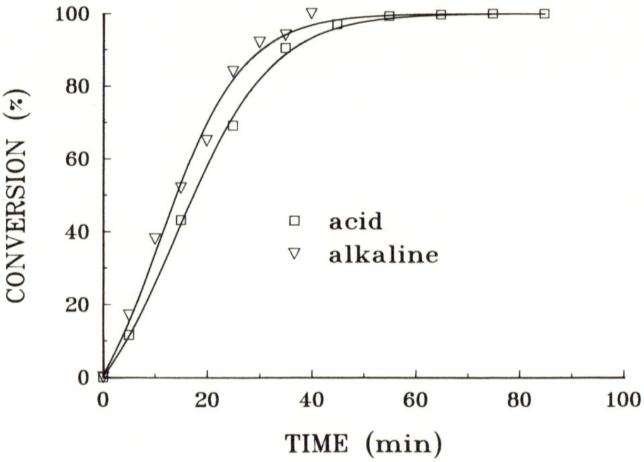

FIG. 3.7. Hydrolysis of poly[MeOx] by acid or base in water; [acid] or [base] = 8 mol L^{-1}, [amide] = 0.6 mol L^{-1}, t is *ca* 100 °C (reflux).

3.3.3.2. Polymers Derived from N-Alkylaziridines

The living character of the polymerization of *N*-alkylaziridines, expressed as the ratio of k_p/k_t, depends dramatically on the nature and number of substituents on the monomer.[27] From kinetic studies as well as from molecular-weight measurements, it was established that the termination reaction is predominantly intramolecular, i.e., with formation of macrocyclic ammonium end groups. It is shown in Table 3.4 that the k_p/k_t

TABLE 3.4. Values of k_p/k_t for the Polymerization of *N*-Alkyl Aziridines[a]

Monomer substituents 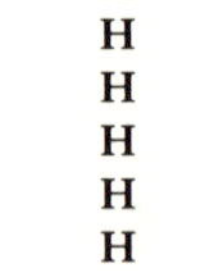				
R_1	R_2	R_3	k_p/k_t ($L\,mol^{-1}$)	Ref.
C_2H_5	H	H	6	27
$CH(CH_3)_2$	H	H	20	27
$C(CH_3)_3$	H	H	12,000	28
$CH_2C_6H_5$	H	H	85	27
$CH_2CH_2C_6H_5$	H	H	14	27
CH_2CH_2CN	H	H	82	27
$CH_2C_6H_5$	CH_3	H	1100	27
$CH_2CH_2C_6H_5$	CH_3	H	10,000	27
CH_2CH_2CN	CH_3	H	~5000	27
$CH_2C_6H_5$	CH_3	CH_3	0[b]	29

[a] Polymerizations at 0 °C in CH_2Cl_2 with Et_3OBF_4 as initiator.
[b] No homopolymerization, copolymerization is possible.

value (in the given reaction conditions) increases from $4\,L\,mol^{-1}$ for *N*-ethylaziridine to $12{,}000\,L\,mol^{-1}$ for *N-tert*-butylaziridine. Also, the introduction of a substituent on one of the carbon atoms of the monomer has a marked effect on the living character. This is explained by assuming that the steric hindrance of the substituents had a greater shielding effect in the polymer than in the monomer due to the cyclic structure of the latter. The termination reaction by intramolecular attack of a polymer amino function on its own active chain end (in contrast to the intermolecular one) does not affect the degree of polymerization which, provided a quantitative initiation, is determined by

$$DP_n = \Delta m/[In]$$

Monomers with a high living character, such as *N-tert*-butylaziridine (TBA) or 1-(phenethyl)-2-methyl aziridine (PEMA), give almost quantitative conversions to polymers with predictable molecular weight and narrow dispersities. As an example, Fig. 3.8 shows the GPC analyses of three poly(PEMA)s obtained with methyl triflate as initiator for different $m_0/[In]_0$ ratios.

The polymerization of TBA is very rapid at room temperature and therefore, to obtain polymers with predictable molecular weight and narrow dispersity, an equally fast initiation is required. Trialkyl oxonium salts and triflate esters are such fast initiators. Proton acids are also fast initiators, but

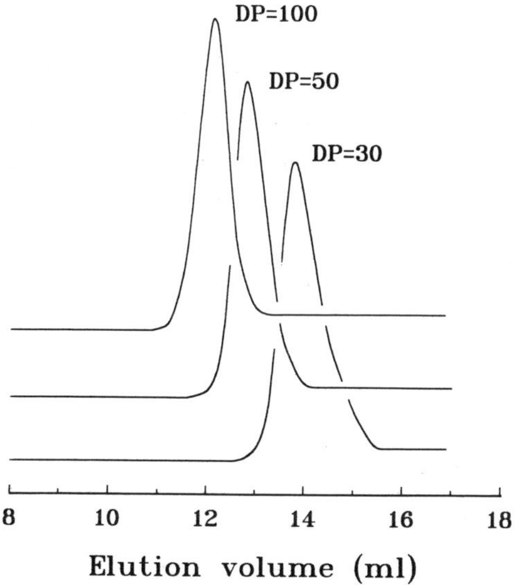

FIG. 3.8. GPC analyses of poly[1-(2-phenylethyl)-2-methyl aziridine]s obtained with different monomer/initiator ratios (in CH_2Cl_2 at 20 °C; initiator: methyl triflate).

they lead to mixtures of high-molecular-weight polymers and cyclic oligomer due to head-to-tail coupling reactions.[30]

A possible problem that can occur with alkylating initiators is that the quaternary aziridinium ion formed by the initiation reaction has limited solubility in the reaction medium and, as a consequence, the efficiency of the initiation is apparently low due to precipitation of a part of the active species formed.

Thus, it was found that polymerization of TBA in THF solution with methyl triflate as initiator leads to broad molecular-weight distribution, but on addition of 10% (vol/vol) of hexamethyl phosphoric acid triamide (HMPA) polymers with $M_w/M_n < 1.2$ are obtained. The effect of the HMPA addition is to keep the methyl *tert*-butyl aziridinium triflate in solution.[31]

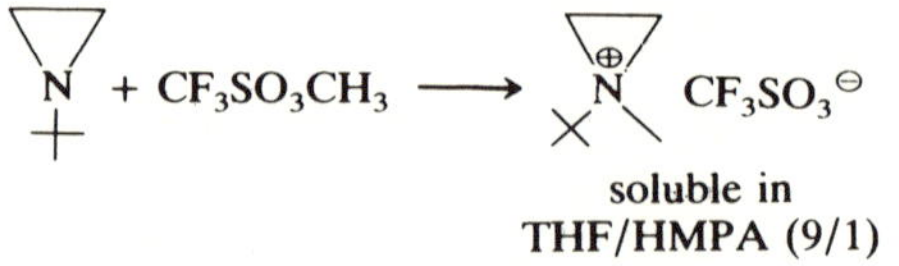

Bifunctionally living polyTBA can be obtained by initiation of the polymerization with a bifunctionally living polyTHF,[32] which in turn is obtained by initiation of THF polymerization with triflic anhydride. If the TBA is added two minutes after the initiation reaction of THF (at 20 °C), the initiating polyTHF segment has a molecular weight of approximately 800.

The aziridinium ion end groups can be used to introduce a number of functional end groups on polyTBA,[33] or to produce block-copolymers or polymer networks.

3.3.3.3. Polymerization of Oxazolines

Unsubstituted as well as substituted 1,3-oxazolines can be polymerized with typical cationic initiators to produce the corresponding poly(*N*-acylaziridine)s. The polymerization is a ring-opening reaction with

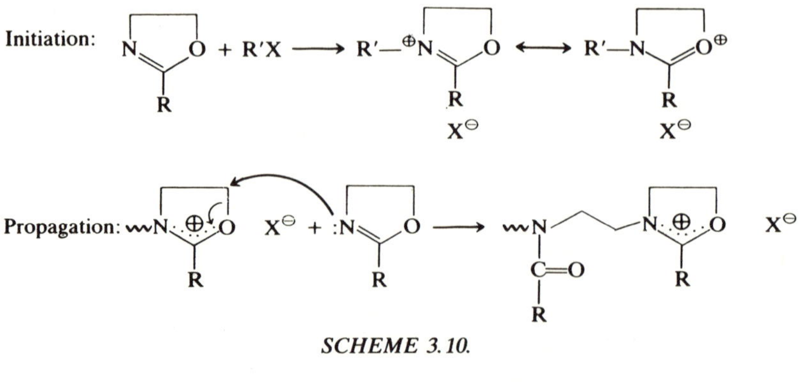

SCHEME 3.10.

isomerization.[34] The mechanism of the polymerization is shown in Scheme 3.10.

A termination reaction by attack of a polymer heteroatom on the active species, which usually occurs in the cationic ROP of heterocyclic monomers, does not occur in the polymerization of oxazolines because the nucleophilicity of the amide functions of the polymer chain is much lower than that of the monomer. As a consequence, the polymerization is expected to be of the living type and the molecular weight to be controlled by varying the ratio of monomer to initiator concentrations. However, in practice, the attainment of high-molecular-weight polymers turns out to be difficult due to some transfer reaction, the nature of which has been discussed by Litt *et al.*[35]

For the mechanism of polymerization of 1,3-oxazoline, two types of propagating species have been identified.[34] One is the oxazolinium salt, which is obtained with initiators producing counterions with low nucleophilicity, such as *p*-toluenesulfonate (tosylate). The other is the covalent species, which is formed if an initiator such as methyl iodide is used. In the latter case, the first-formed oxazolinium iodide is unstable under the reaction conditions used because of the high nucleophilicity of the iodide ion. This causes a rapid ring opening, forming the covalently bound alkyl iodide:

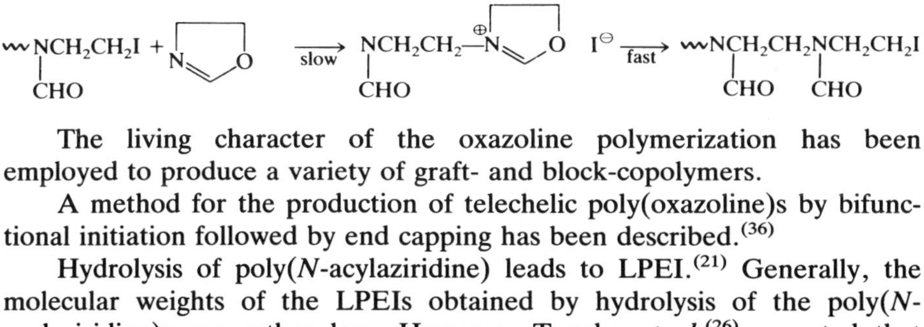

The living character of the oxazoline polymerization has been employed to produce a variety of graft- and block-copolymers.

A method for the production of telechelic poly(oxazoline)s by bifunctional initiation followed by end capping has been described.[36]

Hydrolysis of poly(*N*-acylaziridine) leads to LPEI.[21] Generally, the molecular weights of the LPEIs obtained by hydrolysis of the poly(*N*-acylaziridine)s are rather low. However, Tanaka *et al.*[26] reported that polymerization of 2-phenyl-1,3-oxazoline leads to polymers which, after hydrolysis, yield LPEIs with molecular weights above 100,000.

3.3.4. *Polymerization of Epoxides*

3.3.4.1. *Via the Activated Chain End Mechanism*

The cationic polymerization of three-membered cyclic ethers, the oxiranes, generally leads to a mixture of low-molecular-weight polymer and cyclic oligomers. The nature of the cyclic oligomers varies from dimer to pentamer depending on the substituents of the monomer. Ethylene oxide, for example, produces mainly 1,4-dioxane[37] while propylene oxide produces a mixture of dimer, trimer, and (predominantly) tetramer.[38] In some cases an isomerized dimer, i.e., a five-membered 1,3-dioxolane, is formed. This occurs with monomers which have a tendency to form carbenium ions (e.g., styrene oxide) which then isomerize to the more stable α-oxocarbenium ion by a hydride shift[39] (Scheme 3.11).

The main mechanism of oligomer formation is by back biting. The same reaction products are formed when polymer (obtained, for instance, by an anionic mechanism) is treated with an oxonium-forming initiator.[40]

From these facts it is clear that the "classical" cationic polymerization of oxiranes is not suited for the synthesis of well-defined polymers. Therefore the development of a nonclassical polymerization, in which the cyclic oligomer formation is strongly reduced, is of great interest as shown in the next section.

3.3.4.2. *Via the Activated Monomer Mechanism*

In contrast with the activated chain end (ACE) mechanism, the activated monomer (AM) mechanism occurs between a nucleophilic end group of the growing polymer chain and the activated monomer. The activation of the monomer occurs by interaction with a proton or a Lewis acid which enhances the polarization of the carbon–heteroatom bond and thus its reactivity toward the nucleophile. The AM mechanism has long been known in the anionic ring-opening polymerization of lactams. In cationic polymerization it was also known to occur in the case of cyclic secondary amines as a consequence of proton transfer, which takes place with these monomers (see Section 3.3.3.1). For cyclic ethers, and more particularly for oxiranes, the interest to shift the mechanism from ACE to AM is that the AM mechanism does not lead to cyclics, which is one of the main problems in the ACE mechanism. This stems from the fact that the "active" chain end in AM polymerization is in fact not very active and is therefore not attacked by a heteroatom of its own chain, i.e., formation of cyclics by back biting is absent.

The AM mechanism for oxirane was described in Section 3.2.5. As already mentioned there, the mechanism is useful for the production of

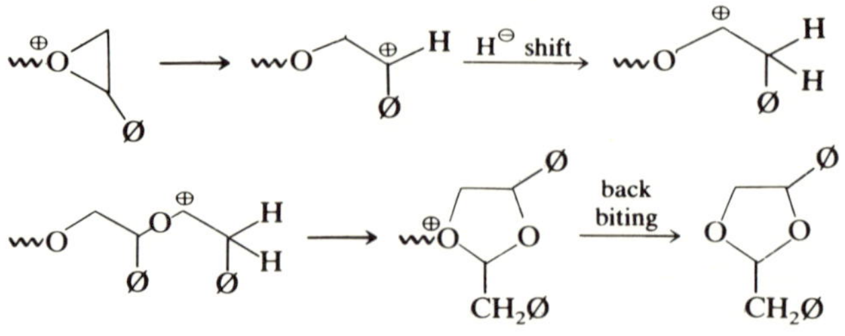

SCHEME 3.11.

low-molecular weight polymers derived from substituted monomers. For example, it works fine for epichlorohydrin but not for ethylene oxide. It is necessary to carry out the polymerization at low monomer concentration by slowly supplying monomer to the reaction mixture.

An additional characteristic of the AM polymerization is that it allows one to produce polymers with functional end groups by choosing the appropriate initiating alcohol. Thus, methacrylate-terminated poly(epichlorohydrin) has been produced using (2-hydroxyethyl)-methacrylate (HEMA) and triflic acid as initiating system.[41] By using a diol as initiator an α,ω-dihydroxy terminated polymer is obtained, the molecular weight of which can be varied from a few hundreds to 2500 by adjusting the [ECH]/[diol] ratio.[42,43] Boron trifluoride was used as the activator in this system.

3.3.5. Block-Copolymers

3.3.5.1. By Sequential Monomer Addition

This method, also referred to as the "blocking from" method, can be used if a first monomer is polymerized to a polymer carrying a functional end group capable of initiating (by bond formation) the polymerization of a second monomer. This second polymerization should show no transfer reactions. The ideal case is when the first as well as the second polymerizations are of the living type. Since various cationic ROPs have a high living character, several of these polymerizations can be utilized to prepare block-copolymers. There are also a number of examples where only the second polymerization is living and where the macroinitiator was obtained by the end-group transformation of a commercial polymer.

Some representative examples are given below.

(a) PolyTHF–PolyTBA. Under well-chosen conditions, the polymerization of THF is living, which means that each polyTHF carries a tertiary oxonium ion end group capable of alkylating tertiary amines to form the corresponding quaternary ammonium salts. If the tertiary amine is cyclic and polymerizable, block-copolymers are obtained.[44] A typical case where such block-copolymers are formed is when TBA is the cyclic amine. Due to the living character of the TBA polymerization, the length of the polyamine blocks can be controlled by the ratio m_0/[oxonium].

When the polymerization of THF is monofunctional, e.g., with methyl triflate as initiator, AB-type block-copolymers are obtained. If the polymerization of THF is bifunctional, e.g., with triflic anhydride as initiator, ABA-type block-copolymers are formed (Scheme 3.12).

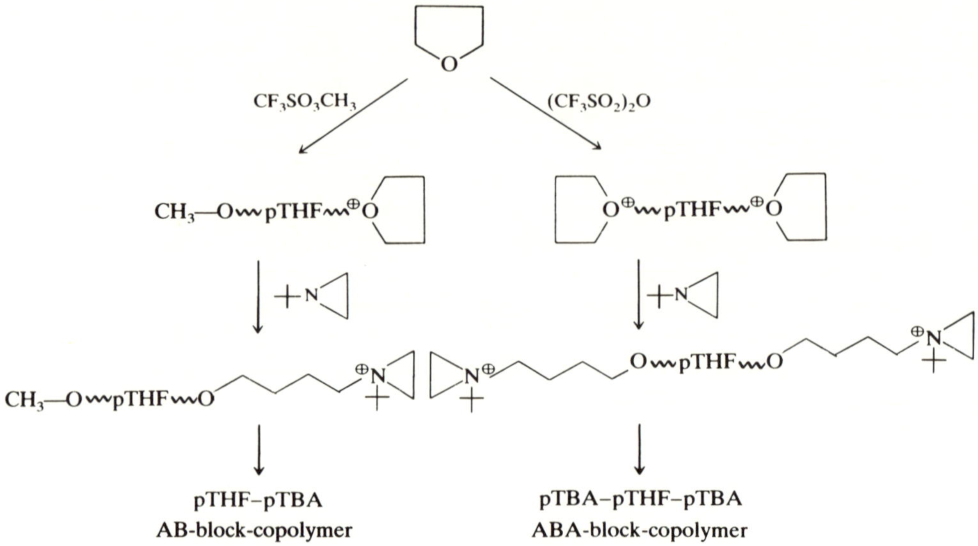

SCHEME 3.12.

(b) PolyTHF–Poly(Vinylmonomers). Another approach of the blocking from method applied to polyTHF is to prepare a polymer containing azo groups in the main chain and use this as a macroinitiator for radical polymerization. For the synthesis of the macroinitiator two approaches were used. Yagci[45] prepared an azo-containing polyTHF by initiation with an azo-containing initiator obtained *in situ* from silver tetrafluoroborate and 4,4′-azobis(4-cyanopentanoyl chloride); see Scheme 3.13.

D'Haese *et al.*[46] reported the preparation of azo-containing polyTHF by end capping of the living polymer (via the thiolanium end-group, see Section 3.3.5.2.a) with an aqueous solution of the sodium salt ACVA (Scheme 3.14). With the thus obtained azo polymers, the free-radical polymerization of styrene or methyl methacrylate can be initiated with the formation of the corresponding block-copolymers.

(c) PolyTHF–Polyoxazolines.[47] Oxonium salts are good initiators for the polymerization of oxazolines and therefore addition of an oxazoline to a

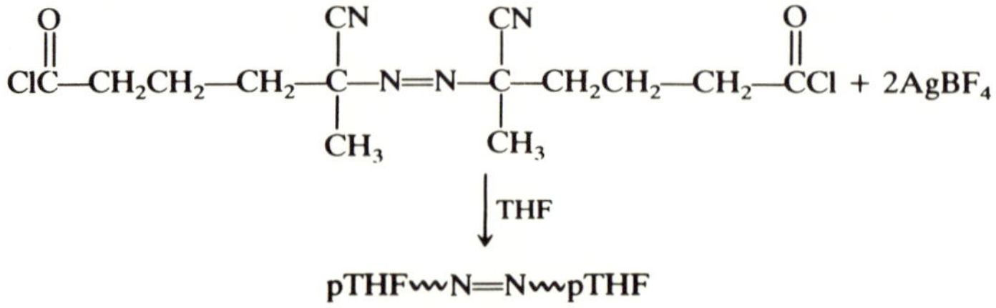

SCHEME 3.13.

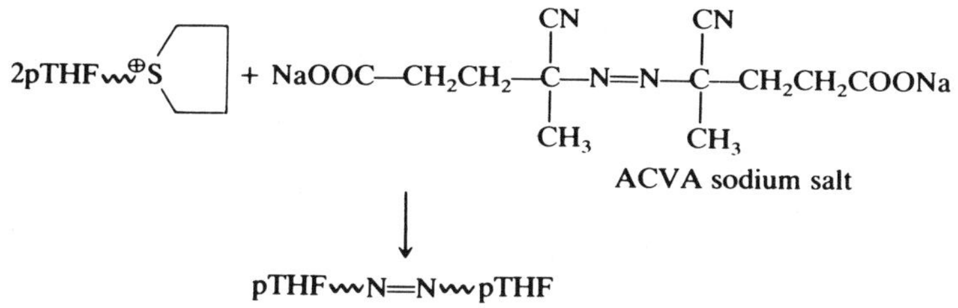

SCHEME 3.14.

living polyTHF should lead to the corresponding block-copolymers. However, the polymerization of oxazolines requires a polar solvent and temperatures above 80 °C. Therefore, the reaction is carried out in two stages: in the first stage the THF polymerization is terminated with an equimolar amount of oxazoline with the formation of an oxazolinium-terminated polyTHF. Then, after evaporation of the unreacted THF a solution of the oxazoline in acetonitrile is added and the mixture is heated at 80 °C.

When 2-methyloxazoline is used, the resulting block-copolymer consists of a hydrophobic polyether segment and a hydrophilic polyamide segment. Such amphiphilic polymers are of interest for applications as dispersion agents.

(d) PolyEO–Polyoxazoline. Tosylate-terminated PEO has been used as macroinitiator for the polymerization of oxazolines leading to poly(oxyethylene)polyoxazoline block-copolymers.[48,49]

(e) Polybutadiene–Polyoxazoline and Polybutadiene–LPEI.[50] Liquid α,ω-dihydroxy butadiene (MW 2840) can be tosylated to give α,ω-polybutadiene ditosylate. This macroinitiator was used for the polymerization of oxazoline in toluene/DMF giving a solid material which, on the basis of NMR analysis and solubility properties, was shown to be an ABA-type block-copolymer. Alkaline hydrolysis in aqueous solution at 100 °C for 4 h lead to the corresponding block-copolymer with linear polyethylenimine segments.

3.3.5.2. By Coupling Reactions

This method, also referred to as "blocking onto," requires two prepolymers carrying complementary functional groups, i.e., which allow the coupling of the two chains by a covalent bond. In principle, this could be achieved by mixing equimolar amounts of an anionically living polymer and a cationically living polymer. That this is possible has indeed been demonstrated for the couple cationic living polyTHF–anionic living polystyrene.[51] However, for practical reasons it is preferred to derivatize at least one of the two kinds of active species into a less reactive form, in order to avoid side reactions during the coupling reaction.

Coupling of two different polymer chains is hampered by two facts, which become more important as the molecular weights of the prepolymers are higher: the low concentration of end groups and the incompatibility of polymers having different structures. The low concentration of reactive groups implies that highly reactive species or long reaction times are necessary to attain high coupling efficiencies. The incompatibility increases the difficulty to react, because each reactive group is buried in its own polymer domain and therefore the probability of encountering the reactive group of the other polymer is further reduced. Therefore, the production of block-copolymers by the coupling method is generally restricted to relatively low molecular weight prepolymers.

A possible way of overcoming the above-mentioned problems is to provide the two reacting polymers with oppositely charged reactive end groups. If the small counterions are removed, the two macroions are forced to stay in each other's vicinity due to Coulombic interaction.

A number of examples of block-copolymer formation based on coupling reactions and involving at least one prepolymer obtained by CROP are given below.

(a) Block-Copolymers with PolyTHF Segments. Combination of a nucleophile-terminated polymer with living polyTHF should lead to the corresponding block-copolymer. Due to the high reactivity of the oxonium group, traces of water in the reaction mixture disturb the reaction. It was found that reaction of the oxonium end group with thiolane leads to a thiolanium end group, which is stable toward water but which is still reactive toward a negatively charged nucleophile, such as a carboxylate anion. If the carboxylate is situated at the end of a polymer chain, the

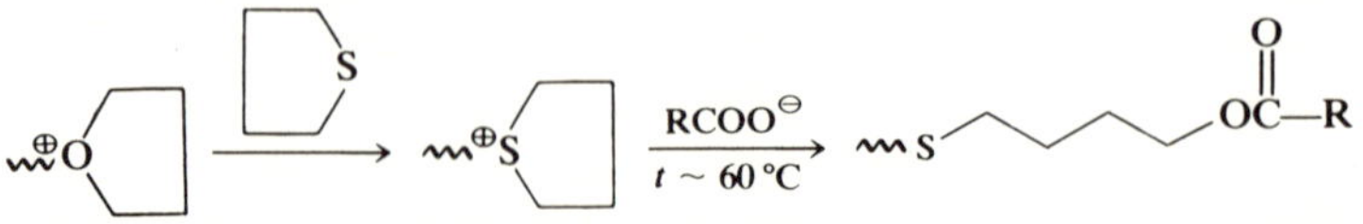

coupling results in a block-copolymer. This reaction has been carried out with α,ω-carboxy terminated polybutadiene (CTpBu) and with α,ω-carboxy terminated polystyrene, and with monofunctionally as well as with bifunctionally living polyTHF;[12] see Scheme 3.15.

(b) Block-Copolymers of Polyacetals. The preparation of block-copolymers of polydioxolane by a "blocking from" method has recently been described,[52] but due to the occurrence of transacetalization reactions in this system, the polymer inevitably must contain a fraction of (cyclic)

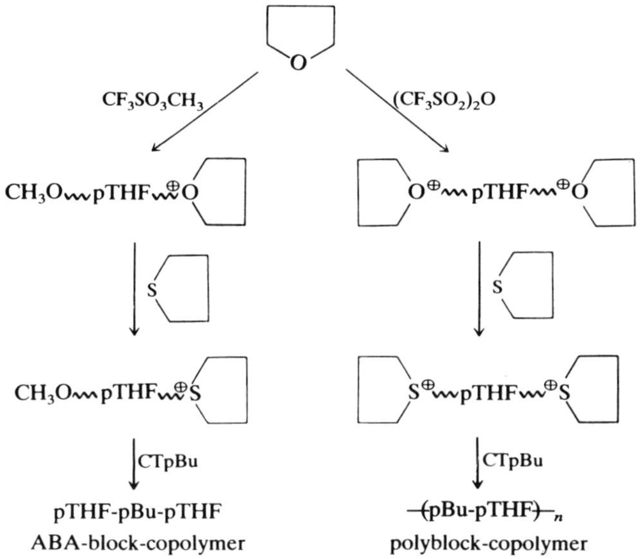

SCHEME 3.15.

homopolymer. A method for the preparation of polyblock-copolymers with polyDXL segments consists in end-capping bifunctionally living polymerization of DXL with a tertiary amine followed by reaction of the resulting reactive α,ω-bis quaternary ammonium salt with a α,ω-dicarboxy-terminated polymer.[18] As initiator for the bifunctional polymerization, terephthaloyl triflate (from terephthaloyl chloride and silver triflate) was used (see Section 3.3.2.2).

3.3.6. *Graft Copolymers*

3.3.6.1. *"Grafting From"*

In this method an initiating group attached to a polymer backbone is used to initiate a polymerization. The method has been used to prepare graft-copolymers in which the grafts are polyTHF or polyoxazolines.

PolyTHF grafts were obtained from reactive halide-containing polymers[53] in combination with silver salts (Scheme 3.16). Chloromethylated polystyrenes (noncrosslinked or crosslinked) were used to initiate the polymerization of 2-methyloxazoline to yield a graft-copolymer.[54] Potassium iodide was added to promote the grafting rate and efficiency (Scheme 3.17). Analogous reactions have been carried out on 1-chloro-1,3-butadiene–butadiene copolymer[55] and on poly(vinyl chloride).[56]

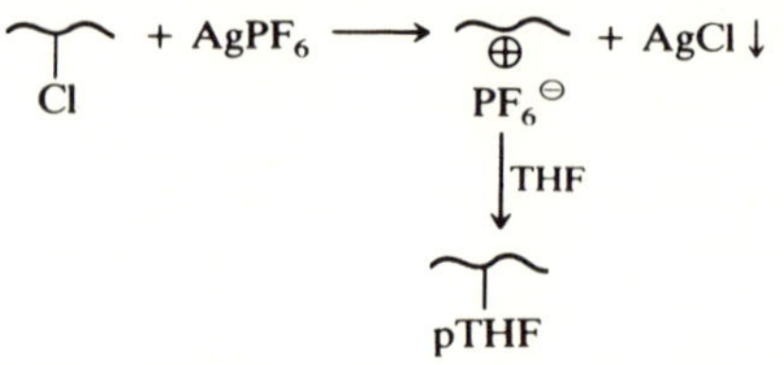

SCHEME 3.16.

3.3.6.2. *"Grafting Onto"*

In this method a polymer with a reactive end group (usually the active chain end of the polymerization) is combined with another polymer which contains "complementary" functional groups along the backbone. Thus, living polyTHF has been grafted on salts of polyhydroxystyrene–styrene copolymers[57] and on polymers containing amino functions.[58]

Living polyTBA was grafted onto silica which had previously been treated with 3-aminopropyl triethoxysilane. Another approach consisted in first end capping the living polymer with a triethoxy silane group and coupling this polymer with silica.[59] Living polyTBA has also been grafted onto cellulose derivatives.[60]

3.3.6.3. *The Macromonomer Procedure*

Macromonomers are polymers (generally of relatively low molecular weight) provided with a (co)polymerizable end group. Mostly this end group is a carbon–carbon double bond, but it may also be a cycle or a difunctional moiety capable of participating in a stepwise polymerization reaction. Copolymerization of a macromonomer with a normal monomer leads to the corresponding graft-copolymer:

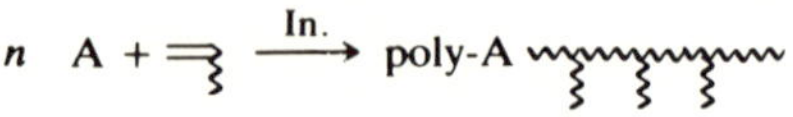

Several macromonomers obtained via CROP have been described. An overview is given in Table 3.5.

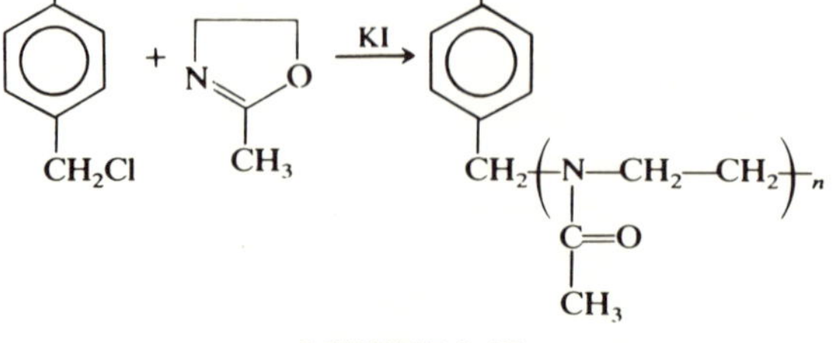

SCHEME 3.17.

TABLE 3.5. Overview of Macromonomers Obtained by CROP

Monomer	End group	Refs.
	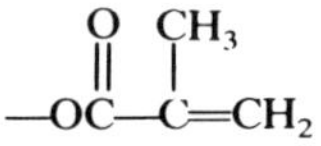	61, 62, 12, 67
	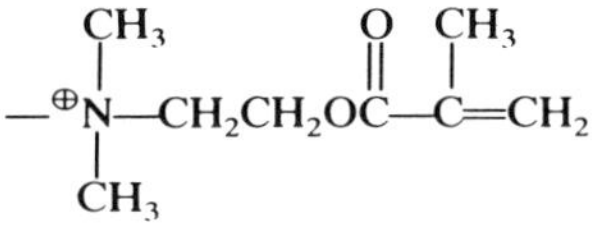	63
	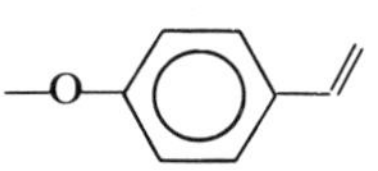	64
	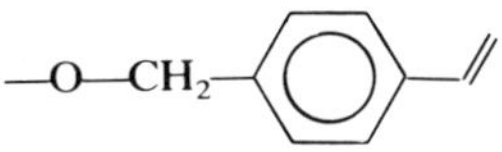	65, 66
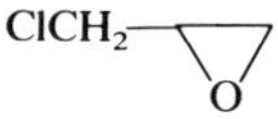	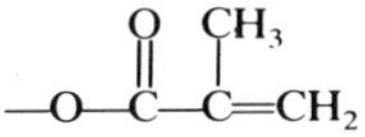	41
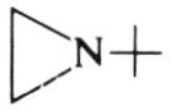	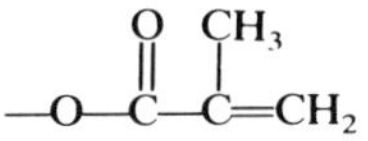	68
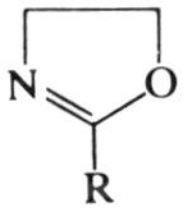	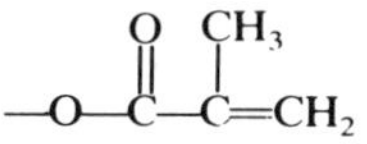	69, 72
	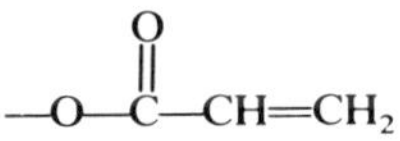	69, 72
	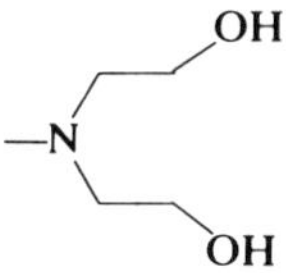	70
	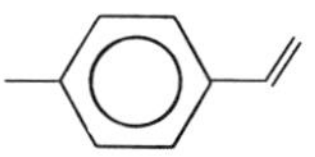	71
	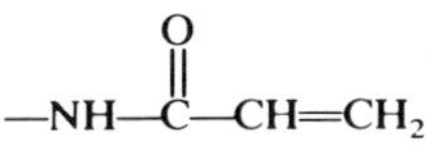	73
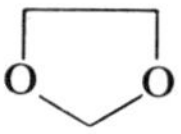	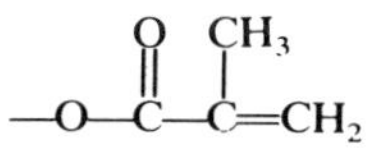	17

Via initiation

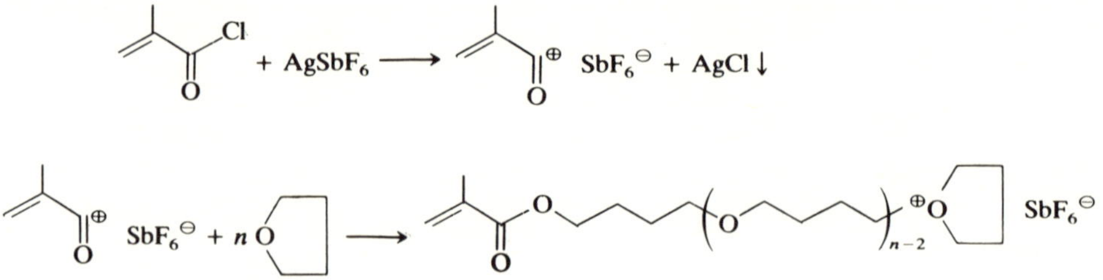

Via end capping

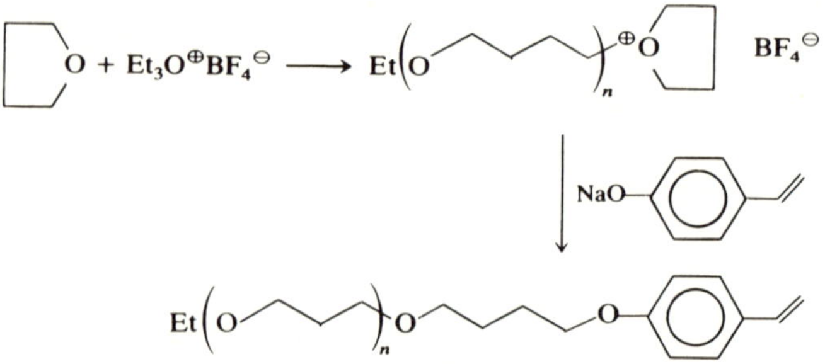

SCHEME 3.18.

The polymerizable end group can be introduced via the initiation reaction or via termination (end capping), as illustrated by the examples of polyTHF macronomers in Scheme 3.18.

A few cases are known where the active species of the polymerization is first transformed into a less reactive end group and subsequently end capped to form a macromonomer. This is, for example, the case for the synthesis of poly(1,3-dioxolane) α,ω-bis methacrylate, where the oxonium end group was first converted into a reactive quaternary ammonium salt and then coupled with the methacrylate anion.[17]

Finally, macromonomers from substituted oxiranes have been prepared by the "activated monomer" method using a hydroxyl group containing (meth)acrylate ester as initiator.[41]

3.4. *EXPERIMENTAL PROCEDURES*

In this section a number of recipes describing the synthesis of polymers described in Section 3.3 are assembled. The examples were chosen from the literature and most of them have been repeated successfully in the authors' laboratory.

3.4.1. α,ω-Dihydroxy PolyTHF[64]

150 mL of THF is distilled directly into a two-necked flask equipped with a magnetic stirrer rod and an inlet for dry nitrogen. The flask is thermostated at 20 °C and 1.21 mL (7×10^{-3} mol) of triflic anhydride is introduced through a rubber septum by means of a hypodermic syringe. After 10.5 min the reaction is terminated by adding 3 mL of water in which 0.6 g (15×10^{-3} mol) of sodium hydroxide is dissolved. Then, the THF solution is concentrated to approximately 50 mL. The residual solution is slowly added to 0.5 L of water cooled at 0 °C. The precipitate is filtered off, washed with cold water, and dried in vacuum yielding 18 g (13.5%) of α,ω-dihydroxy polyTHF with $M_n = 3500\ \mathrm{g\,mol^{-1}}$ and $M_w/M_n = 1.30$.

3.4.2. Polymerization of Trioxane[14]

A four-necked flask, fitted with a nitrogen inlet, thermometer, stirrer, and condenser, is charged with 3.6 g of finely powdered poly(ethylene oxide) ($M_n \approx 15{,}000\ \mathrm{g\,mol^{-1}}$), 120 mL of dry cyclohexane, 144 mL (1.87 mol) of molten, freshly distilled trioxane, and 6.6 mL (0.094 mol) of dry 1,3-dioxolane. The flask is thermostated at 60 °C and 0.14 g (7×10^{-4} mol) of boron trifluoride dibutyl etherate is added under constant stirring. Polymerization ensues after a short induction period, the temperature rises to 70–75 °C, and the polymer precipitates. After 60 min the reaction is terminated by adding 5 mL of a 0.2N solution of NH_3 in CH_3OH. The polymer is filtered off, washed with warm water and methanol, and then dried at 70 °C. A typical yield is 120–135 g (70–80%).

3.4.3. Synthesis of α,ω-Dihydroxy PolyDXL[19]

3 μL (3.4×10^{-5} mol) of triflic acid dissolved in 0.6 mL of dry dichloromethane is placed in a three-necked flask fitted with a mechanical stirring device, dropping funnel, and condenser. The reaction vessel is thermostated in a water bath at 50 °C and kept under a slight argon overpressure. 470 μL (8.4×10^{-3} mol) of dry ethylene glycol is introduced and then 8 mL (0.114 mol) of DXL is added dropwise to the reaction mixture. The polymerization is carried out for 3 hours at 50 °C under argon atmosphere. The polymerization is stopped by adding 1.8 mL (0.017 mol) of tributylamine. The reaction mixture is evaporated in vacuum. The remaining polymer is dissolved in CCl_4. The insoluble tributylammonium triflate is filtered off. The solution is allowed to stand over barium oxide to eliminate the last traces of acid. After filtration, the polyDXL is recovered by lyophilization from its CCl_4 solution yielding 6.5 g of hydroxy-terminated polymer with $M_n = 900\ \mathrm{g\,mol^{-1}}$.

3.4.4. Synthesis of PolyDXL-α,ω-Bis Methacrylate[17]

In a conical reaction flask provided with an inlet for dry argon and a glass neck equipped with a rubber septum, 1.00 g (3.9×10^{-3} mol) of silver triflate dispersed in 10 mL dichloromethane and 381 mg (1.9×10^{-3} mol) of terephthaloylchloride are introduced. The mixture is stirred with a magnetic stirrer for 30 min at room temperature. Then, the mixture is allowed to rest for a few minutes whereby the silver chloride precipitate concentrates itself at the narrow lower part of the flask. A needle connected with a teflon tubing is introduced through the rubber septum in the supernatant solution. The other end of the tubing is also connected to a needle, which is introduced in the polymerization flask through a rubber septum. By putting argon pressure on the flask containing the initiator solution, the solution of terephthaloyl triflate is transferred to the polymerization flask via the teflon tubing. The polymerization flask, containing a magnetic stirrer rod, is cooled to −40 °C and then 9.5 mL of dry dichloromethane and 10.5 mL (0.15 mol) of DXL are introduced through the septum by means of hypodermic syringes. The mixture is stirred at −40 °C for 10 min and then 2 mL (0.014 mol) of triethylamine is introduced. The polymer is isolated and purified by threefold precipitation in cold methanol (−10 °C) from dichloromethane solution and dried in vacuum at room temperature yielding 10 g of telechelic polyDXL with two quaternary ammonium end groups and $M_n = 6500\,\mathrm{g\,mol^{-1}}$ and $M_w/M_n = 1.95$.

For the synthesis of the methacrylate bis-macromonomer, 0.5 g (7.7×10^{-5} mol) of polymer is added to a solution of 0.198 g (7.1×10^{-4} mol) of cetyltrimethylammonium methacrylate in 10 mL of chloroform, and the solution is refluxed during 2 hours. The polymer is isolated by precipitation in cold methanol (−10 °C), washed with diethylether, and dried in vacuum. The cetyltrimethylammonium methacrylate salt is prepared by dissolving 2 g (9.5×10^{-3} mol) of cetyltrimethylammonium hydroxide and 0.644 g (7.5×10^{-3} mol) of methacrylic acid in 300 mL of benzene. The mixture is stirred at room temperature overnight and the formed water removed by azeotropic distillation. The salt is recovered by lyophilization. The very hygroscopic, white solid is stored under dry argon atmosphere.

3.4.5. Methacrylate-Terminated PolyTBA[68]

3.4.5.1. Synthesis of TBA

112 g (0.96 mol) of 2-(*tert*-butylamino)ethanol (Janssen Chimica) is dissolved in 500 mL of dry dichloromethane. The mixture is cooled in an ice bath and, under constant stirring, 65 ml (0.98 mol) of chlorosulfonic acid is added dropwise to the reaction mixture while maintaining the temperature

lower than 10 °C. When the addition of the chlorosulfonic acid had been completed the mixture was stirred for half an hour and the solvent removed on the rotavapor. The product is purified by recrystallization from an 80% ethanol/20% water mixture. 197 g of 2-(*tert*-butylamino)ethyl sulfate is obtained.

50 g (0.27 mol) of the recrystallized product is dissolved in 25 mL of water at approximately 35 °C and this solution is added dropwise to a well-stirred solution of 140 g of potassium hydroxide dissolved in 150 mL of water. The solution is stirred for about half an hour and then TBA is steam distilled. The distillate is cooled in an ice bath and potassium hydroxide added until the appearance of two layers. The upper TBA layer is isolated with a separating funnel and dried over calcium hydride. It is then distilled twice over calcium hydride to obtain 19 g of TBA (yield 70%, bp 91 °C).

3.4.5.2. Synthesis of α,ω-Bis Methacrylate-Terminated PolyTBA

165 mL (2.06 mol) of dry THF is placed in a two-necked flask thermostated at 20 °C and provided with a magnetic stirrer rod, an inlet for dry nitrogen, and a glass neck equipped with a rubber septum. Under vigorous stirring 150 μL (0.89 mmol) of triflic anhydride is added quickly through the septum with a hypodermic syringe. After 1.5 min, 4.5 mL (35 mmol) of TBA is added to the system. The mixture is stirred for 10 min at 20 °C before adding 0.75 mL (8.9 mmol) of methacrylic acid. One hour after the acid is added the polymer is precipitated in 1.5 L of methanol to which 9 mL of an aqueous 1N sodium hydroxide solution has been added. After filtering off the precipitate and threefold washing with methanol and drying in vacuum, 4.0 g (yield 80%) of α,ω-bis methacrylate-terminated polyTBA with $M_n = 7000\ \mathrm{g\,mol^{-1}}$ and $M_w/M_n = 1.12$ is obtained.

3.4.6. Poly(2-Methyl-1,3-Oxazoline) and Hydrolysis to LPEI[47]

3.4.6.1. Synthesis of Poly(2-Methyl-1,3-Oxazoline)

A solution of 4.97 mL (0.058 mol) of 2-methyl-1,3-oxazoline in 20 mL of acetonitrile is placed in a glass tube under argon atmosphere. The reaction vessel is thermostated at 80 °C and 6.6 μL (5.8×10^{-4} mol) of methyl triflate is added to the solution. The mixture is stirred for 24 hours at 80 °C. Then, the solution is poured into 200 mL of cold ether (0 °C) and the precipitate filtered off. After drying in vacuum, 4.9 g (98% yield) of poly(2-methyl-1,3-oxazoline) with $M_n = 7800\ \mathrm{g\,mol^{-1}}$ (by GPC, polystyrene standards) is obtained.

3.4.6.2. Hydrolysis to LPEI

2.5 g (0.53 mol L^{-1}) of the poly(2-methyl-1,3-oxazoline) and 33 mL of a 37% HCl (8 mol L^{-1}) solution are placed in a two-necked flask containing 22 mL of water. The solution is refluxed for 24 hours. The precipitate is filtered off and redissolved in 150 mL of water. By adding sodium hydroxide the pH is increased to 9, the precipitate is filtered off and washed with water until pH 7, yielding 1.2 g (98%) of LPEI with a molecular weight of 3800 g mol^{-1} (based on the molecular weight of the starting material).

3.4.7. α,ω-Dihydroxy-Poly(Epichlorohydrin) by AM Polymerization[74]

62 g (1 mol) of ethylene glycol is placed in a 5-L three-necked flask fitted with a mechanical stirrer, dropping funnel, thermometer, and condenser, thermostated in a water bath at 20 °C. 3 g of BF_3·EG complex (80% mol of BF_3·ethylene glycol complex and 20% of free ethylene glycol) (0.02 mol of pure BF_3) is added and, under constant stirring, epichlorohydrin is slowly added at a rate of approximately 40 g per hour. The addition of ECH is continued for 48 hours, during which time 1940 g (21 mol) of ECH is introduced into the system. As the viscosity of the reaction mixture increases considerably toward the end of the reaction, particular care should be taken to remove effectively the heat of the reaction, so that the temperature of the reaction mixture does not exceed 25 °C. After adding the required amount of ECH the reaction mixture is kept for an additional 2 hours at 25 °C. Then, the catalyst is neutralized by adding 100 g of solid CaO. The temperature of the reaction mixture is raised to 60 °C and the suspension of CaO is stirred vigorously for 2 hours. Then, CaO is filtered off. The filter is heated to maintain the temperature of the solution at approximately 60 °C in order to reduce the viscosity of the mixture and facilitate filtration. The product is further purified by treatment with an ion-exchange resin (Zenolit KMP, BDH Chemicals). To eliminate all remaining volatile parts the product is heated to 90 °C under a vacuum of 1 mm Hg and a stream of nitrogen is passed through from a capillary for 3 hours. α,ω-polyECH with a molecular weight of 2000 g mol^{-1} and $M_w/M_n = 1.13$ is obtained. The yield is close to 100%.

3.4.8. PolyTHF–PolyTBA Block-Copolymer[44]

30 mL (12.3 mmol) of dry THF is placed in a two-necked flask fitted with a magnetic stirrer rod and an inlet for dry argon. The reaction vessel is thermostated in a water bath at 20 °C. 41.7 μL (0.37 mmol) of methyl triflate is introduced under vigorous stirring through a rubber septum by means of a syringe. After 20 min 1.42 mL (11.6 mmol) of TBA is added to

the solution. The reaction mixture is stirred for an additional 10 min at 20 °C and the solution poured into 300 mL of dry cold methanol. The polymeric precipitate is filtered off and dried in vacuum, yielding 2.25 g of the AB-block-copolymer with a molecular weight of 3000 g mol^{-1} for both the polyTHF and polyTBA segment.

3.4.9. PolyTHF–Poly(2-Methyl-1,3-Oxazoline) Block-Copolymer[47]

100 mL of dry THF is placed in a two-necked flask fitted with a magnetic stirrer rod and a nitrogen inlet, thermostated in a water bath at 25 °C. Under continuous stirring, 1.39 mL (12.3 mmol) of methyl triflate is injected. 14 min after the initiator has been added, 0.208 mL (2.46 mmol) of 2-methyl-1,3-oxazoline is added to the reaction mixture. The solution is stirred for 1 hour at 25 °C. Then, the remaining THF is evaporated in vacuum, yielding 38 g of oxazolinium-terminated polyTHF. The resulting polymer is dissolved in 100 mL of dry acetonitrile and 5.2 mL (61.5 mmol) of 2-methyl-1,3-oxazoline is added. The mixture is kept for 2 hours at 80 °C. After cooling to room temperature the triflate ions are exchanged for hydroxyl ions by adding 1 g of $Ca(OH)_2$ powder. After 30 min stirring, calcium triflate and excess $Ca(OH)_2$ are filtered off. The filtrate is concentrated by evaporating half of the solvent on the rotavapor. The mixture is added to 500 mL of water and the resulting emulsion is freeze-dried, yielding 43 g of a polyTHF–poly(2-methyl-1,3-oxazoline) block-copolymer with $M_n = 2740$ g mol^{-1} [2300 g mol^{-1} polyTHF and 440 g mol^{-1} poly(2-methyl-1,3-oxazoline)].

3.4.10. Poly(Ethylene Oxide)–Poly(2-Methyl-1,3-Oxazoline) Block-Copolymers[49]

3.4.10.1. Preparation of Methanesulfonic Acid Diester of Poly(Ethylene Glycol) (MsPEO)

While cooling with ice, 2.20 mL (28.4 mmol) of mesylchloride is added to a stirred solution of 9.92 g of poly(ethylene glycol) ($M_n = 1400$ g mol^{-1}) in 30 mL of dry pyridine. After the addition is completed, the reaction mixture is stirred at 0 °C for 4 hours and kept in a freezer for 10 days. The mixture is then poured into 300 mL of ice-water. This solution is extracted with chloroform three times. The chloroform extracts are combined and dried over anhydrous K_2CO_3. After evaporation of chloroform at room temperature in vacuum, diethylether is added to the remaining pyridine solution until the mixture becomes turbid. Then, the mixture is allowed to rest for a few minutes and a brown liquid precipitate is decanted off. The skimmed solution is cooled in an ice bath and the resulting white precipitate is filtered off. After recrystallizing the product over a chloroform ether solution and lyophilization from benzene solution, 5.97 g (53% yield) of MsPEO is obtained.

3.4.10.2. Preparation of the Block-Copolymer

4.84 g (3.02 mmol) of MsPEO (MW = 1560 $g mol^{-1}$) is transferred to a 200-mL glass tube containing 60 mL of acetonitrile under nitrogen atmosphere. 19.66 mL (229 mmol) of 2-methyl-1,3-oxazoline is added and the tube is sealed and heated to 100 °C. After 20 hours the mixture is poured into 3.5 L of diethylether. The precipitated block-copolymer is isolated by decantation and purified further by reprecipitation from chloroform in diethylether yielding 23.62 g (96%) of the block-copolymer (M_n = 8400 $g mol^{-1}$).

3.4.11. PolyTHF–Polybutadiene Polyblock-Copolymers[12]

3.4.11.1. Sulfonium-Terminated Bifunctional PolyTHF

100 mL (0.186 mol) of freshly distilled polyTHF is placed in a two-necked flask provided with a nitrogen inlet and a magnetic stirrer rod, and thermostated at 25 °C. Under continuous stirring 414 μL (2.46 mmol) of triflic anhydride is added through a septum by means of a hypodermic syringe. After 20 min the polymerization is terminated by introducing 2.12 mL (24 mmol) of thiolane. The reaction mixture is held for 30 min at room temperature and then poured dropwise and under vigorous stirring into 1 L of cold water. The precipitate is isolated and purified by washing with cold water and dried in vacuum at room temperature yielding 20 g (20% yield) of polymer with a molecular weight of 7800 $g mol^{-1}$ and M_w/M_n = 1.4.

3.4.11.2. Carboxylate Ditelechelic Polybutadiene

5 g (2.5 mmol) of carboxylic-terminated polybutadiene (Goodrich Comp) (MW = 2000 $g mol^{-1}$) is dissolved in 100 mL of THF. The solution is then poured into 2 L of water, containing 0.5 g NaOH. The precipitated polymer is isolated by decantation, washed, and dried in vacuum.

3.4.11.3. PolyTHF Polybutadiene Polyblock-Copolymers

A solution of 2 g polyTHF-bis-thiolanium salt (MW = 7800 $g mol^{-1}$) and 1.23 g carboxylate ditelechelic polybutadiene (MW = 2000 $g mol^{-1}$) in 50 mL of THF is added slowly to 2 L cooled water. The precipitate mixture is filtered off, washed with water, and dried under vacuum yielding 2.84 g of polyblock-copolymer polyTHF–polybutadiene, MW = 50,000 g mol^{-1} (by GPC based on polyTHF standards).

3.4.12. PolyTHF Networks using Spontaneous Termination of Thietane Polymerization[63]

25 mL of dry THF (0.31 mol) is placed in a two-necked flask equipped with a nitrogen inlet and a magnetic stirrer rod. The polymerization is initiated by introducing 0.1 mL (0.59 mmol) of triflic anhydride. The reaction vessel is kept for 16 min at 25 °C. Then, the mixture is cooled to 0 °C and 320 μL (3.6 mmol) of thietane added. Immediately after thietane has been added the mixture is shaken and poured on a polypropylene plate. The solution is allowed to warm up slowly to room temperature and within 2 min a gel is formed. The remaining solvent and thietane are removed by heating in the oven to 50 °C, leaving a film of crosslinked polyTHF which can easily be detached from the substrate.

REFERENCES

1. W. J. Bailey *et al.*, in *Ring Opening Polymerization,* ed. by T. Saegusa and E. J. Goethals, *ACS Symp. Ser.* **59**, 38 (1977).
2. E. J. Goethals, *J. Polym. Sci., Polym. Symp.* **56**, 271 (1976).
3. F. R. Jones and P. H. Plesch, *Chem. Commun.* 1231 (1969).
4. S. Penczek and S. Slomkowski, in *Comprehensive Polymer Science* (Pergamon Press, Oxford, 1989), Vol. 3, Part 1, p. 725.
5. I. Yilgör, J. S. Riffle, and J. E. McGrath, in *Reactive Oligomers,* F. W. Harris and H. J. Spinelli, *ACS Symp. Ser.* **282**, 161 (1985).
6. E. J. Goethals and R. De Clercq, Lecture at International Symposium on Mechanisms and Kinetics of Polymerizations, Paris, 9–13 Sept. 1990. *Makromol. Chem., Macromol. Symp.* **47**, 151 (1991).
7. T. Biedron, R. Szymanski, P. Kubisa, and S. Penczek, *Makromol. Chem., Macromol. Symp.* **32**, 155 (1990).
8. S. Penczek and K. Matyjaszewski, *J. Polym. Sci., Polym. Symp.* **56**, 255 (1976).
9. Y. Yamashita and K. Chiba, *Polymer J.* **4**, 200 (1973).
10. E. Franta, L. Reibel, and J. Lehmann, *J. Polym. Sci., Polym. Symp.* **56**, 139 (1976).
11. S. Smith and A. J. Hubin, *J. Macromol. Sci.-Chem.* **A7**, 1399 (1973).
12. F. D'Haese and E. J. Goethals, *Br. Polym. J.* **20**, 103 (1988).
13. Polymer Laboratories Ltd., Church Stretton, Shropshire, U.K.
14. S. Penczek, J. Fijgin, W. Sadowska, and M. Tomaszewicz, *Makromol. Chem.* **116**, 203 (1968).
15. G. Opitz, *Plaste Kautsch.* **22**, 951 (1975).
16. S. Penczek and K. Kubisa, in *Comprehensive Polymer Science* (Pergamon Press, Oxford, 1989), Vol. 3, Part 1, p. 807.
17. D. Van Meirvenne and E. J. Goethals, *Makromol. Chem., Macromol. Suppl.* **15**, 61 (1989).
18. D. Van Meirvenne and E. J. Goethals, *New. Pol. Mat.* **1**(4), 281 (1990).
19. E. Franta, E. Gerard, Y. Gnanou, L. Reibel, and P. Rempp, *Makromol. Chem.* **191**, 1689 (1990).

20. D. Horn, in *Polymeric Amines and Ammonium Salts,* ed. by E. J. Goethals (Pergamon Press, Oxford 1980), p. 333.
21. T. Saegusa, H. Ikeda, and H. Fujii, *Macromolecules* **5**, 108 (1972).
22. P. A. Gembitskii, A. I. Chmarin, N. A. Kheshcheva, and D. S. Zhuk, *Vysokomol. Soedin., Ser. A* **20**, 1505 (1978).
23. K. Weyts and E. J. Goethals, *Polym. Bull.* **19**, 13 (1988).
24. T. Saegusa and S. Matsumoto, *J. Polym. Sci. A*1 **6**, 1559 (1968).
25. S. Kobayashi, H. Uyama, M. Ogaki, T. Yoshida, and T. Saegusa, *Macromolecules* **22**, 4412 (1989).
26. R. Tanaka, I. Ueoka, Y. Takaki, K. Kataoka, and S. Suito, *Macromolecules* **16**, 849 (1983).
27. E. J. Goethals, E. Schacht, P. Bruggeman, and P. Bossaer, in *Ring Opening Polymerization,* ed. by T. Saegusa and E. J. Goethals, *ACS Symp. Ser.* **59**, 1 (1977).
28. P. Bossaer, *Ph.D. Thesis,* University of Ghent (1978).
29. E. J. Goethals, P. Bossaer, and R. Deveux, *Polym. Bull.* **6**, 121 (1981).
30. E. J. Goethals, A. Munir, R. Deveux, and L. Vandenberghe, *Makromol. Chem., Rapid Commun.* **3,** 515 (1982).
31. A. Munir and E. J. Goethals, in *Polymeric Amines and Ammonium Salts,* ed. by E. J. Goethals (Pergamon Press, Oxford, 1980), p. 19.
32. M. Van de Velde and E. J. Goethals, *Makromol. Chem., Macromol. Symp.* **6**, 271 (1986).
33. A. Munir and E. J. Goethals, *J. Polym. Sci., Polym. Chem. Ed.* **19**, 1985 (1981).
34. S. Kobayashi and T. Saegusa, in *Ring Opening Polymerization,* ed. by K. J. Ivin and T. Saegusa (Elsevier, London ,1984), p. 761.
35. M. Litt, A. Levy, and J. Herz, *J. Macromol. Sci.-Chem.* **A9,** 703 (1975).
36. S. Kobayashi, E. Masuda, S. Shoda, and Y. Shimano, *Macromolecules* **22**, 2878 (1989).
37. S. Kobayashi, K. Marikowa, and T. Saegusa, *Macromolecules* **8**, 952 (1975).
38. R. J. Katnik and J. Schaefer, *J. Org. Chem.* **33**, 384 (1968).
39. Y. Yamishita, I. Katsumi, and I. Kiochi, *J. Polym. Sci., Polym. Lett.* **17**, 1 (1979).
40. E. J. Goethals, D. Van Meirvenne, and R. De Clercq, *Makromol. Chem., Macromol. Symp.* **13/14**, 175 (1988).
41. S. Penczek, P. Kubisa, and R. Szymanski, *Makromol. Chem., Macromol. Symp.* **3**, 203 (1986).
42. Y. Okamoto, in *Ring-Opening Polymerization,* ed. by J. E. McGrath, *ACS Symp. Ser.* **286**, 362 (1985).
43. A. Kuzaev and O. Olkova, *Vysokomol. Soedin.,* **24**, 2194 (1982).
44. E. J. Goethals, M. Van de Velde, and A. Munir, in *Cationic Polymerization and Related Processes,* by E. J. Goethals (Academic Press, New York, 1984), p. 387.
45. Y. Yagci, *Polym. Commun.* **26**, 7 (1985).
46. F. D'Haese, E. J. Goethals, Y. Tezuka, and K. Imai, *Makromol. Chem., Rapid Commun.* **7**, 165 (1986).
47. F. Hosteaux and E. J. Goethals, unpublished results.
48. V. Percec, *Polym. Bull.* **5**, 643 (1981).
49. M. Miyamoto, Y. Sano, T. Saegusa, and S. Kobayashi, *Eur. Polym. J.* **19**, 955 (1983).
50. T. Saegusa and H. Ikeda, *Macromolecules* **6**, 805 (1973).
51. D. H. Richards, S. B. Kingston, and T. Souel, *Polymer* **19**, 68 (1978).
52. E. Franta, J. Refai, C. Durand, and L. Reibel, *Makromol. Chem., Macromol. Symp.* **32**, 169 (1990).
53. P. Dreyfuss, T. Adaway, and J. P. Kennedy, *Appl. Polym. Symp.* **30**, 183 (1977).
54. T. Saegusa, S. Kobayashi, and A. Yamada, *Macromolecules* **8**, 390 (1975).
55. T. Saegusa, A. Yamada, and S. Kobayashi, *Polym. J.* **11**, 53 (1979).
56. P. D. Trivedi and D. N. Schulz, *Polym. Bull.* **3**, 37 (1980).
57. R. Asami, M. Takaki, Y. Harima, and S. Tsuzuki, *Polym. J.* **13**, 141 (1981).

58. A. Dondas, P. Lutz, L. Reibel, P. Rempp, and E. Franta, *Makromol. Chem.* **179**, 2549 (1978).
59. A. Munir and E. J. Goethals, *Makromol. Chem., Rapid Commun.* **2**, 693 (1981).
60. A. Munir and W. D. Daly, *Makromol. Chem., Rapid Commun.* **4**, 589 (1983).
61. J. Sierra Vargas, J. G. Zilliox, P. F. Rempp, and E. Franta, *Polym. Bull.* **3**, 83 (1980).
62. M. Takaki, R. Asami, and T. Kuwabara, *Polym. Bull.* **7**, 521 (1982).
63. G. Trossaert and E. J. Goethals, unpublished results.
64. R. Asami, M. Takaki, K. Kyuda, and E. Asakura, *Polym. J.* **15**, 139 (1983).
65. E. J. Burgess, A. V. Cunliffe, D. H. Richards, and D. Thompson, *Polymer* **19**, 334 (1978).
66. R. Asami, M. Takaki, K. Kita, and E. Asakura, *Polym. Bull.* **2**, 713 (1983).
67. H. J. Kress and W. Heitz, *Makromol. Chem., Rapid Commun.* **2**, 427 (1981).
68. E. J. Goethals and M. A. Vlegels, *Polym. Bull.* **4**, 523 (1981).
69. M. Miyamoto, K. Naka, M. Tokumizu, and T. Saegusa, *Macromolecules* **22**, 1604 (1989).
70. S. Kobayashi, H. Uyama, and H. Shirasaka, *Makromol. Chem., Rapid Commun.* **11**, 11 (1990).
71. S. Kobayashi, M. Kaku, S. Sawada, and T. Saegusa, *Polym. Bull.* **13**, 447 (1985).
72. S. Kobayashi, E. Masuda, S. Shoda, and Y. Shimano, *Macromolecules* **22**, 2878 (1989).
73. R. C. Schulz and E. Schwarzenbach, *Makromol. Chem., Macromol. Symp.* **13/14**, 495 (1988).
74. T. Biedron, P. Kubisa, and S. Penczek, *J. Polym. Sci., Polym. Chem. Ed.* **29**, 619 (1991).
75. P. Cohen, M. J. M. Abadie, F. Schué, and D. H. Richards, *Polymer* **23**, 1350 (1982).

4

Cationic Polymerization of Acyclic Monomers

Oskar Nuyken and Stephen D. Pask

4.1. INTRODUCTION

Polymerizations are generally considered to be "cationic" if they involve, formally, a positively charged growing chain end. Such, formally positive, growing chain ends may be identifiable (e.g., by conductivity) or they may be present only as transition states (e.g., pseudocationic polymerizations).

This overview is concerned only with the cationic polymerization of ethylenic monomers and does not consider either heterocyclic or acetal polymerization via cationic mechanisms.

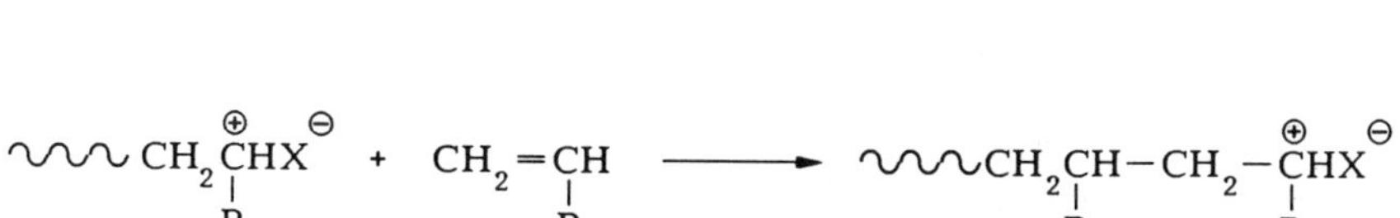

Oskar Nuyken • Macromolecular Chemistry I, University of Bayreuth, D-8580 Bayreuth, Germany. *Stephen D. Pask* • Bayer A. G., KA/F, D-4047 Dormagen, Germany.

New Methods for Polymer Synthesis, edited by W. J. Mijs. Plenum Press, New York, 1992.

Essentially, the type of cationic polymerization considered here involves the following, individual reaction steps:

Initiation

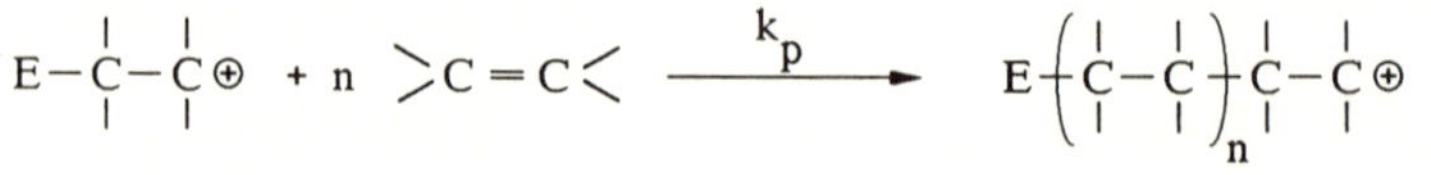

Propagation

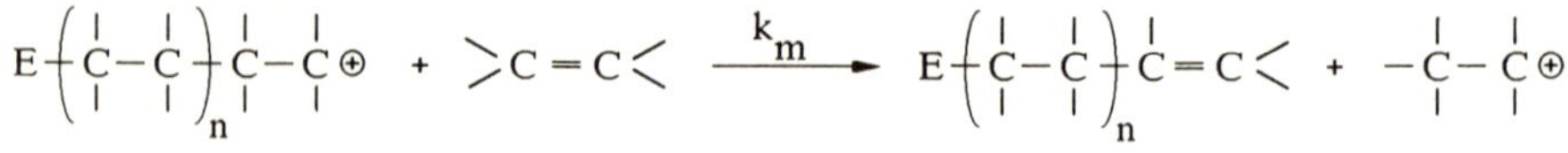

Transfer to monomer

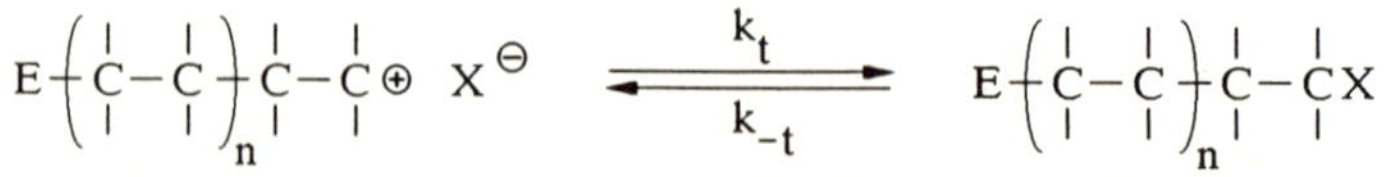

Termination

In some systems (e.g., vinyl ethers/HI/ZnI_2) a direct, simple relationship can be identified between the initiator concentration and the rate of polymerization and the molar mass of the products. However, for many systems involving Brønsted or Lewis acid initiating systems the true initiator (E^+ in the above scheme) remains unclear.

The control of propagation, transfer to monomer and termination, is to a large extent a function of the counterion (gegenion) but it is also a function of temperature and solvent. It is the elucidation of these details in the last 20 years that has led to carbocationic polymerization becoming a realistic technique for the synthesis of well-defined polymers.

Nevertheless, the kinetics and mechanisms remain, in most cases, matter for discussion. The considerable reactivity of carbocations with respect to a number of impurities, but in particular with respect to the ubiquitous water, makes accurate kinetic work a very demanding experimental subject. The different reactivities of solvent-, monomer-, or counterion-complexed growing chains remain to be quantified and indeed are the subject of considerable controversy. The importance of these reactivities for the synthetic chemist lies, in part, in the effect of the various mechanisms on the stereochemistry of the resulting products: an aspect of carbocationic polymerization which has been sadly neglected. The importance also lies in that the synthetic chemist needs to be able to control the number of growing species of different reactivity in order to be able to obtain products having the required molar mass and molar mass distribution under the least demanding experimental conditions.

In this chapter the authors have not attempted an updated review of the cationic polymerization of acyclic monomers nor do they make any claim to an exhaustive description of the synthetic pathways for those polymers described.

General reviews on the cationic polymerization of acyclic monomers have been published recently[1–6] as have more specialized reviews concerning the preparation of telechelic polymers[7] and macromonomers[8] using cationic techniques.

The following is an attempt to indicate the variety of well-defined polymers which have now been prepared via cationic polymerization techniques. An attempt has been made to indicate the experimentally most detailed literature rather than the most recent. Where the monomers are not readily available references containing preparative methods are included. Lastly, the scope and limitations of the methods are described.

4.2. STYRENE

4.2.1. Polymerization

Industrially, styrene is not polymerized using carbocationic techniques. This is not only due to the alternatives available for producing polystyrene, but also to the fact that the physical properties of cationally synthesized polystyrene are unattractive. Nevertheless, the cationic polymerization of this monomer has been the subject of numerous scientific investigations. The most important investigations have been summarized by Kennedy.[10] From the table given in Ref. 10, one can see that styrene has been polymerized with practically all known cationic initiators: Brønsted acids, Lewis acids, alkylmetal systems, carbenium ion salts, iodine, clay, and others. The emphasis of much of this work has been on the kinetics of the polymerization reaction.[9–14]

4.2.2. Initiators

Two significant new areas have been of interest since Mathieson's review[11] appeared: alkylaluminum compounds and pseudocationic polymerization.

(a) Alkylaluminum Initiator System. Styrene can be polymerized by $AlEt_3$ in the presence of suitable protogens such as H_2O or HCl.[15] $AlEt_2Cl$ was found to be inactive for styrene polymerization, while $AlEt_2Cl$/protogens (e.g., H_2O, CH_3COOH, C_6H_5COOH) initiate a

$$(AlEt_3)_2 + H_2O \longrightarrow Et_2Al{-}O{-}AlEt_2 \; (+\; 2\; EtH)$$

$$Et_2Al{-}O{-}AlEt_2 + H_2O \longrightarrow H^+[Et_2Al{-}O{-}AlEt_2]^-$$

SCHEME 4.1.

polymerization.[16] Detailed studies by Saegusa *et al.*[17] with various $AlEt_3/H_2O$ ratios have shown that this ratio has to be kept between 1.0 and 0.5 in order that polymerization takes place. The role of water was explained on the basis of Scheme 4.1.

Another active system described by the same authors is $AlEt_3/CH_3O{-}CH_2Cl$ in CH_2Cl_2 at $-78\,°C$ (Scheme 4.2). Direct evidence for this suggestion comes from endgroup analysis. Kennedy has investigated styrene polymerization initiated by $AlEt_2/Cl/RCl$ systems in which R was n-butyl, *i*-propyl, *sec*-butyl, *i*-butyl, *tert*-butyl, allyl, and others.[18,19] Certain combinations of Et_2AlCl/RCl are efficient initiators for styrene and give rise to products having high molar masses.[20] The initiation is explained according to Eq. (4.1).

$$RCl + AlEt_2Cl \rightleftharpoons R^+[AlEt_2Cl_2]^- \tag{4.1}$$

The unexpected low initiating efficiency of benzyl chloride has been attributed to self-alkylation:

$$C_6H_5CH_2Cl + AlEt_2Cl \longrightarrow C_6H_5CH_2^+ + AlEt_2Cl_2^- \longrightarrow \text{+}C_6H_4CH_2\text{+}_n \tag{4.2}$$

Kennedy also studied the effect of temperature, solvent, and nature of R on styrene polymerization.[19] The most important observations are that the rate of polymerization increases with increasing solvent polarity and is greatest at lower temperature in polar solvents.

(b) Pseudocationic Polymerization. One of the most interesting developments in this field is the polymerization of styrene initiated by water-free $HClO_4$.[21,22] This polymerization reaction involves three stages. In the first stage an orange-red color is observed, electrical conductivity can be detected, and the polymerization is fast. The second stage is characterized by the absence of electrical conductivity and color, conductivity and color reappearing in the third stage. The behavior can be explained in terms of the formation of intermediate perchlorate esters.[23] Macromolecular and

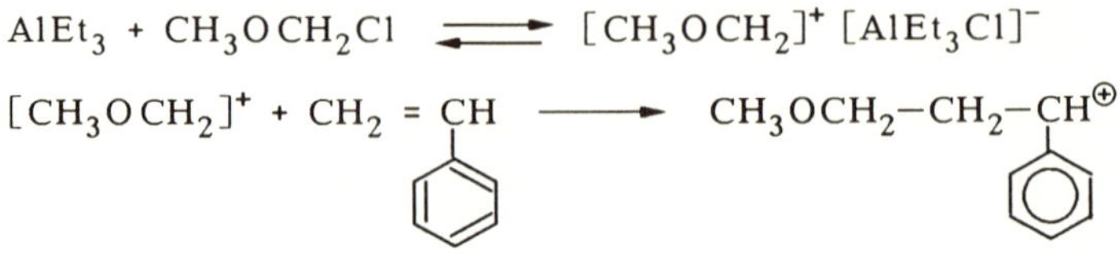

SCHEME 4.2.

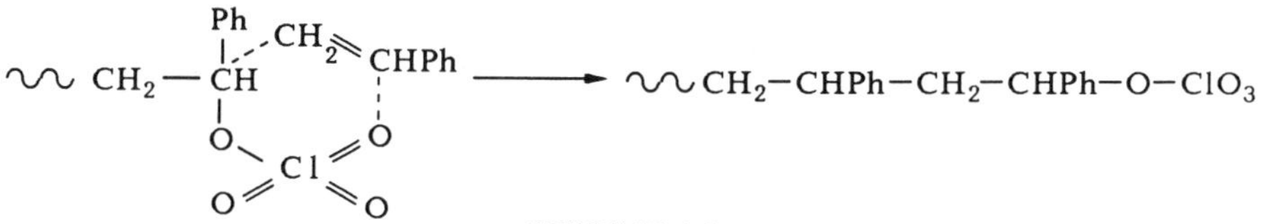

SCHEME 4.3.

model covalent species in the perchloric acid initiated polymerization of styrene have been directly observed by NMR.[23] For the propagation two different pathways have been discussed: a multicenter rearrangement (Scheme 4.3) and an ionic intermediate with preliminary dissociation (Scheme 4.4). The relatively small effect of the solvent polarity on the polymerization rate supports Scheme 4.3, since multicenter rearrangements and a majority of pericyclic reactions are insensitive to solvent polarity.

However, conductivity and color support, at least partly, the mechanism in Scheme 4.4. The exact mechanism is still a matter of controversial discussion.[24,25] The bimodal molar mass distribution observed in the polymerization of styrene initiated by CH_3COClO_4 can be explained by postulating two types of growing species (the mechanisms in Schemes 4.10 and 4.11). This view is supported by the fact that the higher molar mass is sharply reduced if a common salt was added[26] or the polarity of the solvent was reduced.[27]

Another interesting initiating system is trifluoroacetate activated by the acid. It has been shown that 1-phenylethyl trifluoroacetate alone is inactive as an initiator for the polymerization of styrene.[28,29] Indeed, the polymerization initiated by the corresponding acid was retarded by the ester. From mechanistic investigations using deuterated styrene and 1-phenylethyl trifluoroacetate and trifluoroacetic acid it has been shown that the polymer head groups were undeuterated styrene [Eq. (4.4)] moieties rather than singly protonated ones [Eq. (4.3)]. This clearly indicates that the acid

$$CF_3COOH + CD_2{=}CDPh \longrightarrow CF_3COO^{\ominus}\ HCD_2{-}CDPh{\sim\sim}^{\oplus} \qquad (4.3)$$

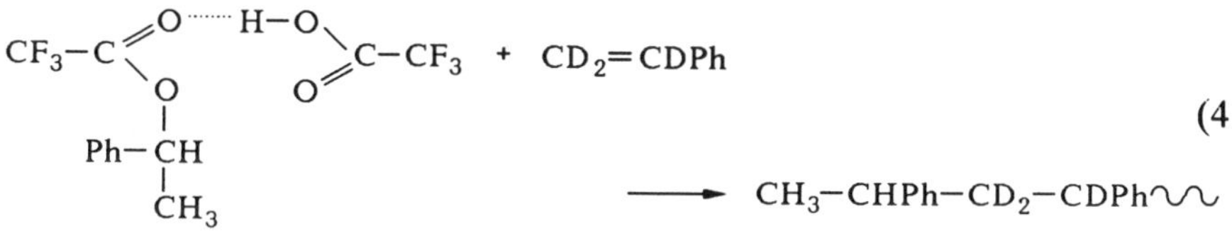

(4.4)

$$\sim\sim CH_2{-}CHPh{-}O{-}ClO_3 \rightleftharpoons \sim\sim CH_2{-}CHPh^{\oplus}ClO_4^{\ominus} \xrightarrow{M}$$

$$\sim\sim CH_2{-}CHPh{-}CH_2{-}CHPh^{\oplus}ClO_4^{\ominus} \rightleftharpoons \sim\sim CH_2{-}CHPh{-}CH_2{-}CHPhO{-}ClO_3$$

$$\longrightarrow \text{polymer}$$

SCHEME 4.4.

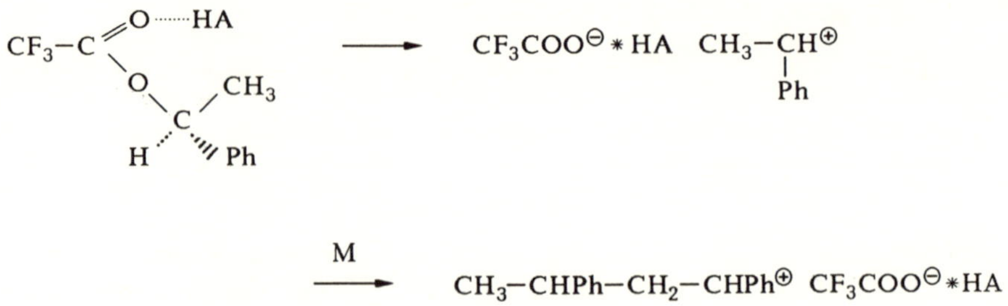

SCHEME 4.5.

functions as an activator and that initiation by the acid itself is relatively slow.[30] Equations (4.3) and (4.4) show an overall reaction rather than a mechanism. From mechanistic studies using an optically active ester the authors concluded that an ionic mechanism (Scheme 4.5) is favored over a multicenter rearrangement. Their main arguments are:

(a) that the rate of racemization is higher than the rate of incorporation of the ester, and
(b) the complete loss in optical activity in the polymer[30] (see Scheme 4.5).

The claim of a patent recently published by Kennedy[31] is the living carbocationic polymerization of styrene. The initiator used is shown in Scheme 4.6.

Further addition of monomer results in an increase in the molar mass of the products. Termination by indane formation could be avoided by addition of monomer in small portions.[32,33] similar activation by Lewis acids was observed for 1-phenylethyltrifluoroacetate and trichloroacetate. The value $\bar{M}_w/\bar{M}_n = 1.15$ was found for the polymerization of styrene if the trichloroacetate was used in combination with BCl_3.[16]

1-phenylethyl-trifluoromethanesulfonate has been shown to be unstable even at low temperatures (−78 °C). Thus, there is no real evidence in favor of the pseudocationic polymerization of styrene in CH_2Cl_2 initiated by CF_3SO_3H and a classical ionic mechanism must be preferred to explain the experimental results.[34]

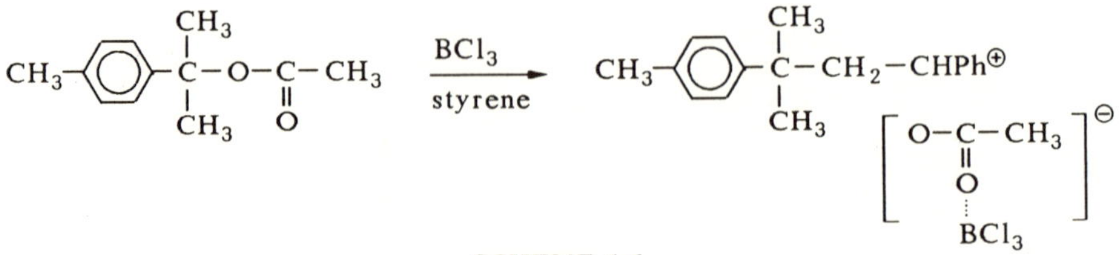

SCHEME 4.6.

4.2.3. Experimental Procedures

A classical cationic-solution polymerization of styrene is described in detail by Overberger[35] using $SnCl_4$ as catalyst.

Freshly distilled styrene was pipetted into a dried screw-cap bottle, then an approx. 1:1 mixture of carbon tetrachloride and nitrobenzene was added to the monomer. This mixture was cooled to 0 °C, then a solution of $SnCl_4$ in carbon tetrachloride was added. After a reaction period of 5 h the polymerization was stopped by dropping the polymerized mixture into an excess of methanol. The precipitated polymer was separated from the solvent by filtration redissolved in methyl ethyl ketone and reprecipitated in methanol. This procedure yields >80% polymer.

A general procedure for metal alkyl–water initiation of styrene is described by Saegusa *et al.*[17]: Freshly distilled $AlEt_3$ ($bp_{7\,Torr}$ 80 °C) was distilled into an organic solvent (toluene, hexane, CH_2Cl_2). Water was added dropwise to this solution with vigorous stirring at 0 °C.

The polymerization was carried out in sealed tubes under a nitrogen atmosphere. In a tube a mixture of monomer, solvent, and catalyst was cooled to the polymerization temperature, then a cocatalyst (acetyl chloride or *t*-butyl chloride) dissolved in the same solvent was added. This mixture was allowed to stand for a desired period of time. After polymerization the reaction mixture was dropped into methanol and the precipitated polymer was isolated by filtration and characterized by the usual methods.

A detailed description of a styrene polymerization initiated by trifluoroacetic acid alone and in combination with 1-phenylethyl trifluoroacetate

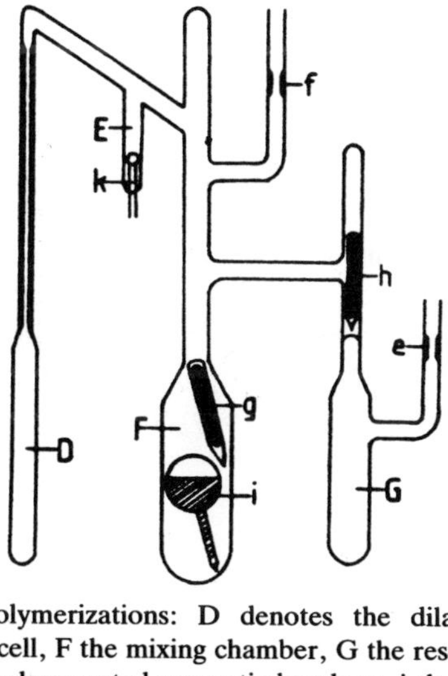

FIG. 4.1. Assembly for polymerizations: D denotes the dilatometer compartment with capillary, E the conductivity cell, F the mixing chamber, G the reservoir for monomer solution, e and f constriction, g and h glass-coated magnetic breakers, i the initiator phial, and k the Pt wires.

is given by Obrecht and Plesch.[28] The polymerization was carried out in the apparatus shown in Fig. 4.1. This apparatus was attached to the vacuum line via f. After evacuation the device was dosed with CH_2Cl_2 and styrene by distillation through f into F and the device sealed off from the vacuum line at f. The whole device was thermostated before the acid phial was crushed with the breaker g. The components were mixed carefully and then tipped into the dilatometer D, in which the polymerization could be followed quantitatively. For the polymerization in the presence of the 1-phenylethyl trifluoroacetate the ester was distilled into G, as were styrene and solvent. By opening the break seal with the breaker h both solutions (in F and G) could be combined by tipping the device.

This device also allows the change of conductivity during polymerization to be followed. This apparatus, as well as many others useful for carrying out cationic polymerizations are described in excellent detail in a recently published book.[36]

4.3. STYRENE DERIVATIVES

4.3.1. Ortho- and Para-Methyl Styrene

4.3.1.1. Polymerization

These monomers can be polymerized via radical, anionic, and cationic mechanisms; however, most of the available literature deals with their cationic polymerization. Kanoh *et al.*[37,38] have studied the polymerization of *p*-methyl styrene in detail. These authors have shown that this monomer polymerizes much faster than unsubstituted styrene under similar conditions. They also observed a considerable effect of solvent polarity, the rate increasing as solvent polarity increases.

According to the investigations of Prishchepa *et al.*,[39,40] poly (*p*-methyl styrene) synthesized in toluene or ethyl chloride with BF_3Et_2O as initiator at 0–130 °C contained up to 35% anomalous structural units. It was proposed that these are formed via a hydride transfer reaction (Scheme 4.7). However, it has not been found possible to reproduce these interesting results, despite considerable effort.[41,42] Thus, Kennedy *et al.* could only identify structures resulting from a conventional 1,2-addition in polymers prepared from both *o*- and *p*-methyl styrene.

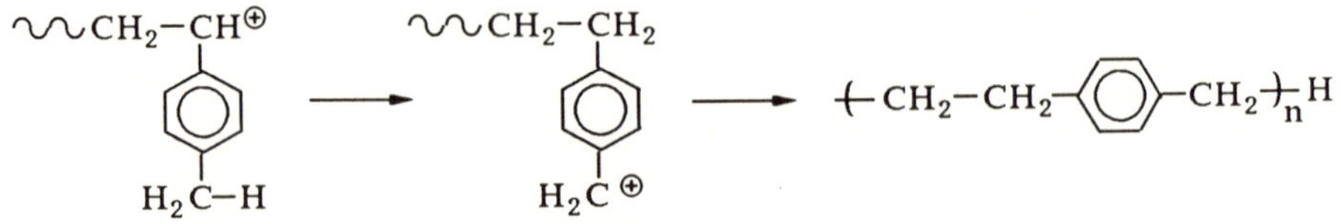

SCHEME 4.7.

A strong solvent effect has also been observed by Heublein and Dawczynski,[43] who report rate constants of propagation in four different solvents.

Polymerization of *p*-methyl styrene initiated with acetyl perchlorate in the presence of $Bu_4N^+ClO_4^-$ in CH_2Cl_2 at −78 °C shows the characteristics of a living polymerization so that this method[44] is probably the most convenient for preparing well-defined polymers. Very narrow molar mass distributions ($\bar{M}_w/\bar{M}_n \approx 1.1$–1.4) are possible using this technique.

For *p*-methyl styrene in $CHCl_3$ at −30 and −50 °C the initiating system, cumyl acetate/BCl_3, has been studied by Kennedy and Faust.[45] Despite finding a relatively broad molar mass distribution ($\bar{M}_w/\bar{M}_n \approx 2$–5) these authors maintain that this system is also living.

4.3.1.2. Experimental Procedure

The monomers are distilled twice over CaH_2 under reduced pressure and stored in ampules under nitrogen. $AcClO_4$ was prepared from acetyl chloride and silver-perchlorate.[46] n-$Bu_4N^+ClO_4^-$ is available in polarographic quality and can be used without further purification except drying. All solvents used should be purified by standard methods.

Polymerizations were carried out under dry nitrogen in baked glass tubes equipped with three-way stopcocks. Each reaction was started by adding an $AcClO_4$ solution to a monomer solution via a dry syringe and, after a certain time, the polymerization was stopped by addition of ammoniacal methanol. The polymer conversion was determined gravimetrically and was in good agreement with that calculated by analyzing the residual monomer gas chromatographically.

The experiments for the living polymerization[45] were carried out in a dry box under dry nitrogen. Otherwise, the techniques used are comparable with those applied for standard organic reactions.

4.3.2. Ortho- and Para-Methoxy Styrene

4.3.2.1. Polymerization

In the same way as other styrene derivatives with donor substituents, *o*- and *p*-methoxy styrene can be polymerized via cationic pathways. As early as 1935, *p*-methoxy styrene was polymerized already by Staudinger and Dreher[47] in benzene at 0 °C using $SnCl_4$ as initiator. Molar masses (determined by cryoscopy) up to $\bar{M}_n = 13{,}000$ were observed. $BF_3 \cdot Et_2O$ was also found to be an active initiator in toluene, methylene chloride, or mixtures of both at temperatures between −78 °C and −20 °C.[48] The latter authors found that molar masses increased with decreasing temperature if toluene was used as solvent, while in methylene chloride no significant change of molar mass with temperature was observed. The samples polymerized in the more polar solvent generally had the higher molar masses. Okamura *et al.*[49] have studied the polymerization kinetics of *p*-methoxy styrene initiated by iodine in methylenchloride and carbon tetrachloride. They found that polymerization rates increase as the polarity of the solvent increases.

Systematic studies of the polymerization of *o*- and *p*-methoxy styrene in several chlorinated solvents at different temperatures have shown that, for the *ortho*-derivative, the ratio of monomer transfer to propagation is not effected by the temperature and polarity of the solvent. In contrast, for the *para*-derivative, this ratio decreases with decreasing temperature and increasing polarity of the solvent.[50]

The polymerization of *p*-methoxy styrene with $SnCl_4/H_2O$ has been studied in detail by Heublein and Dawczynski[51]; molar masses up to $\bar{M}_n = 19{,}000$ could be prepared.

Iodine has also been shown to be an active initiator[52–54] for the polymerization of *p*-methoxy styrene; in CH_2Cl_2 or CCl_4 polymers with $\bar{M}_w/\bar{M}_n = 1.3–1.4$ have been synthesized.

A living polymerization was observed if HI/ZnI_2 was used as initiator even at room temperature.[55,56] Similar to the vinylether/HI/ZnI_2 systems, narrow molar mass distributions ($\bar{M}_w/\bar{M}_n \leq 1.1$) and control of the molar masses by the ratio [Monomer] : [Initiator] were observed. As well as with ZnI_2, it is possible to activate these polymerizations with $Bu_4N^+ClO_4^-$[57] (Scheme 4.8).

The molar mass increases with conversion, however the $\bar{M}_w/\bar{M}_n = 1.5–1.8$ was broader than that found for vinylether polymerizations using an identical initiator system.

Rather unusual initiators have been used by Hall,[58,59] who initiated the polymerization with trialkyl silyl triflate in CH_2Cl_2 and

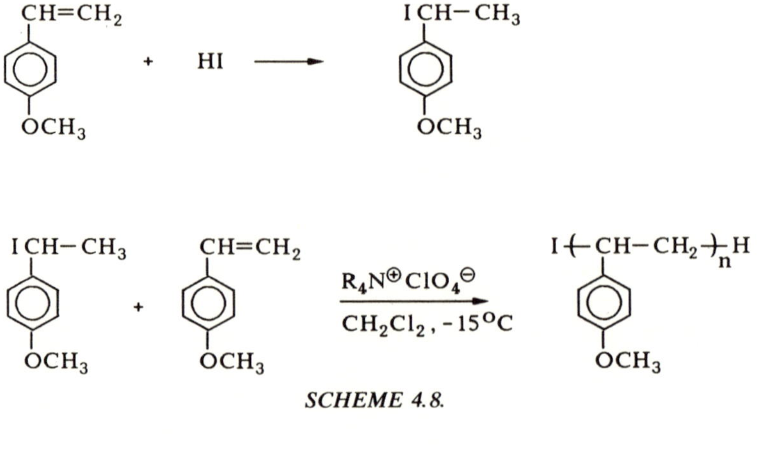

SCHEME 4.8.

bis(trifluoromethane sulfonyl)methane in nitromethane. Also unusual are the initiators which have been used by Heublein such as Ph_3CBr/I_2, Ph_3CSCN/I_2, and triphenyl picrate/picric acid.[60]

The polymerization of *p*-methoxy styrene has also been initiated by aryl-diazonium salts.[61] Using *p*-nitrobenzenediazonium hexafluorophosphate in nitromethane gave complete conversion in less than 1 h at 30 °C. However, whether the initiating cation is the corresponding substituted phenyl-carbenium ion (NO_2-$C_6H_4^+$) or the original diazonium ion (NO_2-C_6H_4- N_2^+) remains uncertain.

4.3.2.2. Experimental Procedure

(a) $HI/NR_4^+\ ClO_4$. The adduct between monomer and HI is synthesized *in situ* and can be stored for months without decomposition if it is kept cool in solution (CH_2Cl_2) and in the dark.[57]

For polymerization a master batch solution of this adduct is injected into a precooled monomer solution in CH_2Cl_2. However, polymerization does not start before the coinitiator $R_4N^+ClO_4^-$ is added. The progress of polymerization can easily be followed either by gas chromatography of the unconverted monomer or gravimetrically. The reaction can be terminated by adding ammonical methanol. The polymer solution is then washed with a 5% thiosulfate solution in water, the solution is then dried and the polymer isolated by solvent evaporation. In order to separate the polymer from the ammonium salt it is more convenient to dissolve the polymer in ethyl ether (this is only possible if the molar mass is not too high), since the salt is not soluble in ether.

(b) NO_2-$C_6H_4N_2^+PF_6^-$. The initiator was recrystallized from diethyl ether, vacuum dried and kept thereafter under nitrogen in the dark at −20 °C. Polymerization was carried out under nitrogen at temperatures of 30 °C or lower. The monomer solution was added to the dry salt with vigorous stirring to ensure rapid dissolution. The polymerization was stopped by adding alcoholic ammonia.

4.3.3. Ortho- and Para-Chlorostyrene

4.3.3.1. Polymerization

The cationic polymerization of *p*-chlorostyrene using CCl_3COOH as initiator has been described by Brown and Mathieson.[62] The *para*-isomer has also been polymerized in 1,2-dichloro-ethane with iodine as initiator.[63] From kinetic investigations the following order of reactivity toward cationic initiators has been established:

$$p\text{-chlorostyrene} < o\text{-chlorostyrene} < \text{styrene}$$

Relatively small molar masses $\bar{M}_n \approx 1400$ are the result of *p*-chlorostyrene polymerization by perchloric acid at 25 °C in dichloroethane, as has been described by Pepper *et al.*[64] In contrast, Heublein *et al.*[(65)] have shown that higher molar masses $\bar{M}_n \approx 15{,}000$ are possible with $SnCl_4/H_2O$ initiating systems.

Bimodal molar mass distribution was observed for poly-*p*-chlorostyrene with $AcClO_4$ as initiator. Addition of common salt $Bu_4N^+ClO_4^-$ suppressed the formation of the higher molar mass product. The polymer is unimodal if the salt concentration was some 50 times that of the catalyst.[(66,67)] The polymerization rate and the molar mass of the product decreased monotonically as the salt concentration was increased. *p*-chlorostyrene has also been polymerized using CF_3SO_3H. In CH_2Cl_2 as solvent a bimodal molar mass distribution was observed.[(68)]

4.3.3.2. Experimental Procedure

(a) CF_3SO_3H.[(68)] A solution of the monomer (1 mol L^{-1}) in CH_2Cl_2 was polymerized at 0 °C under nitrogen, initiated by CF_3SO_3H (10^{-4} mol L^{-1}). The polymerization was slow compared to that of both *p*-methoxy styrene and styrene. The conversion, which was followed by gas chromatography, was less than 20% in 40 min.

(b) CH_3COClO_4.[(69)] The principal experimental procedure has already been described in Section 4.3.2.2. In this procedure, the polymerization solution usually contained a trace of water (0.2–0.3 mmol L^{-1}). The polymerization, yielding a polymer with molar mass ≈30,000 was terminated by addition of methanol containing a small amount of aqueous ammonia solution.

4.3.4. α-Methyl Styrene

4.3.4.1. Polymerization

The polymerization of α-methyl styrene has been thoroughly reviewed by Bywater[(70)] and, more recently, by Kennedy.[(71)] The cationic polymerization of this monomer can yield highly stereoregular polymers provided that the reaction temperature is kept low (<−30 °C). However, the detailed stereochemical structure has been the subject of considerable controversy. Thus, Brownstein *et al.*[(72)] proposed a predominantly syndiotactic configuration for polymers initiated by BF_3, $AlCl_3$, or $TiCl_4$ in toluene at −70 °C and with $SnCl_4$ in nitromethane or ethylene chloride at −35 °C. In contrast, Sakurada *et al.*[(73)] discussed the spectra they obtained in terms of an essentially isotactic configuration despite similar reaction conditions. The latter view has also, more recently, been taken by Higashimura *et al.*[(74,75)]

Additionally, these authors found a considerable solvent effect on the course of the reaction. Thus, atactic polymers were obtained in nonsolvents for the polymer, such as n-hexane, while stereoregular polymers were obtained in toluene or CH_2Cl_2 which are good solvents for these polymers.

More recently still, studies[76–79] using high-field NMR, ^{13}C NMR, and partially deuterated polymers have confirmed the original assumptions of Brownstein. Surprisingly, the type of initiator seems to have little if any effect on the stereoregularity of the polymer which is controlled by the temperature and solvent used.

Kunitake and Aso[80] have used $BF_3{\cdot}Et_2O$, $TiCl_4$, as well as $AlCl_3$ to synthesize 100% syndiotactic polymers in toluene at −75 °C. By increasing the temperature and/or adding a nonsolvent such as methyl-cyclohexane for the polymer, these authors were able to reduce the degree of tacticity of the polymer. More detailed studies using $Ph_3C^+X^-$ ($X = AlCl_4^-$, $SnCl_5^-$, $AlBr_4^-$, or $SbCl_6^-$) as initiator at −78 °C in solvent mixtures have yielded similar results. The most polar solvent or solvent mixture yielded the highest degree of syndiotactic polymer.[81]

Kennedy and his co-workers have investigated the effectiveness of the cationic initiators: $SnCl_4/H_2O$, BCl_3/H_2O, and $SnCl_4$/pentamethyl benzyl chloride. They also studied the effect of 2.6-di-*tert*-butyl pyridine (DtBP) on the polymerization of α-methyl styrene initiated by these systems.[82–88] Their results suggest that the conversion of monomer is drastically decreased by increasing concentrations of DtBP but that molar mass increased and the molar mass distribution narrowed concurrently.

"Quasi-living" polymerization of α-methyl styrene was observed if the monomer concentration was kept low using BCl_3/H_2O or cumyl chloride initiators in CH_2Cl_2/cyclohexane solvent mixtures at −50 °C. With this technique products having molar masses ($\bar{M}_w$) of *ca* 2×10^5 and $\bar{M}_w/\bar{M}_n \simeq 1.5$ could be prepared.[89]

"Quasi-living" cationic polymerization of α-methyl styrene was achieved by slow addition of monomer to an "aged" initiating system containing 2,5-dichloro-2,5-dimethylhexane and BCl_3 in CH_2Cl_2 at temperatures between −20 and −40 °C.[90] Aging is indicated by a brown color—in contrast, the freshly mixed initiator system is colorless. Since the freshly prepared initiator is also able to initiate the polymerization of α-methyl styrene, it is still uncertain whether the chloronium mechanism (Scheme 4.9) suggested by the authors[90] and by Kennedy[91] is correct.

4.3.4.2. *Experimental Procedure*

(a) $BF_3{\cdot}Et_2O$.[80] α-methyl styrene was washed with dilute aqueous solution of alkali and with water, dried over $CaCl_2$ and distilled under nitrogen with use of a 30-cm column packed with glass helices, bp

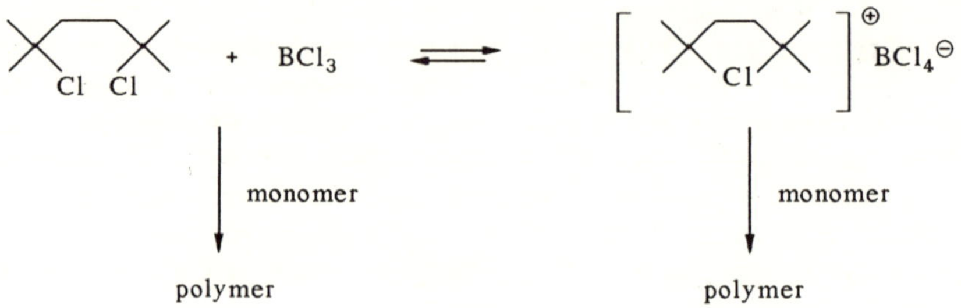

SCHEME 4.9.

59.5 °C/16 mm Hg. Solvents were purified by conventional methods. $BF_3 \cdot Et_2O$ was purified by distillation.

Polymerization was carried out in a Schlenk tube with a side arm for nitrogen inlet; toluene or methyl cyclohexane or their mixtures were used as solvent. The monomer (2 mL) and the solvent were placed in the vessel, cooled to the reaction temperature. The catalyst solution was then added with a syringe through a self-sealing rubber cap [0.05–0.4 mL (0.02–0.1 mol L^{-1})]. The polymerization was homogeneous in toluene and was terminated by adding methanol.

(b) Cumyl Chloride/BCl_3.[89] The purification procedure of the monomer was similar to that described above but the α-methyl styrene was finally distilled from CaH_2.

Polymerization was performed at −50 °C in a stainless-steel dry box under nitrogen in round-bottom, three-neck flasks, equipped with stirrer, teflon plug for monomer addition, and an outlet for sampling. The initiator (cumyl chloride) was dissolved in CH_2Cl_2 and BCl_3 was then added. Lastly, the monomer was added continuously by a pump into solvent charges of 50 mL. During the polymerization, samples were taken with a syringe and were injected into an excess of methanol to ensure complete termination. By this method polymers with $\bar{M}_n > 2 \times 10^5$ can be prepared.

4.3.5. *Cis- and Trans-β-Methyl Styrene*

4.3.5.1. *Polymerization*

The polymerization of *cis*- and *trans*-β-methyl styrene to high polymers is not possible via anionic or radical mechanisms: Only oligomers have been observed.[92,93] However, despite the steric hindrance of the added methyl group, these monomers can be polymerized via cationic mechanisms due to the increased electron density induced by this methyl group.

Staudinger and Dreher[94] have studied the polymerization of β-methyl styrene (the *cis*- or *trans*-isomer is not specified), initiated by BF_3 in toluene

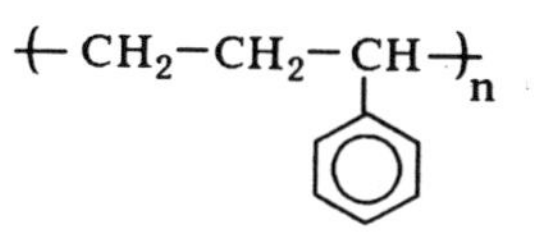

1

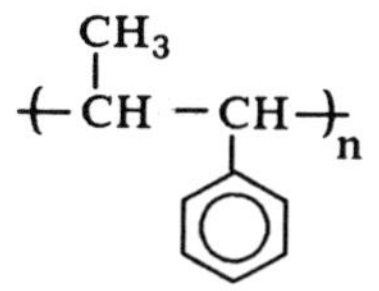

2

at −80 °C. They obtained products having molar masses $\bar{M}_z < 3000$. These authors proposed structure **1** for their polymers rather than the more obvious structure **2**. However, later workers[95–98] have clearly demonstrated that structure **2** predominates in the products of such polymerizations although structure **1** cannot be categorically excluded.

4.3.5.2. *Experimental Procedure*[94]

β-methyl styrene was distilled before use. 10 mL (9.14 g) of monomer in 60 mL CH_3Cl was polymerized by adding a solution of $AlCl_3$ in CH_3Cl (7.2×10^{-2} mol L^{-1}) at −60 °C. A total of 110 mL catalyst solution was introduced gradually over a period of 65 min into the monomer mixture. The reaction was terminated by addition of cold methanol. From the IR spectra a conventional 1,2-head-to-tail polymerization was assumed.

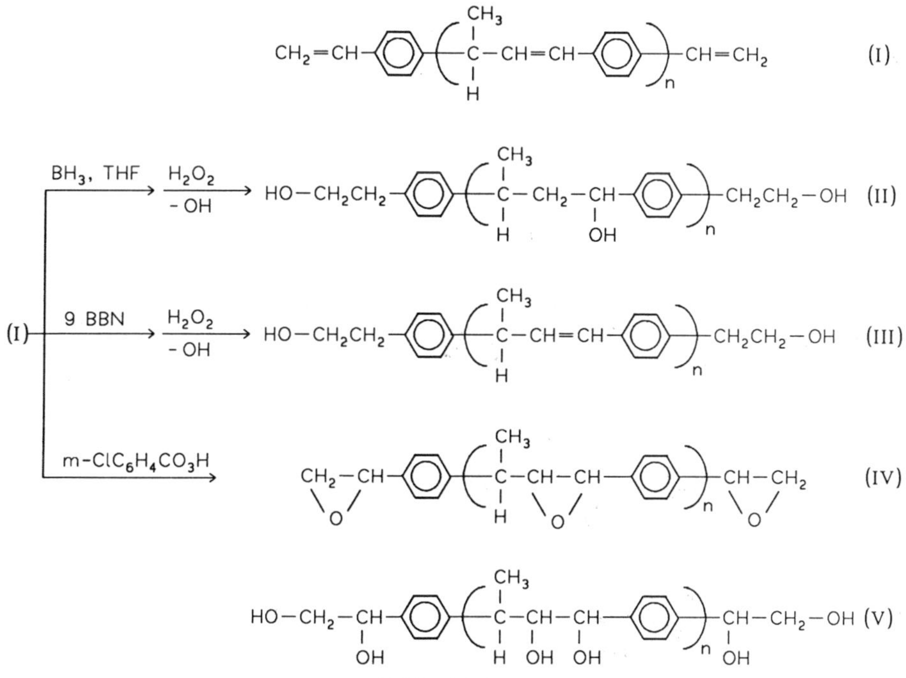

SCHEME 4.10.

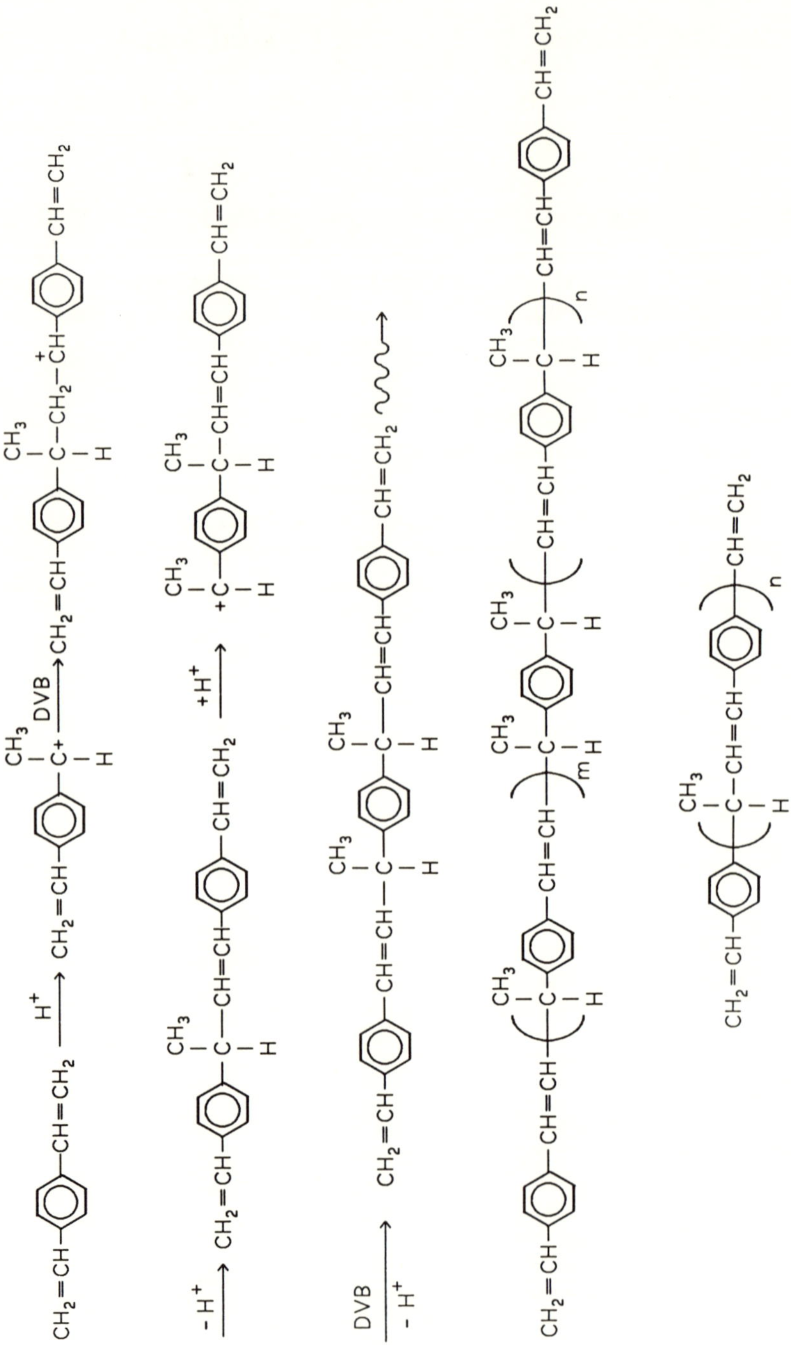

SCHEME 4.11.

4.3.6. Divinylbenzene

4.3.6.1. Polymerization

A mixture of *m*- and *p*-divinyl benzene is freely available and, although the mixture can be separated,(99) simple syntheses of *p*-divinyl benzene(100) and of *o*-divinyl benzene(101) have been described.

Via radical(102) or anionic(103) polymerizations divinyl benzenes (DVB) usually yield insoluble, crosslinked polymers. Via cationic polymerization linear polymers are obtainable. Such linear polymers can be modified to yield a variety of telechelic polymers(104) and indeed completely new polymers (Scheme 4.10(105)). Using triflic acid (CF_3SO_3H) or acylium perchlorate (CH_3COClO_4) initiators,(106) soluble linear polymers can be prepared.

The mechanism put forward by Higashimura and Hasegawa is based on their observations that their oxo acid initiators selectively yield an unsaturated linear dimer from styrene(107–109) and is a type of step-growth polymerization (Scheme 4.11).

In order to achieve higher molar mass ($M_n > 2.5 \times 10^4$) soluble products, it was found necessary to keep the monomer concentration below 0.1 M. Mixtures of *m*- and *p*-divinyl benzene (70/30) using the above method led to soluble polymers in which, using NMR, no cyclic structures could be identified. The intramolecular cyclization (Scheme 4.12) has been studied for *o*-divinyl benzene.(110,111)

By analyzing the concentration of pendant C=C double bonds (IR peak at $1600\,cm^{-1}$ in CCl_4 solution or $1600\,cm^{-1}$ and $1630\,cm^{-1}$ for KBr disks) values for $r_c = k_c/k_p$ could be obtained from

$$\frac{-d[m]}{d[M]} = \frac{2[M]}{2[M] + r_c} \tag{4.5}$$

$$\frac{[m]}{[M]_0 \kappa} = 1 - \frac{r_c}{2[M]_0 \kappa}\left(\ln \frac{2[M]_0 + r_c}{2[M]_0(1-\kappa) + r_c}\right) \tag{4.6}$$

where [m] and [M] are the concentrations of pendant double bonds and monomer, respectively, and κ is the fractional conversion of monomer to

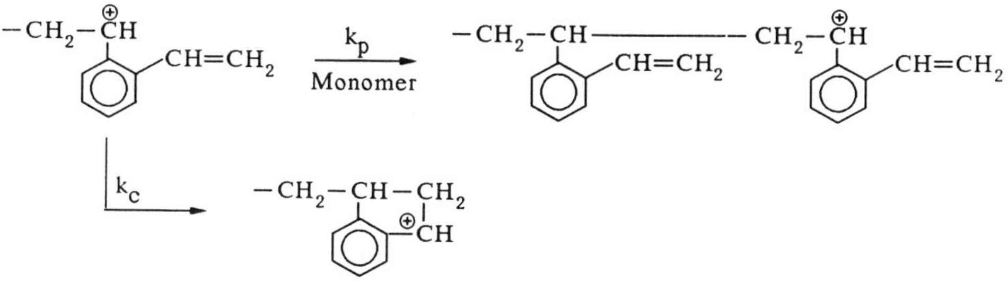

SCHEME 4.12.

polymer. It was found that r_c is increased in less polar solvents and where weaker Lewis acids were employed [i.e., $r_c(BF_3{\cdot}Et_2O$, toluene) > $r_c(BF_3{\cdot}Et_2O, C_2H_5NO_2)$ > $r_c(ZnCl_2/TCA$, toluene)]. These experimental findings corroborate a proposed mechanism involving "tighter" and "looser" ion pairs, the latter leading to an increase in the degree of intramolecular cyclization.

4.3.6.2. Experimental Procedure

Polymerization of the mixture of *meta*- and *para*-divinylbenzene (70/30) was performed in baked-out, Erlenmeyer flasks under dry N_2. The catalyst solution (CH_3COClO_4 in CH_2Cl_2 or CF_3SO_3H in CH_2Cl_2) was added via syringe to the monomer solution (in C_6H_6 or CH_2Cl_2, $c \leq 0.1\ \mathrm{mol\,L^{-1}}$). The water concentration (checked by Karl–Fischer titration) was below 3×10^{-3} M. Additional monomer was added after about 90% conversion, also via syringe, to yield larger molar mass products. Molar masses up to $\bar{M}_n = 2.5 \times 10^4$ were observed.[106]

4.3.7. Diisopropenylbenzene and Related Monomers

4.3.7.1. Polymerization

Monomers 3–6 listed in Scheme 4.13 polymerize at higher temperature to yield soluble, structurally uniform polyindanes.[112] Such polyindanes are

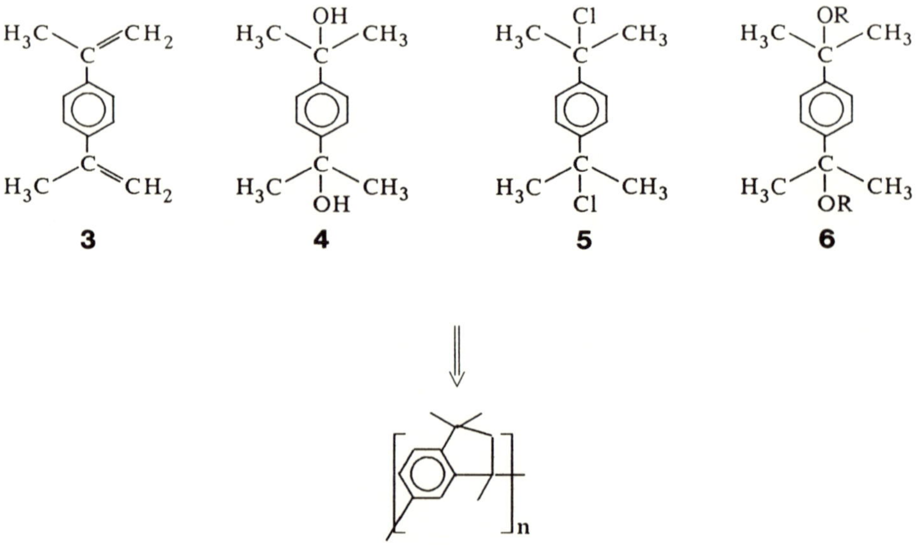

SCHEME 4.13.

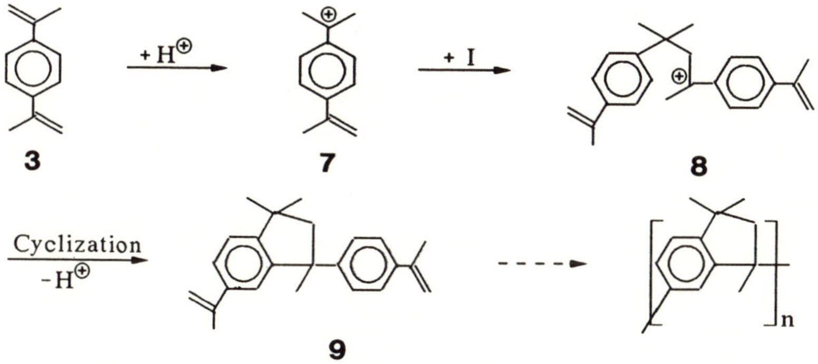

SCHEME 4.14.

interesting due to their excellent thermal stability ($T_g > 220\,°C$, $T_{dec} > 450\,°C$).[113,114] However, the thermal characteristics of these polymers are very sensitive to structural defects. Side reactions leading to unsaturation along the chain should be avoided by using proper synthesis conditions.

The formation of the polyindane and of unsaturated structure units can be explained on the basis of the mechanism in Scheme 4.14, which is discussed for monomer **3**. With a variety of reagents [BCl_3, HCl, H_2SO_4, CF_3COOH $(CF_3CO)O$ · · ·] **3** is postulated to form **7**, which can then add to monomer to form **8**. By intramolecular Friedel–Crafts alkylation followed by deprotonation species **9** is formed. **9** behaves like a monomer again.

A second reaction pathway (Scheme 4.15) leads to unsaturation along the chain. Here, instead of stabilization by an intramolecular Friedel–Crafts reaction, deprotonation occurs directly to form **10**. However, treating polymers having those unsaturated structures with protonic acid yield uniform polyindane again.[114,115] Crosslinking is avoided by keeping the reaction temperature above the ceiling temperature of the monomer (25 °C for I).[115] In view of the present interest in high-temperature thermoplastics these systems deserve more attention.

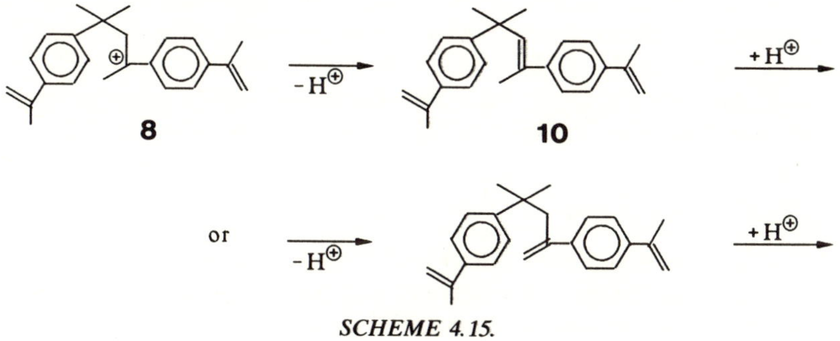

SCHEME 4.15.

4.3.7.2. Experimental Procedure

The polymerization is carried out in a simple two-necked flask equipped with a reflux condenser, a dropping funnel, and a magnetic stirrer. A 1:1 mixture of **3** and **4** were dissolved in 1,2-dichloroethane ([**I**] = [**II**] = 4.55×10^{-2} M) and the mixture heated to reflux. A solution of $AlCl_3$ in 1,2-dichloroethane to give a concentration of 4.6×10^{-3} M was then rapidly added and the mixture refluxed for 4 h. The polymer was then isolated by precipitation into methanol. By this procedure a polymer containing 99% indane units was obtained. Unfortunately, it was not found possible to increase the molar mass above 3.4×10^3 and, despite adding the catalyst at reflux temperatures, bimodal mass distributions could not be avoided.

4.4. VINYLARENES

4.4.1. Polymerization

The cationic polymerization of vinylarenes (styrene is not included) is of minor importance since in most cases only low molar masses have been observed. However, in the following section such will be discussed in more detail.

4.4.1.1. Napthalene Derivatives

1-vinyl-naphthalene, 2-vinyl-naphthalene, and 1-vinyl-4-methoxy naphthalene have been polymerized under various conditions. The observed molar masses were always relatively low.[117,118] It was also found that 1-vinyl-4-methoxy naphthalene does not copolymerize with styrene, indeed, even the homopolymerization of styrene is hindered in its presence.

The quantitative oligomerization of 2-isopropenyl-2-naphthalene by obvious cationic initiators such as $BF_3 \cdot Et_2O$, CF_3SO_3H, and $SnCl_4$ has been described by Schulz *et al.*[119] Transfer and termination of 1-vinyl- and 2-vinyl-naphthalene has been studied by Blin *et al.*[120] Polymerizations were carried out in CH_2Cl_2 initiated by $TiCl_4$. For both monomers increasing termination and transfer were observed with increasing temperature. These values are significantly higher than those for styrene under the same conditions. This seems to be the consequence of highly reactive aromatic moieties, which permit Friedel–Crafts attack by the growing chain.

4.4.1.2. Vinyl Anthracene

According to Bergmann and Katz[12] the cationic polymerization of 9-vinyl anthracene yields conventional 1,2-chain growing. In contrast to these results, it was shown by Michel[122] in 1964 that instead of a normal

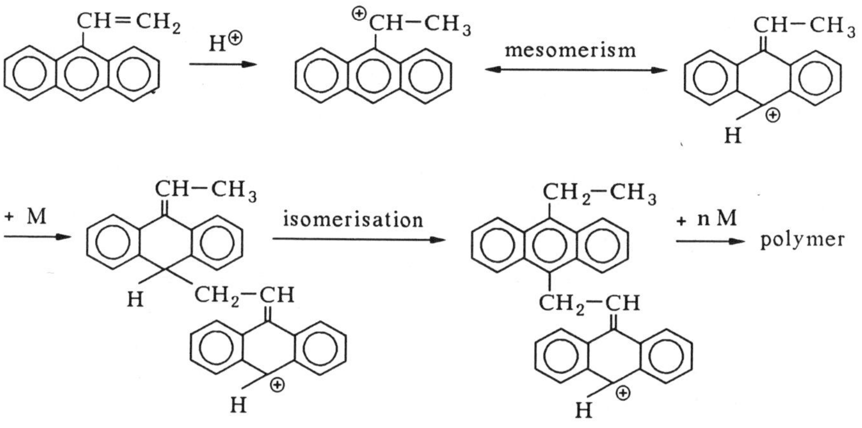

SCHEME 4.16.

1,2-chain growth, polymerization takes place predominantly via a charge transfer across the ring. Bunel *et al.*[117] have shown that this isomerization of the carbocation increases with increasing temperature (Scheme 4.16). The cationic polymerization of 9-vinyl anthracene, 9-vinyl-10-methyl anthracene, 1-vinyl anthracene, and 2-propenyl-1 anthracene have also been studied by Pearson *et al.*[123,124] The latter two monomers failed to polymerize cationically and for the others only oligomers were observed.

4.4.2. Experimental Procedure

2-Isopropenyl-2-Naphthalene.[119] $BF_3 \cdot Et_2O$, CF_3SO_3H, or $SnCl_4$ in CH_2Cl_2 (0.3 to 1 mol% compared to monomer) is added to a monomer solution (1 molL^{-1}) at temperatures between −20 and +20 °C. After 20 h the reaction was terminated by addition of trimethylamine. In all cases only oligomers were observed.

4.5. ACENAPHTHYLENE

4.5.1. Polymerization

Acenaphthylene (ACE) is available as a byproduct of coal-tar distillation[125] but commercial material is generally synthesized by dehydrogenation from acenapththene[126]:

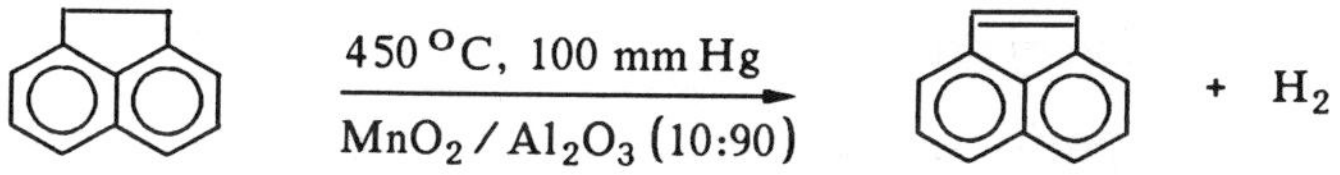

(4.7)

Poly(acenaphthylene) (PACE) is an academically interesting polymer due to the extreme steric hindrance arising from the very bulky naphthalene substituent.[127] PACE is also of industrial interest, because of its fluorescence behavior and also because of its stabilizing effects with respect to radiation[128] and thermal degradation.[129]

ACE can be polymerized radically,[130–132] anionically,[133,134] and by Ziegler–Natta catalysts[135] as well as via cationic techniques. Although using radical initiators and thermal polymerization higher molar mass products are obtained, it seems that both chain branching and double bonds along the main chain are more prevalent in such products than in the products of cationic polymerizations.[130] Thus, the latter tend to be white powders with molar masses up to 10^5 while the former are yellow and have molar masses up to 3×10^6. The limited molar masses obtainable via cationic techniques suggest an efficient transfer to monomer which does not appear to be controllable by reducing the temperature of polymerization.

Aromatic solvents for the cationic polymerization of ACE tend to give better yields and higher molar mass products than aliphatic or chloroalkane solvents.[136] Furthermore, it appears that the different solvents yield polymers of different tacticity. Thus, in CH_2Cl_2 polymers initiated by WCl_6 may contain more *threo*-disyndiotactic structures, while in benzene more *threo*-diisotactic structure are formed. This is interesting since an increase in the isotacticity of the polymer leads to an increase in the occurrence, under appropriate conditions, of the high-energy excimer.

Copolymers of ACE with styrene prepared via cationic polymerization can yield materials with a single T_g up to *ca* 130 °C [T_g(Poly(ACE)) > 210 °C]. Copolymers composed of larger proportions of ACE tend to have a multiphase morphology and multiple T_g. Using $TiCl_4$ as initiator and with ACE/Sty = 0.2 in CH_2Cl_2 at $T = -78$ °C Cohen and Marecha[137] have obtained

$$r_{\text{Sty-ACE}} = 0.36 \qquad \text{and} \qquad r_{\text{ACE-Sty}} = 11$$

The polymerization properties are explained on the basis of ring strain in acenaphthene (25–30 kJ mol^{-1}) compared with that in ACE (*ca* 8 kJ mol^{-1}). Recent studies[138,139] have shown that in nitrobenzene (25 °C) the k_p = 23 L mol^{-1}s^{-1} for the free ion polymerization of ACE. Anasagasti and Leon[140,141] have obtained k_p = 0.88 L mol^{-1}s^{-1} at 18 °C using H_2SO_4 in CH_2Cl_2, although the latter work lacks some of the fundamental details necessary to be able to evaluate the reliability of these numbers.

4.5.2. Experimental Procedure

4.5.2.1. Purification of the Monomer

The major impurity in commercial ACE is (*vide supra*) acenaphthene, but minor impurities such as epoxides are more deleterious for cationic polymerizations. Thus, in addition to sublimation *in vacuo*, recrystallization from ethanol or methanol are typical purification procedures.[130] More exacting work can be preceded by zone refining ACE in evacuated tubes.[138,139] For detailed kinetic work, where the presence of even small amounts of acenaphthene are unacceptable, ACE can be synthesized from acenaphthene by first converting it to acenapthyl acetate in the presence of Pb_3O_4, hydrolyzing this ester to acenaphthenol, and then dehydrating the alcohol by sublimation *in vacuo* in the presence of anhydrous $KHSO_4$.[137] Pure ACE melts at 92.5–93.0 °C and should be stored in the dark.

4.5.2.2. Polymerization Procedure

From the literature it appears that, from a synthetic point of view, in contrast to the purification of monomer, there is little to be gained in terms of molar mass or yield from exacting polymerization conditions. Thus the method given by Cho and Choi[136] can be recommended. These authors took a three-necked flask (250 ml) equipped with a stirrer, thermometer, and nitrogen gas inlet tube, and 30.4 g (0.2 mol) of ACE was dissolved in 180 mL chlorobenzene at 20 °C. To this stirred solution was added, by hypodermic syringe, 0.396 g (10^{-3} mol) WCl_6 homogeneously dissolved in chlorobenzene as 0.05 M solution. the polymerizations were stopped after 12 h of stirring at 20 °C by the addition of 15 mL methanol. Yield *ca* 100%, $\bar{M}_w$ 7.9×10^3. Schmelzer and Springer[130] used higher concentrations of ACE ($1.64\ \mathrm{mol\,L^{-1}}$) in benzene and $BF_3[(C_2H_5)_2O]$ as a catalyst. They stopped their reactions after 1 h but obtained only *ca* 85% yields and molar masses up to *ca* 5×10^3.

4.6. N-VINYL CARBAZOLE

4.6.1. Polymerization

Poly-*N*-vinyl carbazole (Poly-NVC) has been of interest since it was discovered by Hoeg *et al.*[142] that it exhibits photoconductivity. Indeed recent literature[143,144] has described an electrochemical polymerization which yields predoped polymers with electrical conductivity up to $10^{-4}\ \mathrm{S\,cm^{-1}}$. NVC and its polymerization have been covered by the excellent reviews of Pearson and Stolka[145] and Rooney.[146] More recently Nuyken[147] has also reviewed this subject.

Most published synthetic routes to NVC start from carbazole, which is readily available from the distillation of coal tar. However, although a variety of methods for purifying the material have been developed,[148] for detailed investigations it is worthwhile preparing the material from cyclohexanone and phenylhydrazine[149–151] to avoid the sulfur impurities originating from the coal-tar source (see Scheme 4.17).

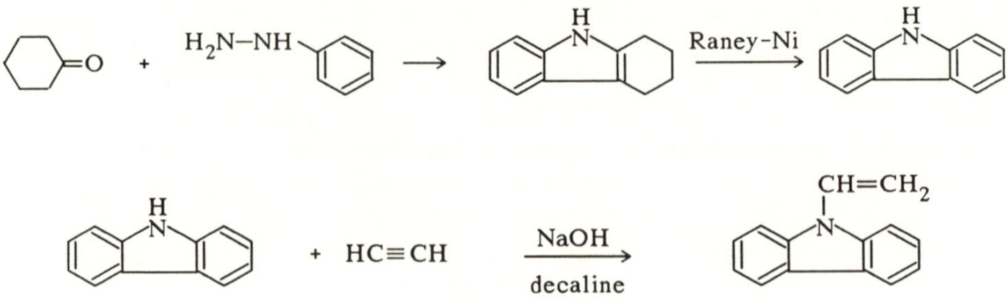

SCHEME 4.17.

As a strong basic monomer, NVC can be easily polymerized by almost any cationic initiating system. Polymerization by Brønsted or Lewis acids is best carried out at low temperatures (*ca* −70 °C) in relatively polar solvents such as CH_2Cl_2 which allow the system to remain homogeneous even to high conversions. Using highly purified and dried solvent and monomer, complete conversion and high molar mass polymers can be obtained.[152]

An interesting alternative to the acid initiation of NVC polymerization, which also exemplifies the use of stable salt initiators, is the work of Ledwith and Sherrington,[153] who have polymerized NVC using amminium salts (Scheme 4.18). With such salts rapid complete conversion to high molar mass polymer ($M_n \approx 10^4$) can be achieved at room temperature.

As is typical of carbocationic polymerization, care must be taken to use carefully purified and dried solvents and monomer.

In recent years a living cationic polymerization of NVC has been realized by using I_2/Bu_4NI in CH_2Cl_2.[154,155] The use of HI/I_2 as an initiating system also leads to more rapid, complete conversion but to slightly broader molar mass distributions. In toluene the monomer concentrations have to be kept low (ca 0.2 M) to maintain homogeneity. In the more polar CH_2Cl_2 solvent higher monomer concentrations can be used, but even at −78 °C the polymerization is very rapid and loses its living character. By adding the common-ion salt Bu_4NI ([Salt]/[HI] ≈ 1/100) the system can be slowed down to yield extremely narrow ($M_w/M_n \approx 1.09$–1.2) distributions and molar masses up to 2×10^4. There seems, however, to be no reason why larger molar masses should not be possible.

The cationic polymerization of *N*-vinylcarbazole (NVC) in the presence of electrophilic olefins such as tetracyanoethylene (TCNE) has been investigated by several research groups.[156–158] Nakamura *et al.*[158] proposed that a cyclobutane adduct was the initiator after proton transfer to monomer. However, Ledwith *et al.*[159] postulated initiation by the cation radical of NVC. Finally, Hall *et al.*[160,161] concluded from intensive mechanistic studies that the reactive species are zwitterions which can combine to form a cyclobutane or initiate cationic polymerization of NVC (Scheme 4.19).

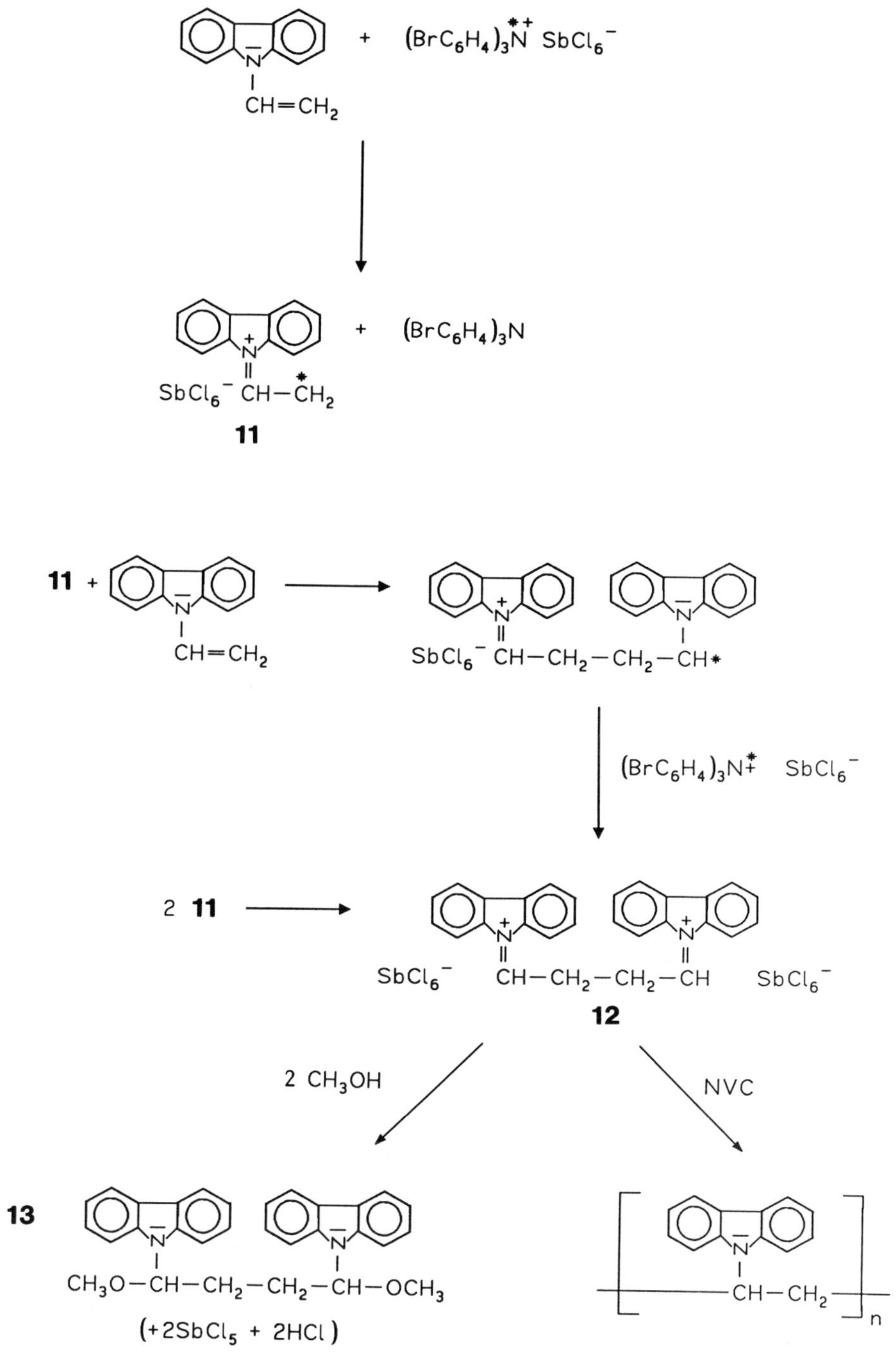
+
(BrC6H4)3N•+ SbCl6−
CH=CH2
+
(BrC6H4)3N
SbCl6− CH−CH2
11
11 +
CH=CH2
SbCl6− CH−CH2−CH2−CH•
(BrC6H4)3N•+ SbCl6−
2 11
SbCl6− CH−CH2−CH2−CH SbCl6−
12
2 CH3OH
NVC
13
CH3O−CH−CH2−CH2−CH−OCH3
(+2SbCl5 + 2HCl)
CH−CH2
n

SCHEME 4.18.

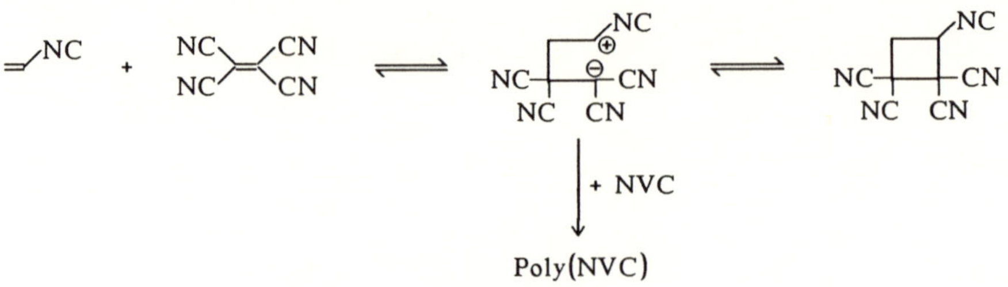

SCHEME 4.19.

Recently Voit *et al.*[162] have found that cyclobutane derivatives formed from TCNE and vinylether can initiate the NVC polymerization at room termperature if polar solvents such as acetonitrile, nitrobenzene, or sulfolane were used. Molar masses up to $\bar{M}_w = 150{,}000$ with $\bar{M}_w/\bar{M}_n = 1.02$ were obtained.

4.6.2. Experimental Procedure

(a) Amminium Salt.[153] The initiator salt (0.0035 g) is dissolved in CH_2Cl_2 (5 mL) and then introduced into a solution of 1 g monomer and 45 mL CH_2Cl_2 at room temperature, open to the air. Amost immediately the deep blue color of the amminium salt is discharged and polymerization is complete within seconds. The polymer is precipitated by the addition of methanol.

(b) HI.[154] The polymerization reactions are carried out in the dark (to avoid disproportionation of HI) and in baked glass tubes filled with dry N_2 and equipped with three-way stopcocks. A solution of HI in CH_2Cl_2 is added into the precooled monomer solution. The polymerization can be terminated with precooled methanol.

(c) TCNE/Vinylether.[162] For a typical experiment NVC (0.0025 mol) was dissolved in 4 mL of solvent (sulfolane or nitro benzene) in a dry polymerization tube. After degassing, 0.00025 mol of the cyclobutane synthesized from TCNE and vinylether, dissolved in 1 mL solvent, were added. If sulfolane was used the reaction temperature had to be increased to 40 °C to induce polymerization. A polymer with $\bar{M}_w \approx 150{,}000$ was isolated by precipitation into methanol.

4.7. VINYL ETHERS

4.7.1. Polymerization

4.7.1.1. Traditional Methods

Vinyl ethers, due to their electron-donating ether group, only yield polymers with large molar masses via a cationic polymerization mechanism, although alternating copolymers with electron-deficient monomers can be prepared by radical polymerization. Poly(vinyl ethers) have been known since 1878.[163]

There are several excellent reviews on the synthesis, reactions, and physical properties of vinyl ether monomers,[164,165] on their polymerization,[166,167] and on the properties of the polymers.[164,168,169]

Due to their basic character vinyl ethers can be polymerized by both Brønsted and Lewis acids, although the Brønsted acid systems rarely lead to high polymers due to side reactions. More recently, however, the kinetics of the polymerization of isobutylvinyl ether with CF_3CO_2H has been studied[170] and also large molar mass ($M_n > 10^5$) polymers have been obtained from 2-chloroethyl vinyl ether using CF_3SO_3H in CH_2Cl_2 at $-78\,°C$.[171]

Among the Lewis acids, the most frequently used initiator for vinyl ether polymerization is $BF_3 \cdot Et_2O$ since it is readily available and easy to handle. However, it remains unclear whether a cocatalyst is essential or whether $BF_3 \cdot Et_2O$ alone can initiate the polymerization via $Et^{\oplus}$.[172–174]

Two types of initiation, which have proved exceedingly useful for determining the rate constants of vinyl ether polymerization but, because of the exacting experimental techniques necessary, have not been developed as preparative methods, are carbocation salts[175–179] and radiation.[180–185] A detailed description of the techniques using carbocation salts can be found in the work of Ledwith *et al.*[178] while those associated with the so-called superdry radiation initiation can be found in the work of Williams.[186]

Much attention has been paid to the problem of stereoisomerism of poly(vinyl ether).[187–192] Although this aspect has been studied extensively, the detailed mechanism has not yet been clearly identified.[193,194] The effects of the reaction conditions on the polymerization structure may be summarized as follows:

1. *Solvent*: Increasing solvent polarity leads to a greater degree of syndio- and heterotacticity. Low-polarity solvents can yield highly isotactic materials. The effect of solvent polarity increases with the volume of the side chains.
2. *Initiator* (*Gegenion*): Although $BF_3 \cdot OEt_2$ in nonpolar solvents at low temperature leads to a crystalline polymer from isobutyl vinyl ether,[187] $SnCl_4$ under the same conditions does not.[195]

3. *Polymerization Temperature*: In homogeneous systems, highly isotactic polymer forms only at low temperatures. In heterogeneous systems (e.g., Cr_2O_3 in toluene), highly stereoregular polymers can be prepared at temperatures up to 80 °C.[196]
4. *Monomer*: The isotacticity of the polymer increases with the volume of the side chain.

In other words, the effects of solvent, initiator, and temperature can be generally interpreted to suggest that "tight" ion pairs lead to isotactic polymers while "loose" pairs or "free" ions tend to yield syndiotactic or heterotactic polymers. This would concur with an insertion mechanism proposed by Sikkema and Angad-Gaur for the isotactic polymerization.[197,198]

4.7.1.2. Photoinitiators

Photoinitated cationic polymerization has been reviewed from several aspects[199–204] and, although not extensively studied with vinyl ethers, Crivello *et al.* have shown that such systems are extremely effective and that particular multifunctional vinyl ethers may prove useful as coating materials.

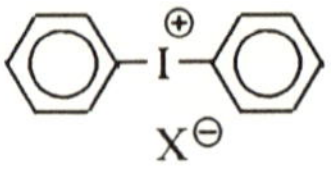

14

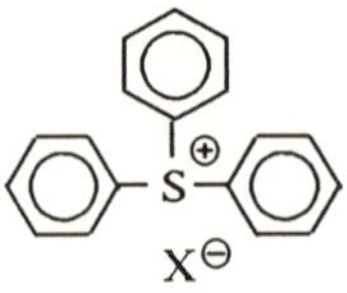

15

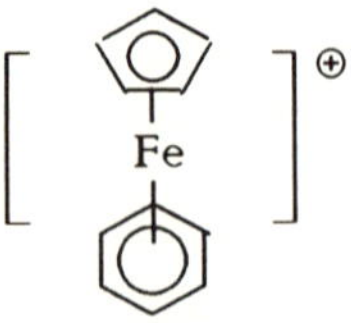

16

Typical structures (**14–16**) of photoinitiators are given. The following equations give the mechanism proposed for diaryl iodonium salts:

$$\begin{aligned} Ar_2I^+X^- &\xrightarrow{h\nu} ArI^{+\cdot} + Ar^{\cdot} \\ ArI^{+\cdot} + \text{solvent-H} &\longrightarrow ArI + \text{solvent}^{\cdot} + HX \end{aligned} \tag{4.8}$$

The Brønsted acid, which is generated by reaction of the aryl iodonium cation-radical with the solvent or monomer, is the initiator. The main advantage of this type of initiator is that such photoinitiators are soluble in these monomers.

4.7.1.3. Recent Developments

The preparation of well-defined vinyl ether polymers is probably most easily achieved using the relatively modern technique using HI/I_2[205–210] or HI/salt[211–213] initiators. These polymerizations are characterized by a fast

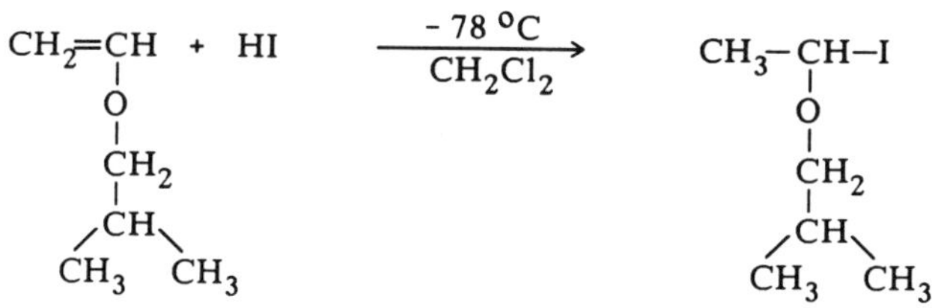

SCHEME 4.20.

initiation ($k_i \gg k_p$) and are living so that polymers having narrow molar mass distributions can be obtained. The combination of HI/I_2 as initiator has been utilized to synthesize well-defined telechelics,[208] macromonomers,[209] and block-copolymers[210] from vinyl ethers. The initial reaction step is the addition of HI across the double bond to yield the corresponding iodide. Thus, from isobutyl vinyl ether 1-(1-iodoethoxy)-2-methylpropane is formed which, under suitable conditions, can be kept as a stable solution (Scheme 4.20).

The addition of I_2 to the iodide in the presence of additional monomer initiates the polymerization. It has been suggested[212] that monomer addition involves the insertion of monomeric units into the terminal C—I bond via a four-center transition state (Scheme 4.21). In such a system an addition of I_2 to monomer yielding the corresponding diiodide[206] is an undesirable side reaction which, depending on the reaction conditions, leads to a broadening of the molar mass distribution.

It has been shown[211,212] that salts, such as ZnI_2 and $ZnCl_2$, can also be used with alkoxy alkyl iodides to initiate a living polymerization of vinyl ethers. Common ion salts, such as $Bu_4N^{\oplus}I^{\ominus}$, retard the polymerization using I_2, presumably due to a consumption of I_2 by a reaction with $Bu_4N^{\oplus}I^{\ominus}$ which yields polyiodide salts, such as $Bu_4N^{\oplus}I_n^{\ominus}$.

Recently it has been shown that $Bu_4N^{\oplus}ClO_4^{\ominus}$ is a very efficient activator for the isobutyl vinyl ether polymerization[213,214] (Scheme 4.22).

The advantages of using an alkoxy alkyl iodide and a salt such as $Bu_4N^{\oplus}ClO_4^{\ominus}$ rather than HI/I_2 are that, using former:

- there are no side reactions;
- there reaction appears to be highly selective so that, e.g., acryl- or 1,2-disubstituted double bonds are not attacked;

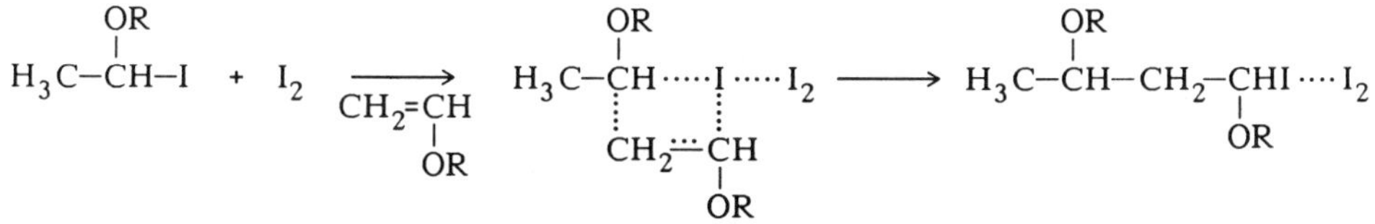

SCHEME 4.21.

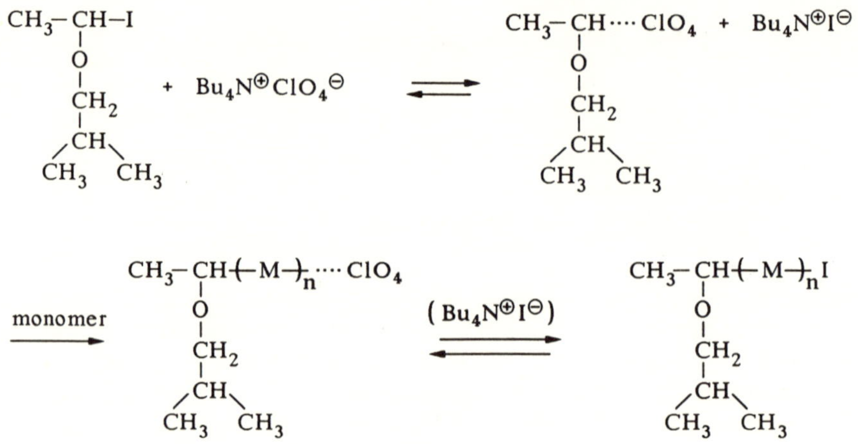

SCHEME 4.22.

- there seems to be no limit to the molar mass that can be achieved while retaining a narrow ($M_w/M_n < 1.2$) molar mass distribution;
- the living character is retained even up to room temperature;
- the reaction is relatively insensitive to impurities so that such polymerizations can also be carried out, e.g., in THF or acetone—although these are not recommended solvents.

A number of monoacidic phosphate esters [$R_2'P(O)(OH)$] have been shown to initiate living cationic polymerization of vinyl ethers in the presence of zinc halide as activator.[215]

The polymerization of isobutyl vinyl ether in toluene below 0 °C initiated by diphenylphosphate/ZnI_2 is controlled by the ratio of [M]:[phosphate] and $M_w/M_n < 1.1$ was observed. The living character was demonstrated by addition of a fresh portion of monomer to polymerized reaction mixture.

The mechanism in Scheme 4.23 was suggested.[216] This mechanism

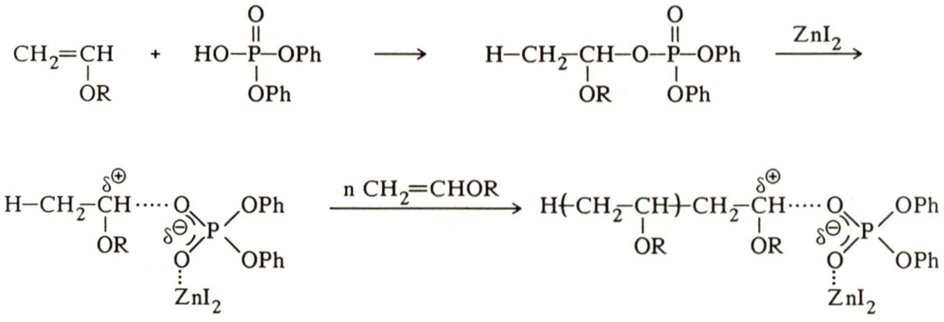

SCHEME 4.23.

suggests that the vinyl ether and the phosphate form an adduct. The subsequent process involves activation of the carbon–phosphate linkage by zinc halide and then propagation starts.

Table 4.1 summarizes some of the most frequently used initiators.

4.7.2. *Experimental Procedure*

(a) CF_3COOH.[170] Monomer and initiator were distilled into a calibrated sidearm of a dilatometer and then into phials having break-seals. The dilatometer was thermostated and the reaction was started by breaking the internal seals. Typical concentrations were $CF_3COOH = 1.2 \times 10^{-3}\,\mathrm{mol\,L^{-1}}$, Monomer $= 0.1$–$0.5\,\mathrm{mol\,L^{-1}}$. These polymerizations can be carried out at room temperature.

(b) Photopolymerization. A detailed description of the initiator synthesis is given in Ref. 228. The polymerization was conducted by placing vials filled with monomer and initiator solution ($Ph_3S^+SbF_6^-$) in a merry-go-round apparatus and irradiating them with a 450-W Hanovia lamp. The samples were quenched at various irradiation times by adding methanol. The polymers were filtered and washed with methanol before characterization.

(c) Et_2AlCl/H_2O.[229] The monomer was dissolved in heptane, which contained traces of water (10 ppm). then a 0.1 wt% solution of Et_2AlCl in

TABLE 4.1. Cationic Polymerization of Vinyl Ethers

Initiator	Monomer R—O—CH=CH$_2$ R	Solvent	Temp (°C)	Ref.
Protonic acids				
$Al(OR)_3/HF$	CH_3	$CH_2Cl/_2$heptane	0–25	217
$Al_2(SO_4)/3H_2SO_4$	i-C_4H_9	hydrocarbons	10–120	218
Lewis acids				
$BF_3{\cdot}Et_2O$	i-C_4H_9	Bulk	−78	219
		Propane	−78	220
$BF_3{\cdot}2H_2O$	Various	Bulk	3–5	221
$SnCl_4$	i-C_4H_9	Benzene	12	222
$EtAlCl_2$	i-C_4H_9	Toluene	−80	223
Miscellaneous				
I_2	i-C_4H_9	Diethyl ether	25	224
Tropylium$^+SbCl_6^-$	i-C_4H_9	CH_2Cl_2	0	225
Trityl$^+SbCl_6^-$	i-C_4H_9	CH_2Cl_2	0	226
$(C_6H_5)_2I^+BF_4^-$ (PF_6^-, AsF_6^-, SbF_6^-)	2-ClC_2H_4	CH_2Cl_2	25	227

heptane was added via syringe at 15 °C. The reaction is almost complete after 1 h. The polymer was isolated by distilling the solvent.

(d) HI/I_2,[206] *$HI/NR_4{}^+ClO_4$.*[213] A convenient method for the preparation of poly(vinyl ethers) is to first prepare a master batch of the corresponding iodide. To this end an aqueous solution of HI can be dried by P_2O_5 kept at 0 °C and then purified by passing it through a column fitted with a mixture of red phosphorous, P_2O_5, and silica gel (1:20:50 w/w) kept at −20 °C before dissolving in a predried hexane or CH_2Cl_2 solution (*ca* 0.5 M) of the desired vinyl ether cooled to −78 °C. Excess HI is then removed *in vacuo* before the concentration is determined by titration with NaOH aq. Such solutions can be kept in a cool, dark place for several months without noticeable decomposition. Polymerization is carried out in predried apparatus under dry N_2 simply by preparing a solution of vinyl ether (up to *ca* 2.0 M) in either CH_2Cl_2 or hexane (the more polar solvent leads to faster reaction) adding the prepared solution of the iodide and finally a solution of I_2 or $Bu_4N^+ClO_4^-$ in CH_2Cl_2. The rate of the polymerization is directly proportional to the concentration of alkoxy alkyl iodide and the concentration of I_2 or added salt. The rate also increases with increasing temperature and although, using added salt solutions, the systems remain living at room temperature it is advisalbe to carry out the reactions at lower temperatures (e.g. <0 °C) to reduce the possibility of side reactions.

4.8. ISOBUTYLENE

4.8.1. Polymerization

4.8.1.1. Traditional Methods

The literature, and in particular the patent literature, is extremely rich in disclosures of isobutylene (methylpropene, IB) polymerization processes. This monomer, which can only be polymerized via a cationic mechanism, is probably the most investigated of those monomers which polymerize cationically. A good overview of early developments of polymers and copolymers of IB is given by Schildknecht.[230] Thomas *et al.* have described the early work on polymerization initiated by Lewis acids.[231] The properties of the homo- and copolymers prepared using IB have been reviewed more recently by Kennedy[232] and extensive critical reviews of the polymerization of IB have been written by Plesch[233] and by Gandini and Cheradame.[234]

IB can be polymerized by a wide range of Lewis acids, and most are well documented in the literature. In contrast simple Brønsted acids, such as CF_3COOH and $HClO_4$[235,236] do not yield polymers but instead simply the corresponding esters. With H_2SO_4, IB oligomerizes and indeed this is an industrial process.[237,238] The preparation of Poly-IB with extremely large molar masses ($\geq 10^5$), although rather broad distribution, can easily be achieved at low temperatures ($< -60\,°C$) with minimum drying procedures and using $BF_3 \cdot Et_2O$ as a catalyst.[239]

PIB is of great technical interest—in principal four different types of PIB are on the market (Table 4.2).

In view of the complexity of the Lewis acid polymerization of IB[234,241] attempts have also been made to use the apparently more simple initiating systems involving carbocation salts[242] and also high-energy radiation.[243,244]

4.8.1.2. *Inifer Technique*

This development of the idea of using carbocation salts prepared *in situ* led Kennedy and his school to develop what has come to be known as the inifer technique—a simple and efficient method for preparing well-defined poly-IB telechelics with molar masses up to *ca* 2×10^4.

The inifer mechanism was initially proposed by Kennedy and Smith[245] and has been studied in some detail by Nuyken and his co-workers.[246–248]

In order to synthesize telechelics from IB, or indeed to produce any well-defined polymers, it is essential that the individual polymerization steps (Scheme 4.24) can be controlled.

The termination and transfer reactions can be utilized to introduce the desired head-groups while initiation and transfer determine the nature of the tail-groups. The principle of this type of scheme was first realized by VerStrate and Baldwin[249,250] who used alkyl bromides as transfer agents in

TABLE 4.2. Technical Types of PIB[240]

Type	$\overline{DP}$	Starting material	T (°C)	Solvent	Transfer	Initiator
A	6–50	IB, C_4-fraction containing small amounts of IB	0 to −20	Propene butene	n-Butene	$AlCl_3$, BF_3 $BF_3 \cdot MeOH$
B	140–360	IB, C_4-fraction enriched with IB	−40 to −103.7	Propene butane	n-Butene C_8-olefines	$AlCl_3$, BF_3
C	530–1400	IB	−60 to −103.7	Ethane ethene dichloro-methane	n-Butene C_8-olefines	AlCl in $CHCl_3$
D	1800–4000	IB	−80 to −103.7	Ethane ethene dichloro-methane	n-Butene C_8-olefines	$AlCl_3$ in CHCl BF_3/ethene

RX	+	MtX_n	⇌	$R^{\oplus}$	+ $MtX^{\ominus}_{n+1}$	Cationation
$R^{\oplus}$	+	M	⇌	$R\text{-}M^{\oplus}$		Initiation
$R\sim M^{\oplus}$	+	n M	⇌	$R\sim M^{\oplus}_{n+1}$		Propagation
$R\sim M^{\oplus}$	+	$MtX^{\ominus}_{n+1}$	⇌	$R\sim X$	+ MtX_n	Termination
$R\sim M^{\oplus}$	+	RX	⇌	$R\sim X$	+ $R^{\oplus}$	Transfer

SCHEME 4.24.

conjunction with $AlBr_3$. However, the transfer in their system was thermodynamically unfavorable $[\Delta G_f^{\ominus}(R^{\oplus}) \approx \Delta G_f^{\ominus}(R \sim M^{\oplus})]$ so that well-defined polymers were not obtained. Thus, the discovery that the inifers (**17–19**) in conjunction with BCl_3 under controlled conditions lead to a relatively ideal system in terms of Scheme 4.24 was a real breakthrough.

In these systems, the transfer reaction is driven (thermodynamically) by the greater stability of the benzylic cation vis-à-vis the *tert*-butyl cation of the growing polymer chain. With **18** and **19**, dichloro- and trichloro-telechelics with $M_n \leq 20{,}000$ can be synthesized and, although not such a perfect system, monochloro-telechelic Poly-IB can be prepared using **17**.

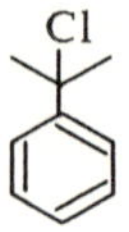

17

18

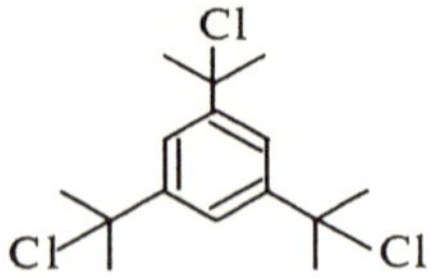

19

The mechanism proposed by Kennedy and Smith does explain the high initiator efficiency and the near-perfect functionality obtainable with these systems. However, the control of molar mass by the ratio [M]/[I] needed a modification to the original mechanism, i.e., that the termination reaction (Scheme 4.24) is reversible. That this is indeed so has since been confirmed by simple chain-extension experiments with prepared α,ω-dichloro-Poly-IB and BCl_3. It has also been confirmed by model experiments with $CH_3C(CH_3)_2CH_2C(CH_3)_2Cl/BCl_3$.[251] Thus, according to this mechanism. BCl_3 reacts with a Cl-terminated molecule via a binary ionogenic equilibrium to yield an active carbocation. This cation then adds monomer until it returns to a dormant (Cl-terminated) form via reaction with a $BCl_4^{\ominus}$ anion. This activation/deactivation process continues, provided no side reactions occur, until no monomer remains.

One of the serious problems with the inifer technique is the termination by indane ring formation as described in Scheme 4.25.

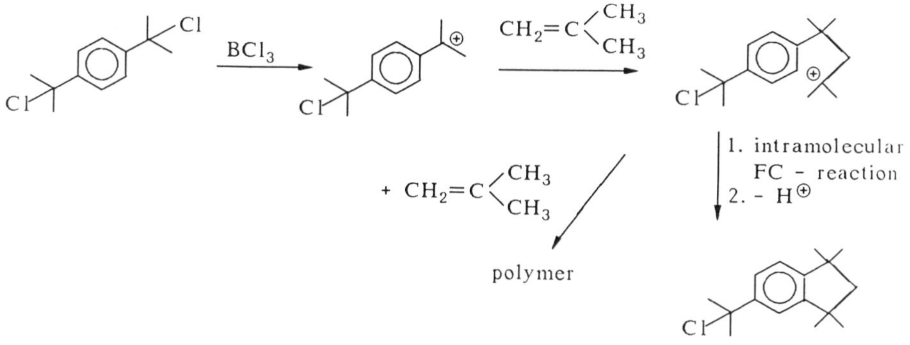

SCHEME 4.25.

More recently, simple nonaromatic inifers (**20**, Scheme 4.26) have been developed which exclude any possibility of indane or polyindane formation and are also more readily available.[(252)]

A further development has been the discovery that the corresponding esters and ethers derived from 2-phenyl-2-propanol can also lead to a polymerization of IB as well as styrene and alkyl-substituted styrenes in the presence of BCl_3 or $TiCl_4$. Since the initial finding, a wide range of esters and ethers have been shown to be effective initiators of living IB polymerizations.[(253–256)]

R^1 = $-C_6H_5$, $-C(CH_3)_3$, $-C(CH_3)_2-CH_2-C(CH_3)_3$,

R^2 = -H, $-CH_3$, $-CH_2CH_3$, $-CH(CH_3)_2$, $-C(CH_3)_3$, $-CH_2Cl$, $-CHCl_2$

R^3 = $-CH_3$

Although these ester/BCl_3 or ether/BCl_3 polymerizations were heralded as living polymerizations, it has since been shown that there are several possible side reactions which lead to difficulties in preparing polymers with molar masses greater than about 5000. Furthermore, the initiation reaction appears to be slower than the propagation in most cases, so that the molar mass distributions are broader than would be expected from living polymerizations. In addition, it has been shown that triphenylmethyl acetate, itself not an initiator, can be added to these polymerizations leading to a reduction in the rate of polymerization, yield, and also narrower molar mass distributions. This "common ion salt" effect suggests

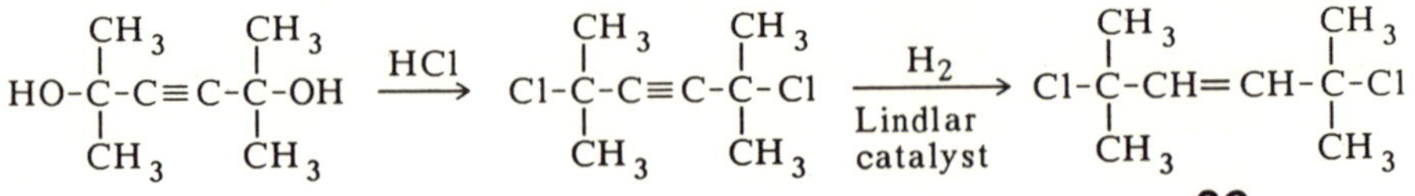

20

SCHEME 4.26.

that the originally proposed "pseudocationic" insertion mechanism is, at least, not entirely correct.[254] It seems at least possible that the true propagating species is of ionic character.

4.8.2. Experimental Procedure

4.8.2.1. Traditional Initiation[257]

To a cooled three-necked 1-L round-bottomed flask equipped with stirrer, nitrogen inlet–outlet tube, and funnel is added a solution of 31.9 g of isobutylene in 250 g of n-pentane. Then 32 mL of 4.13×10^{-2} mol L^{-1} $AlCl_3$ in methyl chloride in slowly added while keeping the temperature at −78 °C. After 1 h the polymer is isolated by evaporating the solvent. Polymers with $10{,}000 < DP < 100{,}000$ are obtained, depending on the purity of the reagents. The better the purity, the lower the yield and the higher the *DP*.

4.8.2.2. Inifer/BCl_3[245, 248]

For more exacting work commercial IB must be purified and dried. This is most easily carried out by distilling IB under its own vapor pressure through molten sodium. In this way finely divided sodium is carried with the dry IB vapor and aids in keeping the distilled material dry.[239]

The polymerizations are best carried out under high vacuum conditions, although with careful use of a dry box larger-scale preparations are possible and perhaps more convenient. Monomer, inifer, and solvent (usually CH_3Cl or CH_2Cl_2) must all be carefully dried. Provided the system is dry a quiescent mixture of IB (*ca* 1.0 M) and BCl_3 (tenfold excess with respect to inifer) in the solvent can be prepared and precooled before initiation by addition of the inifer. Inifer + BCl_3 solutions are not stable and form polyindane (compare Section 4.3.7). Reactions are best carried out at low temperature (−30 °C) to reduce side reactions and, in contrast to the more usual polymerizations, terminating the polymerizations with, e.g., methanol is not advisable. Instead the solvent and BCl_3 mixture is simply removed *in vacuo* after which the polymer can be purified by the usual solution/precipitation techniques. This method has the advantage that the BCl_3/solvent mixtures can be reused,[251] which is not only convenient for small-scale preparation but is vital if this method is to become of commercial importance.

Before carrying out such polymerizations a useful test of the purity of the solvent is to carry out a "blank" without initiator. The formation of significant amounts (>5%) of polymer from IB, BCl_3, and solvent alone indicates that more care must be exercised. Using CH_3Cl as solvent, the reactions can be carried out conveniently at −35 °C by preparing a solution of monomer (*ca* 1.0 M) and initiator (*ca* 10^{-3} M) and finally adding the BCl_3 (*ca* 0.3 M). The reaction rates depend on the concentration of initiator [and triphenylmethyl acetate (*vide supra*) if present].

REFERENCES

1. G. Sauvet, in *Comprehensive Polymer Science,* ed. G. Allen and J. C. Bevington (Pergamon Press, Oxford, 1989), Vol. 3, p. 579.
2. O. Nuyken and S. Pask, in *Comprehensive Polymer Science,* ed. by G. Allen and J. C. Bevington (Pergamon Press, Oxford, 1989), vol. 3, p. 619.
3. K. Matyjaszewski, in *Comprehensive Polymer Science,* eds. by G. Allen and J. C. Bevington (Pergamon Press, Oxford, 1989), vol. 3, p. 639.
4. T. Higashimura and M. Sawamoto, in *Comprehensive Polymer Science,* ed. by G. Allen and J. C. Bevington (Pergamon Press, Oxford, 1989), Vol. 3, p. 673.
5. J. M. Rooney, in *Comprehensive Polymer Science,* ed. by G. Allen and J. C. Bevington (Pergamon Press, Oxford, 1989), Vol. 3, p. 697.
6. J. P. Kennedy, in *Comprehensive Polymer Science,* ed. by G. Allen and J. C. Bevington (Pergamon Press, Oxford, 1989), Vol. 3, p. 705.
7. O. Nuyken and S. Pask, in *Encyclopedia of Polymer Science and Engineering* **16**, 2nd ed. (Wiley, New York, 1989), p. 495.
8. V. Percec, C. Pugh, O. Nuyken, and S. Pask, in *Comprehensive Polymer Science,* ed. by G. Allen and J. C. Bevington (Pergamon Press, Oxford, 1989), Vol. 6, p. 281.
9. J. P. Kennedy, *Cationic Polymerization of Olefins: A Critical Inventory* (Wiley, New York, 1975), p. 228.
10. J. P. Kennedy, *Cationic Polymerization of Olefins: A Critical Inventory* (Wiley, New York, 1975), Table 17.
11. A. R. Mathieson, in *The Chemsitry of Cationic Polymerization,* ed. by P. H. Plesch (Macmillan, New York, 1963), p. 235.
12. D. J. Dunn, in *Developments in Polymerization,* ed. R. N. Haward (Elsevier Appl. Sci. Publ., Barking, 1979), Vol. 1, p. 46.
13. A. Ledwith and D. C. Sherrington, in *Reactivity, Mechanism and Structure in Polymer Chemistry,* ed. by A. D. Jenkins and A. Ledwith (Wiley Interscience, New York, 1974), p. 252.
14. A. Gandini and H. Cheradame, *Adv. Polym. Sci.* **34/35**, 202 (1976).
15. H.-J. Sinn, H. Winter, and W. v. Tirpitz, *Makromol. Chem.* **48**, 59 (1961).
16. E. I. Tinyakova, T. G. Zhuravleva, T. N. Kurengina, N. S. Kirikova, and B. A. Dolgoplosk, *Dokl. Akad. Nauk SSSR* **144**, 592 (1962).
17. T. Saegusa, H. Imai, and J. Furukawa, *Makromol. Chem.* **79**, 207 (1964).
18. J. P. Kennedy, *J. Macromol. Sci., Chem.* **A3**, 861 (1969).
19. J. P. Kennedy, *J. Macromol. Sci., Chem.* **A3**, 885 (1969).
20. J. P. Kennedy, *Cationic Polymerization of Olefins: A Critical Inventory* (Wiley, New York, 1975), Table 18.

21. A. Gandini and P. H. Plesch, *Proc. Chem. Soc.* 240 (1964).
22. P. H. Plesch, *Adv. Polym. Sci.* **8**, 139 (1971).
23. K. Matyjaszewski, *Makromol. Chem. Macromol. Symp.* **13/14**, 433 (1988).
24. K. Matyjaszweski, *Makromol. Chem., Macromol. Symp.* **13/14**, 389 (1988).
25. P. H. Plesch, *Makromol. Chem., Macromol. Symp.* **13/14**, 393 (1988).
26. T. Masuda and T. Higashimura, *J. Polym. Sci., Part B, Polym. Lett.* **9**, 783 (1971).
27. M. Sawamoto, A. Furukawa, and T. Higashimura, *Macromolecules* **16**, 518 (1983).
28. W. Obrecht and P. H. Plesch, *Makromol. Chem.* **182**, 1459 (1981).
29. M. Sawamoto, T. Masuda, T. Higashimura, S. Kobayashi, and T. Saegusa, *Makromol. Chem.* **178**, 389 (1977).
30. K. Matyjaszewski and C. H. Lin, *J. Polym. Sci., Polym. Chem. Ed.* **26**, 3031 (1988).
31. J. P. Kennedy and M. K. Mishra, EP 265053 (27.4.1988); *Chem. Abstr.* **110**, 155 017 z (1988).
32. T. Puskas, G. Kaszas, J. P. Kennedy, T. Kelen, and F. Tüdös, *J. Macromol. Sci., Chem.* **18**, 1315 (1982/83).
33. R. Faust, A. Fehervari, and J. P. Kennedy, *J. Macromol. Sci., Chem.* **18**, 1209 (1982/83).
34. K. Matyjaszewski and P. Sigwalt, *Makromol. Chem.* **187**, 2299 (1986).
35. C. G. Overberger in *Macromolecular Syntheses,* ed. by J. A. Moore (Wiley, New York, 1978), vol. 1, p. 6.
36. P. H. Plesch, *High Vacuum Techniques for Chemical Syntheses and Measurements* (Cambridge University Press, Cambridge, 1989).
37. N. Kanoh, A. Gotoh, T. Higashimura, and S. Okamura, *Makromol. Chem.* **63**, 106 (1963).
38. N. Kanoh, A. Gotoh, T. Higashimura, and S. Okamura, *Makromol. Chem.* **63**, 115 (1963).
39. N. D. Prishchepa, Y. Y. Goldfarb, B. A. Krentsel, and M. V. Shishkina, *Polym. Sci. USSR* **9**, 2743 (1967).
40. N. D. Prishchepa, Y. Y. Goldfarb, and B. A. Krentsel, *Polym. Sci. USSR* **8**, 1829 (1966).
41. J. P. Kennedy, P. L. Magagnini, and P. H. Plesch, *J. Polym. Sci. A*1, **9**, 1635 (1971).
42. P. L. Magagnini, P. H. Plesch, and J. P. Kennedy, *Eur. Polym. J.* **7**, 1161 (1971).
43. G. Heublein and H. Dawczysnki, *J. Prakt. Chem.* **314**, 557 (1972).
44. A. Tanizaki, M. Sawamoto, and T. Higashimura, *J. Polym. Sci., Polym. Chem. Ed.* **24**, 87 (1986).
45. R. Faust and J. P. Kennedy, *Polym. Bull.* **19**, 29 (1988).
46. T. Masuda and T. Higashimura, *J. Macromol. Sci., Chem.* **A5**, 547 (1971).
47. H. Staudinger and E. Dreher, *Justus Liebigs Ann. Chem.* **517**, 73 (1935).
48. S. Matsushita, T. Higashimura, and S. Okamura, *Kobunski Kagaku* **17**, 456 (1960).
49. S. Okamura, N. Kanoh, and T. Higashimura, *Makromol. Chem.* **47**, 19 (1961).
50. Y. Imanishi, S. Matsushita, T. Higashimura, and S. Okamura, *Makromol. Chem.* **70**, 68 (1964).
51. G. Heublein and H. Dawezynski, *J. Prakt. Chem.* **314**, 557 (1972).
52. T. Higashimura and O. Kishiro, *Polym. J.* **9**, 87 (1977).
53. T. Higashimura, M. Mitsuhashi, and M. Sawamoto, *Macromolecules* **12**, 178 (1979).
54. T. Higashimura and M. Sawamoto, *Adv. Polym. Sci.* **62**, 49 (1984).
55. T. Higashimura, K. Kojima, and M. Sawamoto, *Polym. Bull.* **19**, 7 (1988).
56. T. Higashimura, T. Aoshima, and M. Sawamoto, *Makromol. Chem., Macromol. Symp.* **13/14**, 457 (1988).
57. O. Nuyken and H. Kröner, *Makromol. Chem.* **191**, 1 (1990).
58. M. S. Gong and H. K. Hall, *Macromolecules* **19**, 3011 (1986).
59. H. K. Hall and M. Atsumi, *Polym. Bull.* **19**, 319 (1988).
60. G. Heublein, S. Spange, and A. Ulbricht, *Makromol. Chem., Macromol. Symp.* **13/14**, 235 (1988).

61. J. Roudet and A. Gandini, *Makromol. Chem., Rapid Commun.* **10**, 277 (1989).
62. C. P. Brown and A. R. Mathieson, *J. Chem. Soc.* 3608 (1957).
63. N. Kanoh, A. Gotoh, T. Higashimura, and S. Okamura, *Makromol. Chem.* **63**, 106 (1963).
64. G. R. Brown and D. C. Pepper, *Polymer* **6**, 497 (1965).
65. G. Heublein and H. Dawczynski, *J. Prakt. Chem.* **314**, 557 (1972).
66. T. Higashimura, T. Takeda, M. Sawamoto, K. Matsuzaki, and T. Uryu, *J. Polym. Sci., Polym. Chem. Ed.* **16**, 503 (1978).
67. T. Higashimura, *J. Polym. Sci., Polym. Symp.* **56**, 71 (1977).
68. M. Sawamoto, T. Masuda, and T. Higashimura, *Makromol. Chem.* **178**, 1497 (1977).
69. T. Higashimura and O. Kishiro, *J. Polym. Sci., Poly. Chem. Ed.* **12**, 967 (1974).
70. S. Bywater, in *The Chemistry of Cationic Polymerization,* ed. by P. H. Plesch (Macmillan, New York, 1963), p. 305.
71. J. P. Kennedy, *Cationic Polymerization of Olefins: A Critical Inventory* (Wiley Interscience, New York, 1975), p. 270.
72. S. Brownstein, S. Bywater, and D. J. Worsfold, *Makromol. Chem.* **48**, 127 (1961).
73. Y. Sakurada, M. Matsumoto, K. Imai, A. Nishioka, and Y. Kato, *J. Polym. Sci., Part B, Polym. Lett.* **1**, 633 (1963).
74. Y. Ohsumi, T. Higashimura, and S. Okamura, *J. Polym. Sci.* **A3**, 3729 (1965).
75. Y. Ohsumi, T. Higashimura, and S. Okamura, *J. Polym. Sci.* **A4**, 923 (1966).
76. K. F. Elgert, E. Seiler, G. Puschendorf, W. Ziemann, and H. J. Cantow, *Makromol. Chem.* **144**, 73 (1971).
77. K. F. Elgert and E. Seiler, G. *Makromol. Chem.* **145**, 95 (1971).
78. Y. Inoue, A. Nishioka, and R. Chujo, *Makromol. Chem.* **156**, 207 (1972).
79. K. F. Elgert, R. Wicke, B. Stützel, and W. Ritter, *Polymer* **16**, 465 (1975).
80. T. Kunitake and C. Aso, *J. Polym. Sci.,* A1, **8**, 665 (1970).
81. Y. Matsuguma and T. Kunitake, *Polym. J.* **2**, 353 (1971).
82. S. C. Guhaniyogi and J. P. Kennedy, *Polym. Bull.* **4**, 267 (1981).
83. J. P. Kennedy and R. T. Chou, *J. Macromol. Sci., Chem.* **A18**, 3 (1982).
84. J. P. Kennedy and R. T. Chou, *J. Macromol. Sci., Chem.* **A18**, 11 (1982).
85. J. P. Kennedy and R. T. Chou, *J. Macromol. Sci., Chem.* **A18**, 17 (1982).
86. S. C. Guhaniyogi, J. P. Kennedy, and T. Kelen, *J. Macromol. Sci., Chem.* **A18**, 77 (1982).
87. S. C. Guhaniyogi and J. P. Kennedy, *J. Macromol. Sci., Chem.* **A18**, 87 (1982).
88. J. P. Kennedy, T. Kelen, S. C. Guhaniyogi, and R. T. Chou, *J. Macromol. Sci., Chem.* **A18**, 129 (1982).
89. R. Faust, A. Fehervari, and J. P. Kennedy, *J. Macromol. Sci., Chem.* **A18**, 1209 (1982).
90. L. Toman, S. Pokorny, and J. Spevacek, *J. Polym. Sci., Polym. Chem. Ed.* **27**, 2229 (1989).
91. M. Biswas and J. P. Kennedy, *Makromol. Chem. Macromol. Symp.* **3**, 113 (1986).
92. T. Alfrey, J. J. Bohrer, and H. Mark, *Copolymerization* (Interscience, New York, 1952), p. 49.
93. Y. Okamoto, H. Yuki, and S. Murahashi, *Bull. Chem. Soc. Jpn.* **41**, 197 (1968).
94. H. Staudinger and E. Dreher, *Justus Liebigs Ann. Chem.* **517**, 73 (1935).
95. J. P. Kennedy, *J. Polymer Sci.* **A2**, 5171 (1964).
96. J. P. Kennedy and A. W. Langer, *Adv. Polym. Sci.* **3**, 508 (1964).
97. S. Murahashi, S. Nozakura, K. Tsuboshima, and Y. Kotake, *Bull. Chem. Soc. Jpn.* **38**, 156 (1965).
98. A. Shimizu, T. Otsu, and M. Imoto, *Bull. Chem. Soc. Jpn.* **41**, 953 (1968).
99. B. T. Storey, *J. Polym. Sci.* **A3**, 265 (1965).
100. O. Nuyken and F. Siebzehnrübl, *Makromol. Chem.* **189**, 541 (1988).
101. J. O. Halford and B. Weissmann, *J. Org. Chem.* **17**, 1646 (1952).

102. H. Staudinger, W. Heuer, and E. Husemann, *Trans. Faraday Soc.* **32**, 323 (1936).
103. G. F. D'Alelio and T. R. Hoffend, *J. Polym. Sci., A-1* **5**, 1245 (1967).
104. T. Higashimura, S. Aoshima, and H. Hasegawa, *Macromolecules* **15**, 1221 (1982).
105. S. Aoshima and T. Higashimura, *J. Polym. Sci., Polym. Chem. Ed.* **22**, 2443 (1984).
106. H. Hasegawa and T. Higashimura, *Macromolecules* **13**, 1350 (1980).
107. M. Sawamoto, T. Masuda, H. Nishii, and T. Higashimura, *J. Polym. Sci., Polym. Lett. Ed.* **13**, 279 (1975).
108. T. Higashimura and H. Nishii, *J. Polym. Sci., Polym. Chem. Ed.* **15**, 329 (1977).
109. T. Higashimura, M. Hiza, and H. Hasegawa, *Macromolecules* **12,** 217 (1979).
110. C. Aso, T. Kunitake, Y. Matsuguma, and Y. Imaizumi, *J. Polym. Sci., A-1* **6**, 3049 (1968).
111. C. Aso, T. Kunitake, and R. Kita, *Makromol. Chem.* **97**, 31 (1966).
112. T. Dittmer, F. Gruber, and O. Nuyken, *Makromol. Chem.* **190**, 1755 (1989).
113. T. Dittmer, F. Gruber, and O. Nuyken, *Makromol. Chem.* **190**, 1771 (1989).
114. M. Leitner, Diplom Thesis, Universität Bayreuth (1990).
115. O. Nuyken, D. Yang, G. Maier, and F. Gruber, *Makromol. Chem.* **192**, 1969 (1991).
116. P. Lutz, G. Beinert, and P. Rempp, *Makromol. chem.* **183**, 2787 (1982).
117. C. Bunel, S. Cohen, J. P. Laguerre, and E. Marechal, *Polym. J.* **7**, 320 (1975).
118. J. Coudane, M. Brigodiot, and E. Marechal, *Polym. Bull.* **1**, 465 (1979).
119. D. Engel and R. C. Schulz, *Makromol. Chem.* **180**, 2991 (1979).
120. P. Blin, C. Bunel, and E. Marechal, *J. Polym. Sci., Polym. Chem. Ed.* **19**, 891 (1981).
121. E. D. Bergmann and D. Katz, *J. Chem. Soc.* 3216 (1958).
122. R. H. Michel, *J. Polym. Sci., Part A* **2**, 2533 (1964).
123. M. Stolka, J. F. Yanus, and J. M. Pearson, *Macromolecules* **9**, 710 (1976).
124. M. Stolka, J. F. Yanus, and J. M. Pearson, *Macromolecules* **9**, 715 (1976).
125. M. Kaufman and A. F. Williams, *J. Appl. Chem.* **1**, 489 (1951).
126. I. Jones, *J. Appl. Chem.* **1**, 568 (1951).
127. H. Agbaji and J. Springer, *Eur. Polym. J.* **23**, 283 (1987).
128. See, e.g., DE3106954 to Japan Atomic Energy Research Inst. (1985).
129. G. Bald, H. Markert, and R. Wiedenmann, *Siemens Forsch.-Entwicklungsber,* **4**(2), 116 (195).
130. J. Schmelzer and J. Springer, *Eur. Polym. J.* **23**, 243 (1987).
131. J. Ballesteros, G. J. Howard, and L. Teasdale, *J. Macromol. Sci. Chem.* **A11**(1), 29 (1977).
132. G. Galli, R. Solaro, E. Chiellini, and A. Ledwith, *Macromolecules* **16**, 497 (1983).
133. M. Storey and G. Canty, *J. Res. Natl. Bur. Stand. Sect. A* **68**, 165 (1964).
134. A. Rembaum, J. Moacanin, and R. Haack, *J. Macromol. Chem.* **1**, 657 (1966).
135. M. Imoto and I. Soematzu, *Bull. Chem. Soc. Jpn.* **34**, 26 (1961).
136. H.-N. Cho and S.-K. Choi, *J. Polym. Sci., Polym. Chem. Ed.* **25**, 15 (1987).
137. S. Cohen and E. Marechal, *J. Polum. Sci., Symp. Ser.* **52**, 83 (1975).
138. S. D. Pask, P. H. Plesch, and S. B. Kingston, *Makromol. Chem.* **182**, 3031 (1981).
139. G. E. Holdcroft and P. H. Plesch, *Makromol. Chem.* **185**, 27 (1984).
140. S. Anasagasti and L. M. Leon, *J. Polym. Sci., Polym. Lett. Ed.* **21**, 979 (1983).
141. S. Anasagasti and L. M. Leon, *Polym. Bull. (Berlin)* **11**, 13 (1984).
142. H. Hoegl, O. Süs, and W. Neugebauer, Ger. Pat. 1068115 (29 Oct. 1959) to Kalle AG; *Chem. Abstr.* **55**, 20742a (1961).
143. H. Kanega, Y. Shirota, and H. Mikawa, *J. Chem. Soc., Chem. Commun.* 158 (1984).
144. Y. Shirota, N. Noma, H. Kanega, and H. Mikawa, *J. Chem. Soc., Chem. Commun.* 470 (1984).
145. J. M. Pearson and M. Stolka, in *Polymer Monographs, Vol. 6: Poly(N-vinylcarbazole)* (Gordon and Breach, New York, 1981).
146. J. M. Rooney, in *Comprehensive Polymer Science,* ed. by G. Allen and J. C. Bevington (Pergamon Press, Oxford, 1989), Vol. 3, p. 697.

147. O. Nuyken, in *Handbook of Polymer Synthesis,* ed. by H. Kricheldorf (Marcel Dekker, New York, 1991), p. 108.
148. H. Davidge, *J. Appl. Chem.* **9**, 241 (1959).
149. C. U. Rogers and B. B. Corson, *J. Am. Chem. Soc.* **69**, 2910 (1947).
150. G. R. Clemo and W. H. Perkin, *J. Chem. Soc.* **125**, 1804 (1924).
151. V. P. Lopatinskii, and I. P. Zherebtsov, *Izv. Tomskogo Politekhn Inst.* **136**, 23 (1965)' *Chem. Abstr.* **65**, 18550b (1966).
152. B. Bömer, in *Houben-Weyl: Methoden der Organischen Chemie* (G. Thieme Verlag, Stuttgart, 1987), Vol. E 20, p. 1277.
153. A. Ledwith and D. C. Sherrington, in *Macromolecular Synthesis,* ed. by J. A. Moore (Wiley, New York, 1978), p. 571.
154. T. Higashimura, H. Teranishi, and M. Sawamoto, *Polym J. (Tokyo)* **12**, 393 (1980).
155. T. Higashimura, Y.-X. Deng, and M. Sawamoto, *Polym. J. (Tokyo)* **15**, 385 (1983).
156. H. Scott, T. P. Konen, and M. M. Labes, *J. Polym. Sci.* **B2**, 689 (1964).
157. L. P. Ellinger, *Polymer* **6**, 549 (1965).
158. T. Nakamura, M. Soma, T. Onishi, and K. Tamaru, *Makromol. Chem.* **135**, 241 (1970).
159. C. E. H. Bawn, A. Ledwith, and M. Sambhi, *Polymer* **12**, 209 (1971).
160. H. K. Hall jr., *Angew. Chem. Int. Ed. Engl.* **22**, 440 (1983).
161. T. Gotoh, A. B. Padias, and H. K. Hall jr., *J. Am. Chem. Soc.* **108**, 4920 (1986).
162. B. Voit, A. Padias, and H. K. Hall jr., *Polym. Bull.* **24**, 45 (1990).
163. J. Wislicenus, *Justus Liebigs Ann. Chem.* **192**, 106 (1878).
164. C. E. Schildknecht, *Vinyl and Related Polymers* (Wiley, New York, 1952), p. 593.
165. D. H. Lorenz, in *Encylcopedia of Polymer Science and Technology* (Wiley–Interscience, New York, 1971), Vol. 14, p. 504.
166. T. Higashimura and M. Sawamoto, in *Comprehensive Polymer Science,* ed. by G. Allen and J. C. Bevington (Pergamon Press, Oxford, 1989), Vol. 3, p. 673.
167. J. Crivello and O. Nuyken, in *Handbook of Polymer Synthesis,* ed. by H. R. Kricheldorf (Marcel Dekker, New York, 1991), p. 155.
168. L. Lal, in *Polymer Chemistry of Synthetic Elastomers,* ed. by J. P. Kennedy and E. Törnquist (Wiley–Interscience, New York, 1968), Part 1, p. 331.
169. N. M. Bikales, in *Encyclopedia of Polymer Science and Technology* (Wiley–Interscience, New York, 1971), Vol. 14, p. 511.
170. F. Bolza and F. E. Treloar, *Makromol. Chem.* **181**, 839 (1980).
171. S. Aoshima and T. Higashimura, unpublished results (1983).
172. J. D. Coombes and D. D. Eley, *J. Chem. Soc.* 3700 (1957).
173. J. P. Kennedy, *J. Polym. Sci.* **38**, 263 (1959).
174. V. A. Kruglova, V. G. Lipovich, and N. A. Tyukavkina, *Vysokomol. Soedin., Ser. A* **9**, 932 (1967); *Chem. Abstr.* **67**, 64 765 u (1967).
175. A. Ledwith, E. Lockett, and D. C. Sherrington, *Polymer* **16**, 31 (1975).
176. Y. J. Chung, J. M. Rooney, D. R. Squire, and V. Stannett, *Polymer* **16**, 527 (1975).
177. C. C. Ma, H. Kubota, J. M. Rooney, D. R. Squire, and V. Stannett, *Polymer* **20**, 317 (1979).
178. C. E. H. Bawn, C. Fitzsimmons, A. Ledwith, J. Penfold, D. C. Sherrington, and J. A. Weightman, *Polymer* **12**, 119 (1971).
179. F. Subira, G. Sauvet, J. P. Vairon, and P. Sigwalt, *J. Polym. Sci., Polym. Symp.,* **56**, 221 (1976).
180. H. Kubota, V. Y. Kabanov, D. R. Squire, and V. Stannett, *J. Macromol. Sci., Chem.* **A12,** 1299 (1978).
181. W. C. Hsieh, H. Kubota, D. R. Squire, and V. Stannett, *J. Polym. Sci., Polym. Chem. Ed.* **18**, 2773 (1980).
182. A. Deffieux, W. C. Hsieh, D. R. Squire, and V. Stannett, *Polymer* **22**, 1575 (1981).
183. A. Deffieux, W. C. Hsieh, D. R. Squire, and V. Stannett, *Polymer* **23**, 65 (1982).
184. W. C. Hsieh, A. Deffieux, D. R. Squire, and V. Stannett, *Polymer* **23**, 427 (1982).

185. A. Deffieux, J. A. Young, W. C. Hsieh, D. R. Squire, and V. Stannett, *Polymer* **24**, 573 (1983).
186. F. Williams, in *Fundamental Process in Radiation Chemistry,* ed. by P. Ausloos (Interscience, New York, 1968).
187. C. E. Schildknecht, A. O. Zoos, and C. McKinley, *Ind. Eng. Chem.* **39**, 180 (1974).
188. Y. Ohsumi, T. Higashimura, and S. Okamura, *J. Polym. Sci., Part A-1* **5**, 849 (1967).
189. M. Sumi, S. Nozakura, and S. Murahashi, *Kobunshi Kagaku* **24**, 424 (1967); *Chem. Abstr.* **68**, 50 226c (1968).
190. M. Delfini, F. Conti, M. Paci, A. L. Segre, R. Solaro, and E. Chiellini, *Macromolecules* **16**, 1212 (1983).
191. G. Natta, G. Dall'Asta, G. Mazzanti, U. Giannini, and S. Cesca, *Angew. Chem.* **71**, 205 (1959).
192. K. Hatada, T. Kitayama, N. Matsuo, and H. Yuki, *Polym. J.* **15**, 719 (1983).
193. T. Kunitake and C. Aso, *J. Polym. Sci., Part A-1* **8**, 665 (1970).
194. Y. Ohsumi, T. Higashimura, and S. Okamura, *J. Polym. Sci., Part A-1* **5**, 849 (1967).
195. S. Okamura, T. Higahimura, and I. Sakurada, *J. Polym. Sci.* **39**, 507 (1959).
196. K. Iwasaki, *J. Polym. Sci.* **56**, 27 (1962).
197. D. J. Sikkema and H. Angad-Gaur, *Makromol. Chem.* **181**, 2259 (1980).
198. H. Angad-Gaur and D. J. Sikkema, *Makromol. Chem.* **181**, 2385 (1980).
199. J. Hutchison and A. Ledwith, in *Advances in Polymer Science* (Springer, Berlin, 1974), Vol. 14, p. 49.
200. A. Ledwith, M. Szwarc, K. Hayashi, M. Marek, and N. G. Gaylord, *J. Polym. Sci., Polym. Symp.* **56**, 483 (1976).
201. J. P. Kennedy and E. Marechal, in *Carbocationic Polymerization* (Wiley–Interscience, New York, 1982), p. 131.
202. A. Gandini and H. Cheradame, in *Advances in Polymer Science* (Springer, Berlin, 1980), Vol. 34/34, p. 230.
203. J. V. Crivello, in *UV Curing: Science and Technology,* ed. by S. P. Pappas (Technology Marketing Corp., Stamford, USA, 1978), p. 24.
204. J. V. Crivello, in *Advances Polymer Science* (Springer, Berlin, 1984), Vol. 62, p. 1.
205. M. Miyamoto, M. Sawamoto, and T. Higashimura, *Macromolecules* **17**, 265 (1984).
206. T. Higashimura, M. Miyamoto, and M. Sawamoto, *Macromolecules* **18**, 611 (1985).
207. T. Higashimura and M. Sawamoto, *Macromol. Chem. Suppl.* **12**, 153 (1985).
208. M. Miyamoto, M. Sawamoto, and T. Higashimura, *Macromolecules* **18**, 123 (1985).
209. S. Aoshima, K. Ebara, and T. Higashimura, *Polym. Bull. (Berlin)* **14**, 425 (1985).
210. M. Miyamoto, M. Sawamoto, and T. Higashimura, *Macromolecules* **17**, 2228 (1984).
211. M. Sawamoto, C. Okamoto, and T. Higashimura, *Macromolecules,* **20**, 2693 (1987).
212. M. Sawamoto, M. Kamigaito, K. Kojima, and T. Higashimura, *Polym. Bull. (Berlin)* **19**, 359 (1988).
213. O. Nuyken and H. Kröner, *Makromol. Chem.* **191**, 1 (1990).
214. O. Nuyken, H. Kröner, and S. Aechtner, *Makromol. Chem., Macromol. Symp.* **32**, 181 (1990).
215. M. Sawamoto, M. Kamigaito, and T. Higashimura, *Polym. Bull.* **20**, 407 (1988).
216. M. Sawamoto and T. Higashimura, *Makromol. Chem., Macromol. Symp.* (Strasbourg meeting), in press.
217. R. Chang, US 3.025.281 (13.3.1962) to Hercules Powder; *Chem. Abstr.* **57**, 4884 (1962).
218. S.A. Mosley, US 2.549.921 (24.4.1951) to Union Carbide; *Chem. Abstr.,* **45**, 5972 (1951).
219. A. D. Ketley, *J. Polym. Sci.* **62**, S 81 (1962).
220. C. E. Schildknecht and P. H. Dunn, *J. Polym. Sci.* **20**, 597 (1956).
221. J. W. Copenhaver and M. H. Bigelow, *Acetylene and Carbon Monoxide Chemistry* (Van Nostrand–Rheinhold, New Jersey, 1949), p. 49.

222. I. P. Losev, O. Y. Fedotova, and M. F. Shostokovskii, *J. Gen. Chem. (USSR)* **14**, 889 (1944).
223. G. Dall'Asta and N. Oddo, *Chim. Ind. (Milan)* **42**, 1234 (1960).
224. O. Nuyken and H. Kröner, *Polym. Prepr.* **29**(2), 87 (1988).
225. T. Higashimura, K. Suzuoki and S. Okamura, *Makromol. Chem.* **86**, 259 (1965).
226. C. E. H. Bawn, C. Fitzsimmons, and A. Ledwith, *Proc. Chem. Soc.* 391 (1964).
227. J. V. Crivello and J. H. W. Lam, *Macromolecules* **10**, 1307 (1977).
228. J. V. Crivello and J. H. W. Lam, *J. Polym. Sci., Polym. Chem. Ed.* **18**, 2677 (1980).
229. H. S. Schultz, US 3.718.634 (1973) to GAF Corp.; *Chem. Abstr.* **78**, 125 219 (1973).
230. C. E. Schildknecht, *Vinyl and Related Polymers* (Wiley, New York, 1952), Chap. 10.
231. R. M. Thomas, W. J. Sparks, and P. K. Frolich, *J. Am. Chem. Soc.* **62**, 276 (1940).
232. J. P. Kennedy, *Cationic Polymerization of Olefins: A Critical Inventory* (Wiley–Interscience, New York, 1975), p. 86 ff.
233. P. H. Plesch, in *The Chemistry of Cationic Polymerization,* ed. by P. H. Plesch (Macmillan, New York, 1963), Chap. 4.
234. A. Gandini and H. Cheradame, *Adv. Polym. Sci.* **34/35**, 1 (1980).
235. G. A. Latrémouille and A. M. Eastham, *Can. J. Chem.* **45**, 11 (1967).
236. P. H. Plesch, *J. Polym. Sci., Polym. Symp.* **56**, 373 (1976).
237. H. Güterbock, *Polyisobutylen* (Springer, Berlin, 1959).
238. Y. T. Eidus and B. K. Nefedov, *Usp. Khim.* **29**, 833 (1960); *Chem. Abstr.* **54**, 24 329e (1960).
239. R. H. Biddulph and P. H. Plesch, *J. Chem. Soc.* 3913 (1960).
240. H. Güterbock, in *Polyolefine,* Kunststoffhandbuch, Vol. 4, ed. by R. Vieweg, A. Schley, and A. Schwarz (Carl Hanser, München, 1969), p. 112.
241. O. Nuyken and S. D. Pask, in *Comprehensive Polymer Science,* ed. by G. Allen and J. C. Bevington (Pergamon Press, Oxford, 1989), Vol. 3, p. 619.
242. Ref. 3, p. 206; P. H. Plesch and S. D. Pask, unpublished results.
243. G. Akulov, N. Geller, V. Kropachev, V. Nefodov, E. Sinotova, S. Skorokhodov, V. Stepanov, and M. Toropova, *Makromol. Chem.* **179**, 2775 (1978).
244. V. Stannett and J. Silverman, *ACS Symp. Ser.* **212**, 435 (1983).
245. J. P. Kennedy and R. A. Smith, *J. Polym. Sci., Polym. Chem. Ed.* **18**, 1523 (1980).
246. O. Nuyken, S. D. Pask, and A. Vischer, *Makromol. Chem.* **184**, 553 (1983).
247. O. Nuyken, S. D. Pask, A. Vischer, and M. Walter, *Makromol. Chem.* **186**, 173 (1985).
248. O. Nuyken, S. D. Pask, A. Vischer, and M. Walter, *Makromol. Chem., Macromol. Symp.* **3**, 129 (1986).
249. G. VerStrate and F. P. Baldwin, *Polym. Prepr.* **17**, 808 (1976).
250. G. VerStrate and F. P. Baldwin, U.S. Pat. 4,278,822 (1981), to Exxon; *Chem. Abstr.* **95**, 187 880 p (1980).
251. C. V. Freyer, H. P. Mühlbauer, and O. Nuyken, *Angew. Makromol. Chem.* **145/146**, 69 (1986).
252. S. Mah, R. Faust, M. Zsuga, and J. P. Kennedy, *Polym. Bull.* **18**, 433 (1987).
253. M. K. Mishra and J. P. Kennedy, *J. Macromol. Sci., Chem.* **A24**, 933 (1987).
254. R. Faust and J. P. Kennedy, *J. Polym. Sci., Polym. Part A, Chem. Ed.* **25**, 1847 (1987).
255. G. Kaszás, J. Puskás, and J. P. Kennedy, *Makromol. Chem., Macromol. Symp.* **13/14**, 473 (1988).
256. R. Faust and J. P. Kennedy, *Polym. Prepr.* **29**(2), 69 (1988).
257. J. P. Kennedy and R. M. Thomas, *J. Polym. Sci.* **46**, 481 (1960).

5

Free Radical Ring-Opening Polymerization

Takeshi Endo and Tsutomu Yokozawa

5.1. INTRODUCTION*

Ring-opening polymerization has been exploited for incorporation of functional groups such as ethers, esters, amides, and carbonates into the backbone of polymers. However, it is difficult to use commercially available vinyl monomers for copolymerization in the ionic ring-opening polymerization investigated much thus far. Radical ring-opening polymerization would afford a general solution to copolymerization of ring-opening monomers with various kinds of vinyl monomers. Thus, this process makes possible the synthesis of commercially available vinyl polymers containing functional groups such as ether, ester, and ketone groups in the backbone of polymers.[1]

Although conventional cyclic monomers that undergo ionic ring-opening polymerization such as epoxides, lactones, and lactams do not undergo radical ring-opening polymerization, cyclic monomers having structural characteristics as follows can undergo radical ring-opening polymerization: (1) they have a terminal carbon–carbon double bond for attack by radical species; (2) they can undergo release of strain energy of the ring or enable the formation of thermodynamically stable carbon–carbon or carbon–heteroatom double bonds during the ring-opening process (ring-opening isomerization); (3) they have a functional group for stabilization of

* This chapter is dedicated to the memory of Professor W. J. Bailey.

Takeshi Endo and Tsutomu Yokozawa • Research Laboratory of Resources Utilization, Tokyo Institute of Technology, 4259 Nagatsuta, Midori-ku, Yokohama 227, Japan.

New Methods for Polymer Synthesis, edited by W. J. Mijs. Plenum Press, New York, 1992.

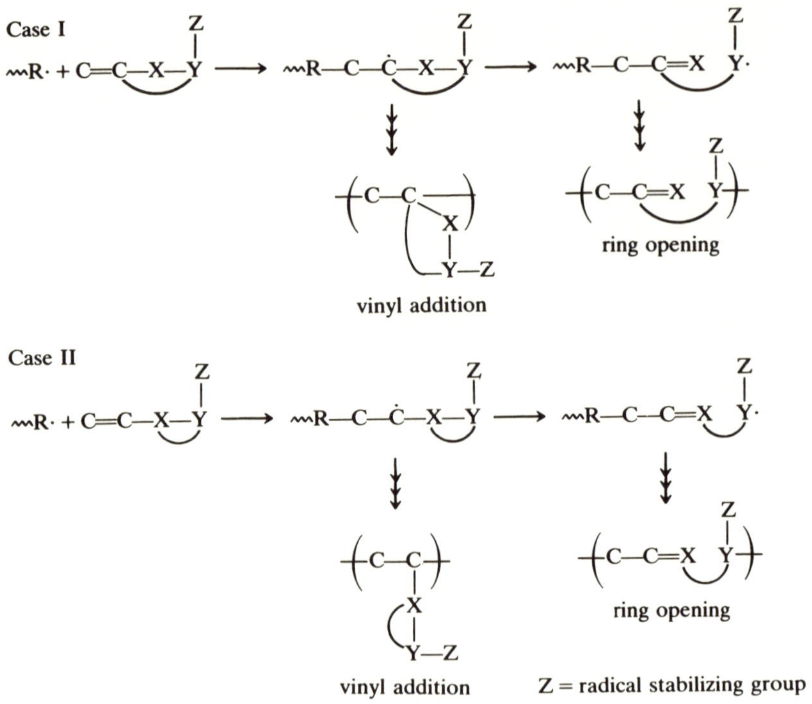

SCHEME 5.1.

the radical generated after the ring-opening process. Thus, the monomers that undergo radical ring-opening polymerization consist of a terminal carbon–carbon double bond and a cyclic compound, and may be classified as case I or II according to the two types of terminal carbon–carbon double bond: the exomethylene group (case I) or the vinyl group (case II). In what follows, monomers that undergo radical ring-opening polymerization will be described according to the respective cyclic compounds including case I and II. Generally ring opening would be promoted by dilution and higher temperatures, as will be elaborated in the reaction described in Scheme 5.1.

5.2. CARBOCYCLES

5.2.1. *Cyclopropanes*

Although the polymerization of methylenecyclopropanes in case I has not been reported yet, vinylcyclopropanes with various kinds of radical-stabilizing groups in case II have been reported to undergo radical ring-opening polymerization. Vinylcyclopropanes **1** and **2** bearing two halogen atoms as the radical stabilizing group selectively ring-opened to give

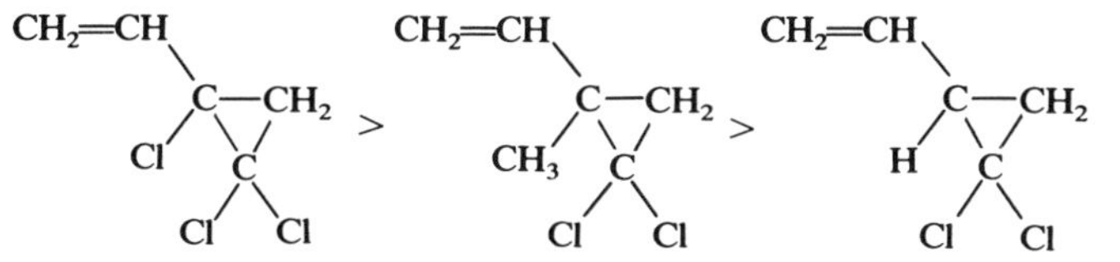

1

SCHEME 5.2.

the corresponding polymers containing only the 1,5-units [Eq. (5.1)],[2–4] while fluorine-substituted **3** has no selectivity of the ring-opening process and affords a polymer consisting of two different units with quantitative ring opening [Eq. (5.2)].[5] Vinyldichlorocyclopropanes substituted on the

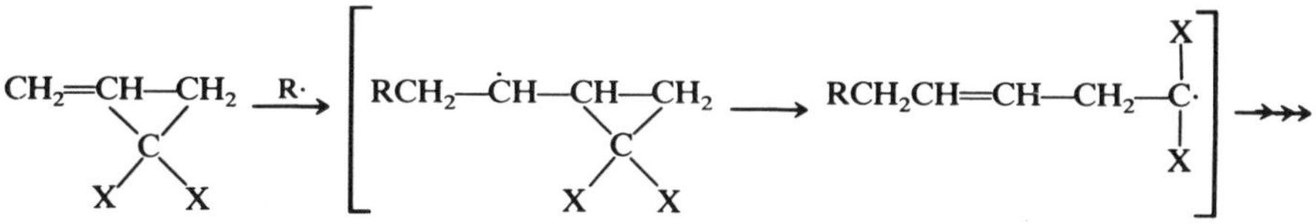

1: X = Cl
2: X = Br

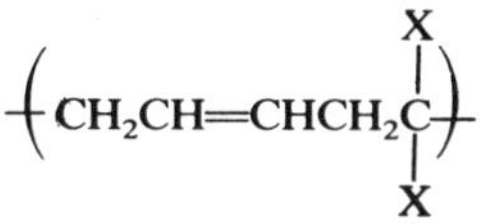

(5.1)

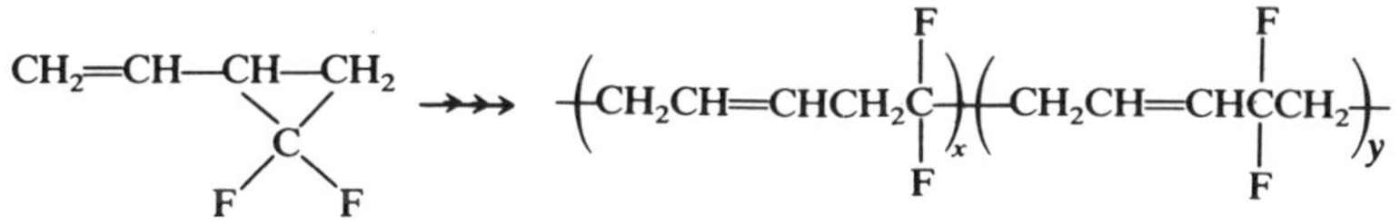

(5.2)

3

cyclopropyl ring were readily prepared from the corresponding substituted diene and chloroform, and polymerized with essentially 100% ring opening; polymerization of higher substituted cyclopropane derivatives proceeds relatively faster than **1** (see Scheme 5.2).[6]

Vinylcyclopropanes **4**,[7] **5**,[8] and **6**[9] bearing an ester, two electron-withdrawing groups, and a phenyl group as the radical-stabilizing group, respectively also underwent quantitative ring opening to afford a polymer containing only the 1,5-units.

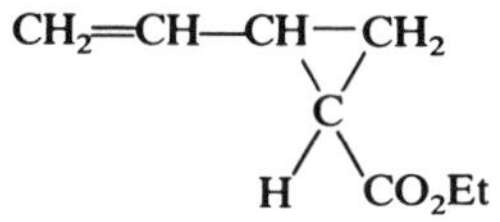

4

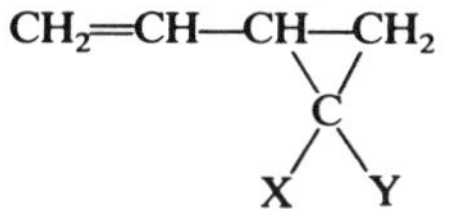

5

X = Y = CO_2Et
X = Y = CN
X = CN, Y = CO_2Et

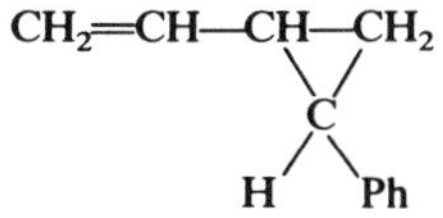

6

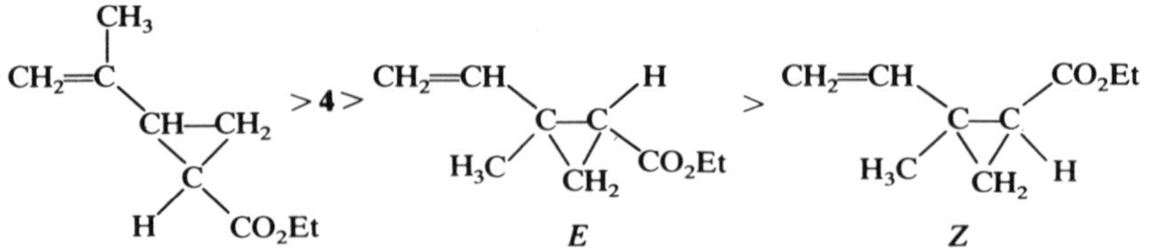

SCHEME 5.3.

Substituent effects of **4** on the relative rate of the radical ring-opening polymerization are shown in Scheme 5.3.[10,11] Furthermore, the effects of radical stabilizing groups on the polymerization rate indicate that monomers with strongly stabilizing groups polymerize slowly, therefore the rate-determining step of the polymerization of vinylcyclopropanes as in Scheme 5.4 is assumed to be not the ring-opening step but the radical addition step.[12,13] Even the polymerization of **7**[4] and **8**[14] carrying no radical

stabilizing group gave a copolymer in which *ca* 90% of rings were opened, so that vinylcyclopropanes seem to have a high potential of radical ring-opening, and the introduction of at least one radical stabilizing group into the cyclopropyl ring may make vinylcyclopropanes undergo quantitative ring opening.

When vinylcyclopropane **1** was copolymerized with vinyl monomers such as methyl acrylate (MA), methyl methacrylate (MMA), and styrene (St), the corresponding copolymers containing carbon–carbon double bonds in the backbone were obtained. However, the content of double bonds in the obtained copolymer decreased to *ca* 40–90% less than the theoretical amount calculated from the polymer composition, depending on the feed

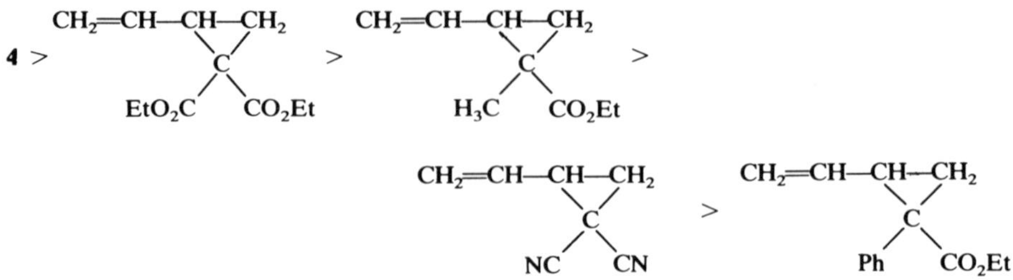

SCHEME 5.4.

ratio of **1** and vinyl monomers. These results have been explained by intramolecular cyclization to afford cyclohexyl rings during polymerization [Eq. (5.3)].[15] On the other hand, copolymerizations of **5** with acrylonitrile

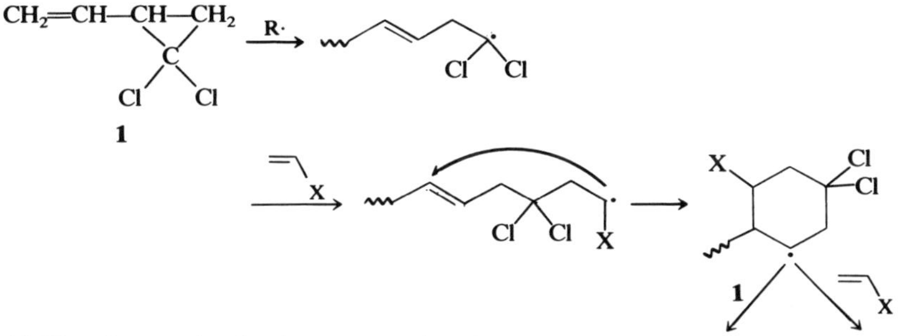

(5.3)

(AN), St, and vinyl acetate (VA) gave the corresponding copolymers without loss of carbon–carbon double bonds in the backbone. In addition, the vinylcyclopropane **9** that does not undergo homopolymerization can also be copolymerized with AN, St, and maleic anhydride (MAn) [Eq. (5.4)].[8b]

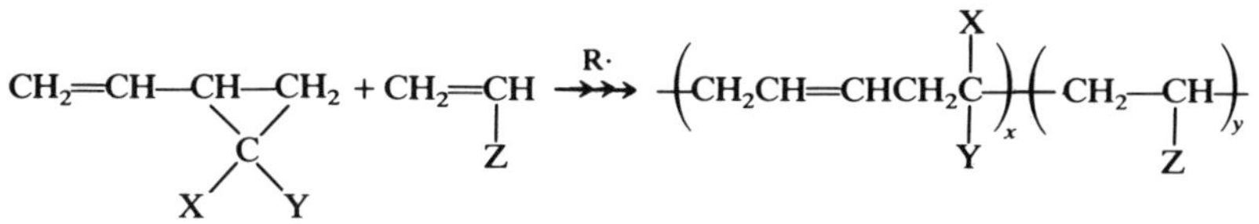

(5.4)

5: $X = Y = CO_2Et$
$X = Y = CN$
$X = CN, Y = CO_2Et$
9: $X = Y = Ph$

5.2.2. Cyclobutanes

Cyclobutanes are potential monomers for radical ring-opening polymerization, because they have high strain energy similar to the cyclopropyl ring. However, examples of radical ring-opening polymerization of cyclobutane derivatives are quite rare, and recently methylenecyclobutane in case I and vinylcyclobutane in case II have been reported.

Methylenecyclobutane **10** with various radical stabilizing groups underwent only vinyl polymerization [Eq. (5.5)],[16] while vinylcyclobutane **11** polymerized with 94% ring opening in solution polymerization at 60 °C (Eq. (5.6)].[17] The observed difference may be accounted for by the possibility of

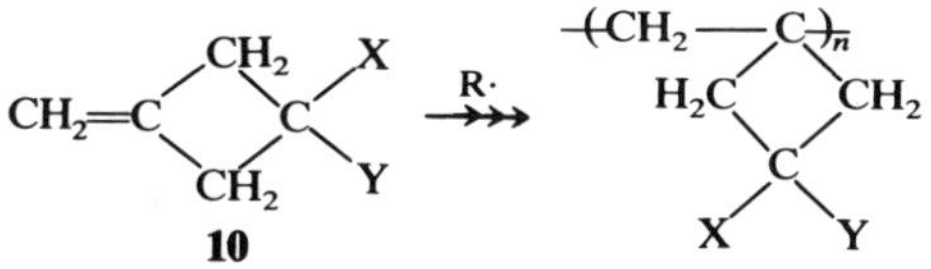

(5.5)

10

$X = CN, Y = H$
$X = CN, Y = CO_2Me$
$X = Ph, Y = H$

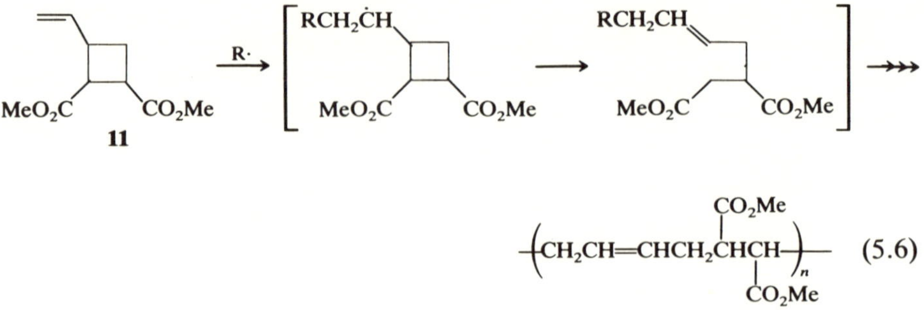

11

(5.6)

overlapping of the p-orbital of the radical generated by radical addition to carbon–carbon double bond with the sp^3-orbital of the cyclobutyl ring. Thus, in the polymerization of **10**, the p-orbital of the radical is perpendicular to the sp^3-orbital of the cyclobutyl ring so that **10** does not undergo ring opening. However, it is possible to achieve coplanarity between the radical p-orbital and the cyclobutyl sp^3-orbital in the monomer **11** and then the ring-opening polymerization may proceed smoothly.

5.2.3. Spiro Compounds

The hydrocarbon spiro compound, the dimer of *o*-xylylene **12**, underwent radical ring-opening polymerization to give the corresponding poly-*o*-xylylene [Eq. (5.7)].[18] The driving force for the ring-opening step appears to be the formation of an aromatic ring (aromatization) and generation of a stable benzyl radical.

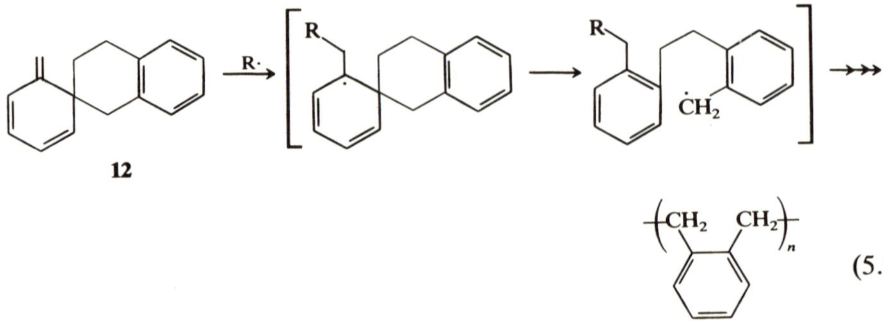

12

(5.7)

5.3. CYCLIC ETHERS

5.3.1. Oxiranes

Vinyloxiranes have been reported as examples for case II in analogy with cyclopropane derivatives as mentioned in 5.2.1. Vinyloxiranes **13** with a radical stabilizing aryl group underwent complete ring opening with selective *scission* of the carbon–carbon bond of the oxirane ring, contrary to the ionic ring-opening polymerization of oxiranes, to afford the corresponding polymers containing vinyl ether groups in the backbone [Eq. (5.8)].[19,20]

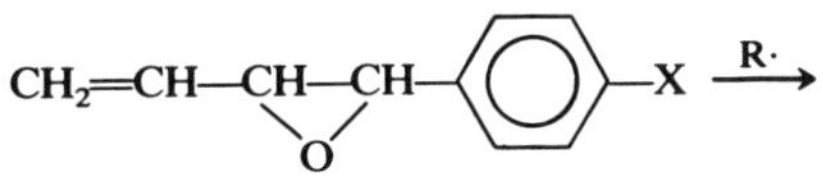

13

X = H, CH_3, OCH_3, Cl, CN

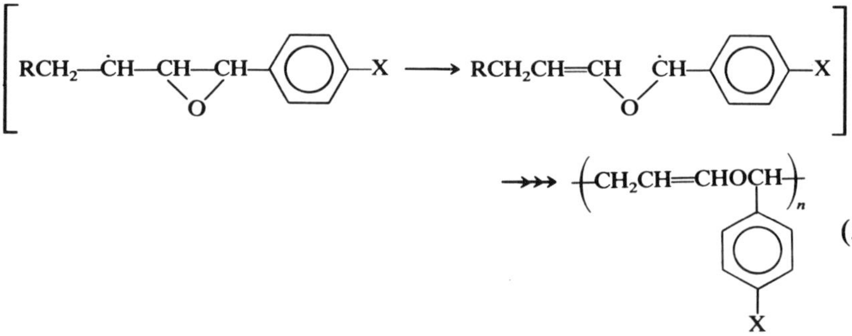

(5.8)

However, vinyloxirane **14** containing a furyl group retards the extent of ring opening and further promotes carbon–oxygen scission in the ring-opening process [Eq. (5.9)].[21] Consequently, vinyloxiranes seem to have a lower potential of radical ring-opening than vinylcyclopropanes.

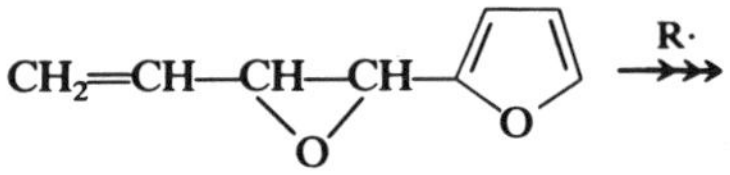

14

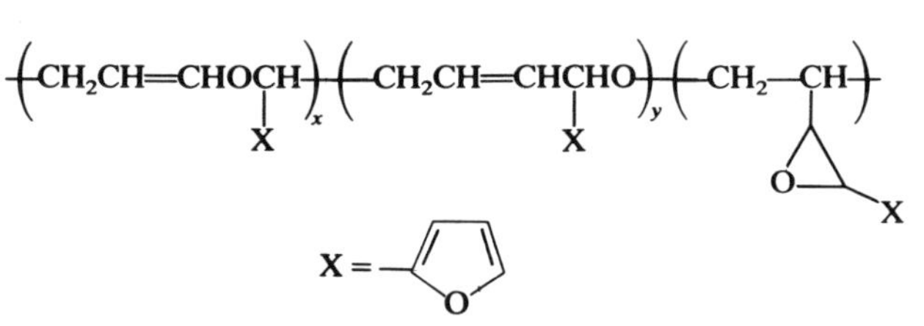

(5.9)

When **13** (X = H) was copolymerized with ethylene at high pressure, the carbon–carbon double bonds of the initially formed vinyl ether groups in the backbone from **13** functioned as grafting sites for ethylene to yield the structure depicted in Eq. (5.10). Also ethylene units appear in the polymer backbone.[22]

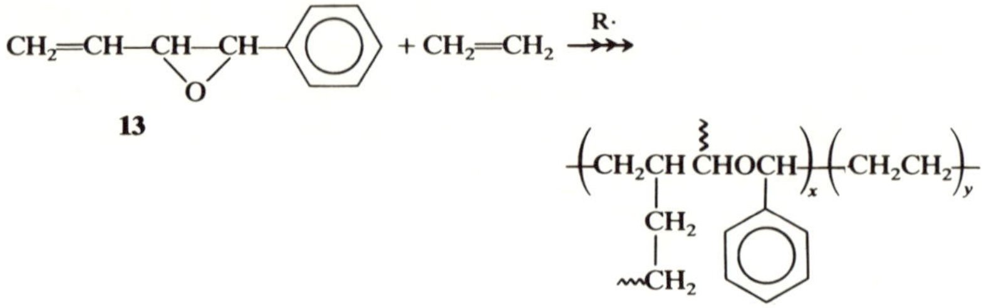

13

(5.10)

5.3.2. Oxetanes

Oxetane monomers in case II have not been reported but only 2-methyleneoxetane in case I has been described. Although the oxetane ring possesses considerable ring strain, the polymerization of 2-methyleneoxetane **15** at 120 °C gave a copolymer in which only about 40% of the rings were opened [Eq. (5.11)].[23]

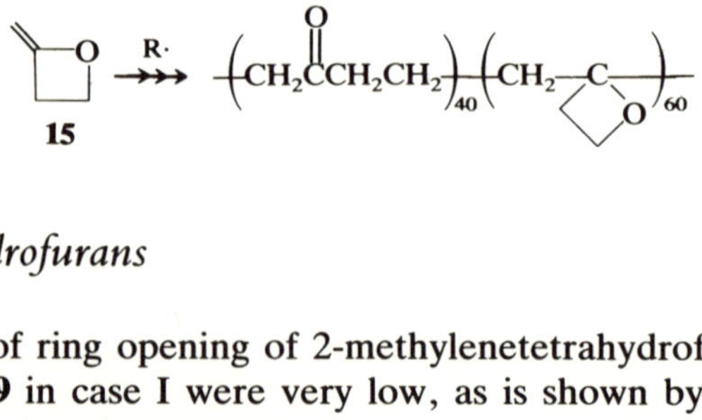

15

(5.11)

5.3.3. Tetrahydrofurans

The extents of ring opening of 2-methylenetetrahydrofuran derivatives **16**, **17**, **18**, and **19** in case I were very low, as is shown by the numbers in parentheses indicating the extent of ring opening at 120 °C in the presence of di-*tert*-butyl peroxide (DTBP).[24]

The polymerization of 2-vinyltetrahydrofuran **20** in case II was also studied to obtain low-molecular-weight oligomer ($M_n < 500$) containing ring-opening and addition units; the exact structure is unclear at the present time.[25]

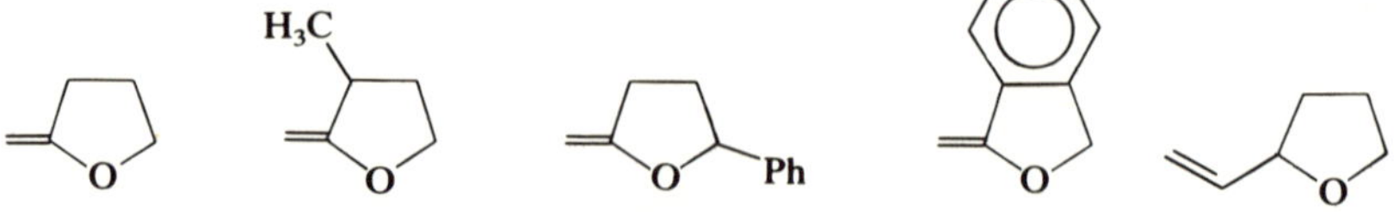

16 (5%) **17** (15–20%) **18** (50%) **19** (0%) **20**

5.3.4. Tetrahydropyrans

2-methylenetetrahydropyran **21** in case I underwent only 4–8% ring opening even at 120 °C with DTBP [Eq. (5.12)].[24] In the cyclic ether series, three-membered cyclic ethers bearing a radical stabilizing group such as **13** undergo quantitative ring opening, but the ring opening of five- and six-membered ones even containing a radical stabilizing group such as **18** or **19** is not complete. Apparently these results suggest that ring strain of monomers effects the extent of ring opening.

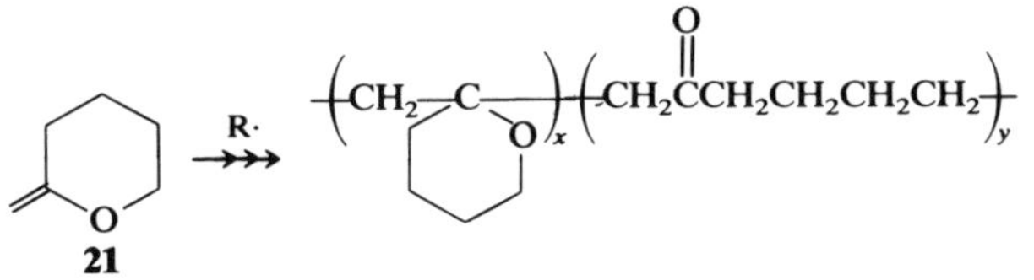

(5.12)

21

5.3.5. Acyclic Vinyl Ethers

Although cyclic vinyl ethers give a copolymer in which the ring opening is not complete, α-benzyloxystyrene **22** undergoes quantitative radical rearrangement of the ketone **23** [Eq. (5.13)].[26] Thus, **22** can be expected to

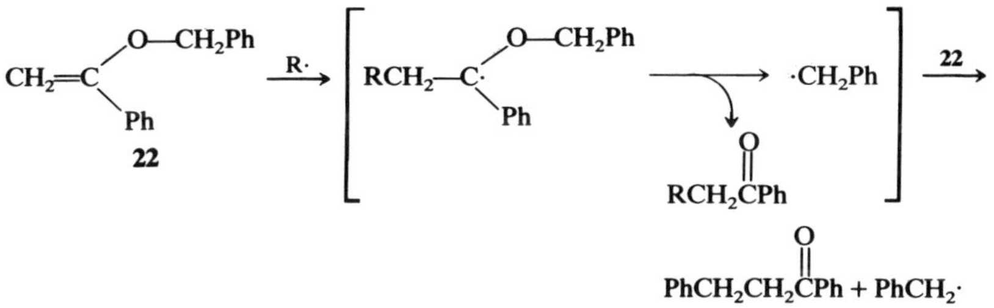

(5.13)

serve as an efficient chain-transfer agent. Actually, when **22**, and St or MMA were heated at 60 °C in the presence of azobis(isobutyrolitrile) (AIBN), the corresponding oligomers capped with a benzyl group at one end and a benzoylmethylene group at the other could be obtained [Eq. (5.14)].[27]

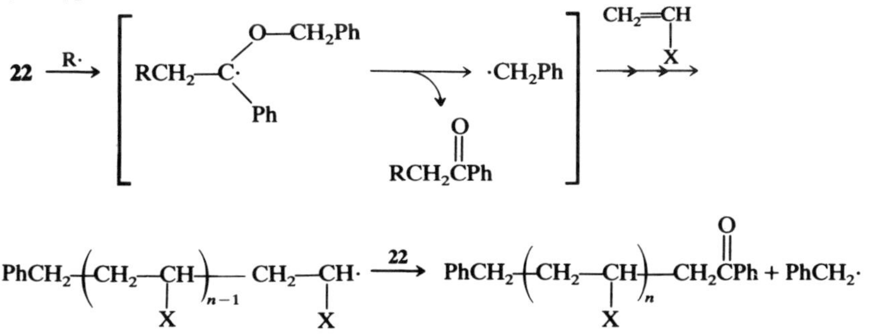

(5.14)

5.4. CYCLIC ACETALS

Only monomers in case I have been reported.

5.4.1. Cyclic Ketene Acetals

Although the polymerization of a five-membered cyclic ketene acetal without the radical stabilizing group, 2-methylene-1,3-dioxolane, gave a copolymer in which only 83% of the rings were opened even at 125 °C,[28] introduction of a phenyl group (**24**)[29,30] or a long alkyl chain (**25**)[31] into the five-membered cyclic ketene acetal makes quantitative ring opening possible to obtain the corresponding polyesters; the substituent R in **25** does not control the selectivity of the carbon–oxygen cleavage [Eq. (5.15)].

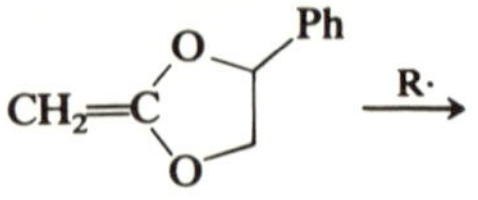

24

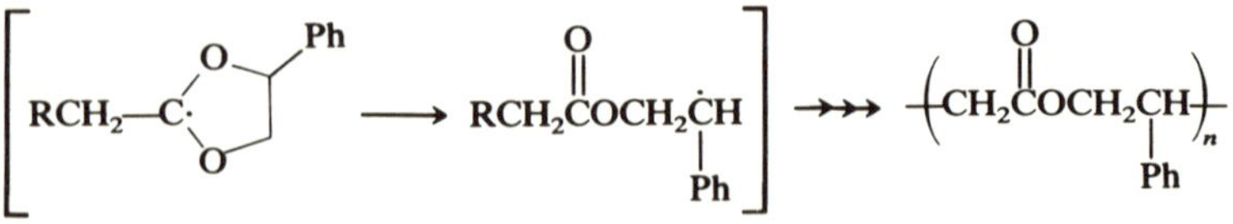

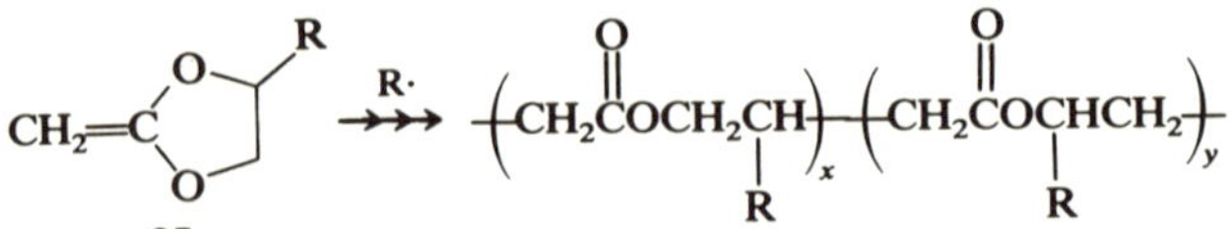

25

R = n-C_6H_{13}
n-$C_{10}H_{21}$

(5.15)

However, **24** polymerized with 100% selective ring opening, even at room temperature, by photoirradiation in the presence of photosensitizer.[32] Apparently, the ring-opening step of **24** is greatly enhanced by the formation of a relatively stable benzyl radical, and the long alkyl chain of **25** increases steric hindrance so that direct addition can be effectively eliminated. Furthermore, the *exo*-methylene-substituted five-membered cyclic ketene acetal **26** underwent ring-opening polymerization at 120 °C in dimethylformamide (DMF) through a stable allyl radical [Eq. (5.16)].[33]

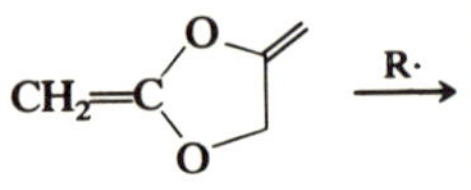

26

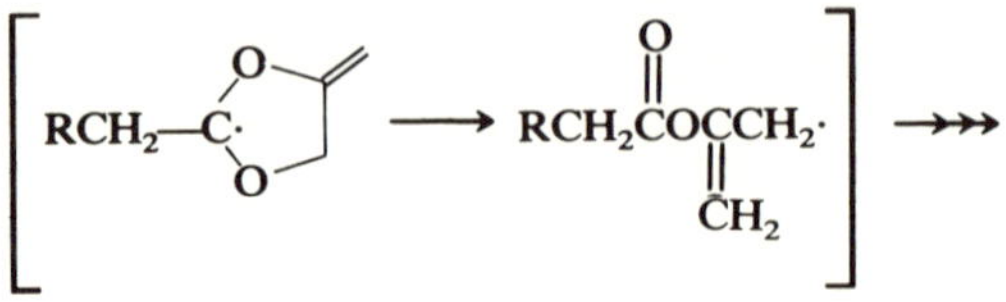

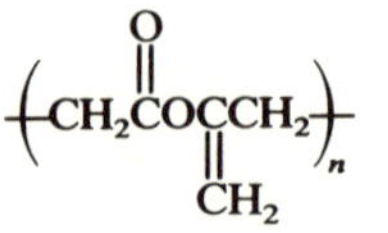

(5.16)

The unsubstituted six-membered cyclic ketene acetal **27** polymerized with 85% ring opening,[28] while the phenyl-substituted one, 4-phenyl-2-methylene-1,3-dioxane (**28**), underwent complete ring opening.[34] These results suggest that it is necessary to introduce a radical stabilizing group such as the phenyl group for five- or six-membered cyclic ketene acetals to undergo quantitative ring-opening polymerization.

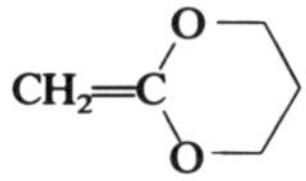

27

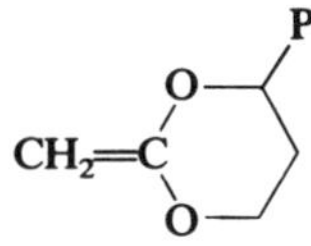

28

However, seven- and eight-membered cyclic ketene acetals **29**[28,35] and **30**[34] even without a radical stabilizing substituent polymerized with 100% ring opening, and also in the polymerization of substituted seven-membered cyclic ketene acetals **31** and **32** the rings are completely opened.[36] Apparently the seven- and eight-membered rings increase steric hindrance in the addition reaction of the monomers and also introduce a small amount of strain so that ring opening may be accelerated.

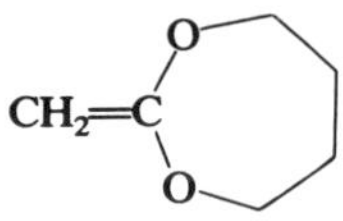

29

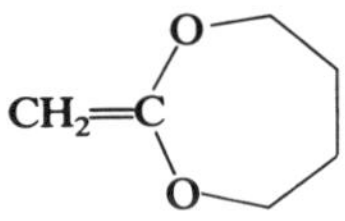

30

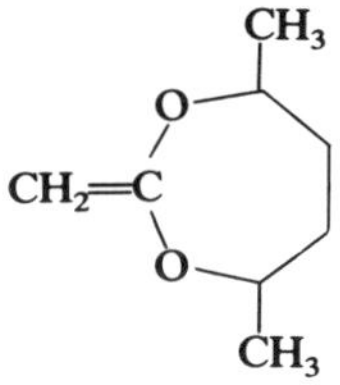

31

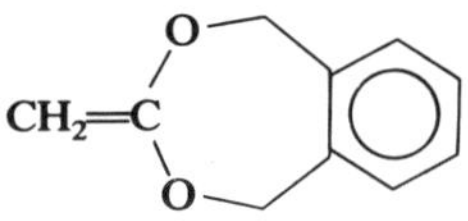

32

In this series of the cyclic ketene acetals that undergo complete ring opening, the relative polymerization rates indicated that ketene acetals with radical stabilizing substituents polymerize more slowly as shown below.[34] Consequently, the rate-determining step of the ring-opening polymerization of cyclic ketene acetals is assumed to be the radical addition step, in analogy with vinylcyclopropanes mentioned in Section 5.2.1.

	29	$>$	**30**	$\gg$	**24**		**28**
Relative rate	46		32		1.0	$>$	0.94

The ease of ring opening of cyclic ketene acetals compared to cyclic ethers is accounted for by suggesting that the ester group formed by the ring-opening isomerization of cyclic ketene acetals is more stable than the ketene group formed by the ring-opening isomerization of cyclic ethers. Furthermore, in contrast to 83% ring opening at 125 °C of the unsubstituted five-membered cyclic ketene acetal, 2-methylene-1,3-dioxolane, as mentioned above, the sulfur analog **33** polymerized with 45% ring opening at 120 °C,[23] while the nitrogen analog **34** polymerized with essentially 100%

ring opening even at 80 °C [Eq. (5.17)].[24,37] This order of ring-opening ability can also be interpreted by the order of thermodynamic stability of the thioester, ester, and amide group formed during polymerization.

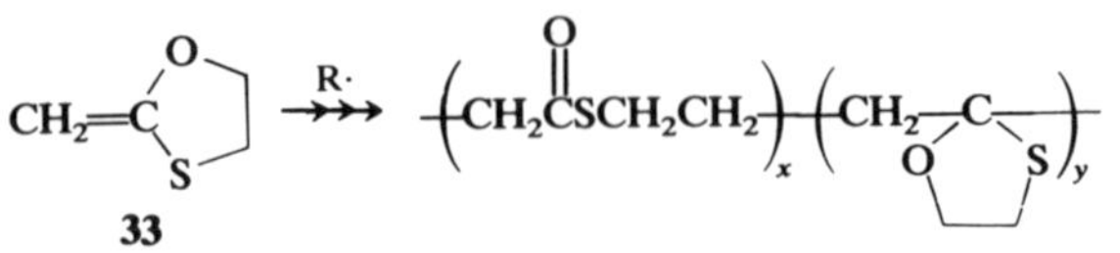

33

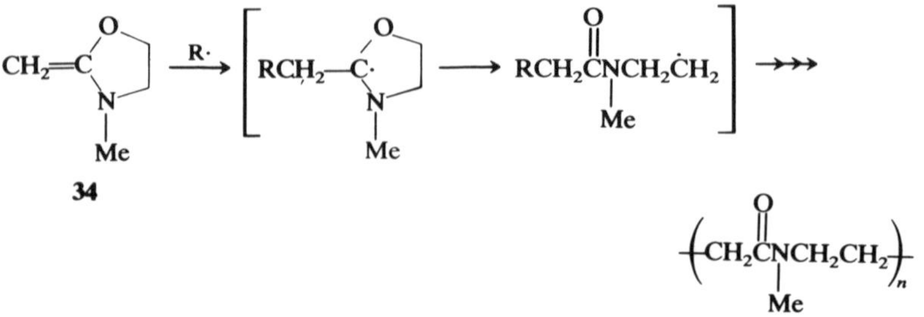

34

(5.17)

This radical ring-opening process makes possible the quantitative introduction of an ester or amide moiety in the backbone of commercially available vinyl polymers. Especially, such polymers containing ester groups can be expected to be biodegradable. The five-membered cyclic ketene acetal **24** could be copolymerized with St, MMA,[32] VA, and 4-vinylpyridine in the presence of DTBP at 120 °C, to afford the corresponding copolymers containing ester groups in the backbone [Eq. (5.18)].[29] Additionally, the five-membered **25** was copolymerized with St.[31] When the seven-membered **29** was copolymerized with St, 4-vinylanisole, MMA, and VA in the presence of DTBP at 120 °C, the corresponding copolymer **35** containing ester groups in the backbone could be obtained [Eq. (5.19)].[28] The cyclic *O*,*N*-ketene acetal **34** could also be copolymerized

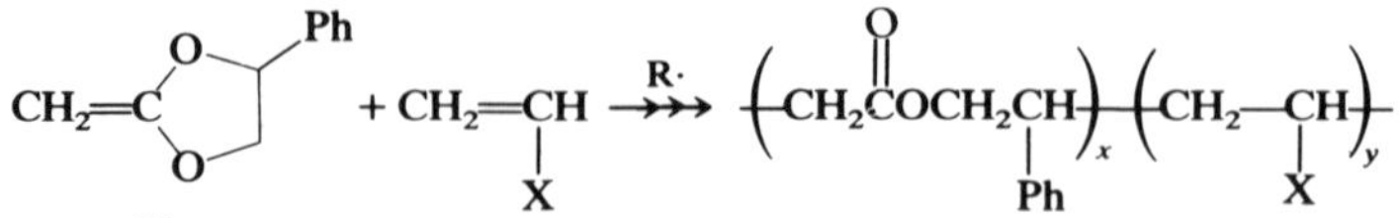

(5.18)

24

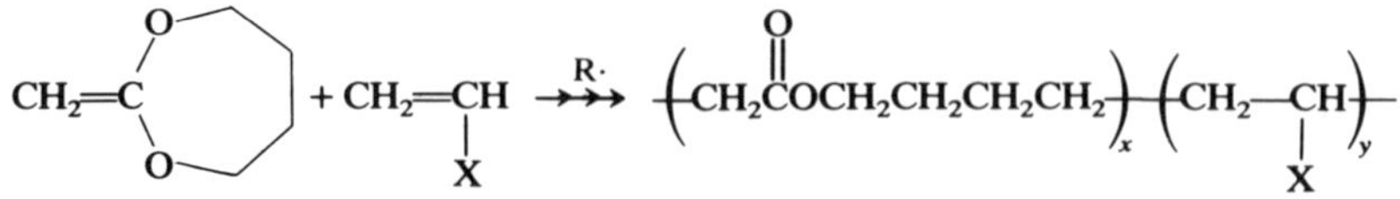

(5.19)

29 **35**

with St in the presence of DTBP at 120 °C to obtain a styrene copolymer **36** containing amide moieties in the backbone [Eq. (5.20)].[24,37]

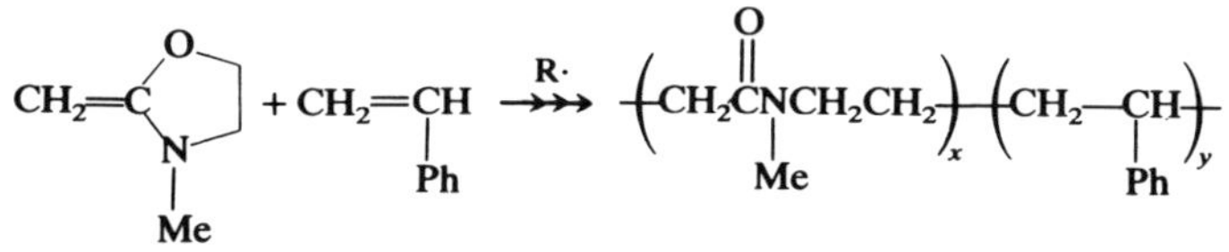

(5.20)

34 **36**

Hydrolysis of these copolymers makes possible the synthesis of oligomers capped with the functional groups on both ends. Thus, the hydrolysis of the copolymer of **35** containing ester groups in the backbone gave a vinyl oligomer capped with a hydroxyl on one end and a carboxylic acid group on the other end [Eq. (5.21)].[38] The hydrolysis of the copolymer of **36** containing amide groups in the backbone gave an oligomer capped with a methylamino group on one end and a carboxylic acid group on the other end [Eq. (5.22)].[38]

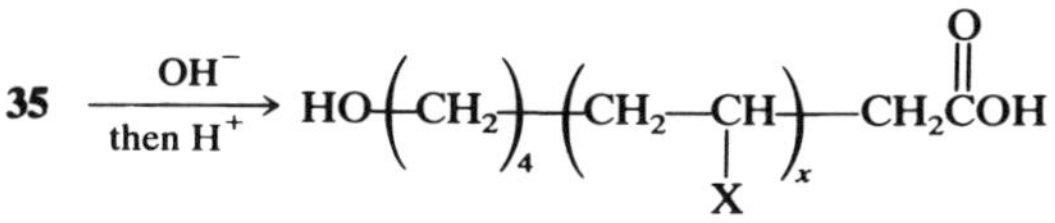

(5.21)

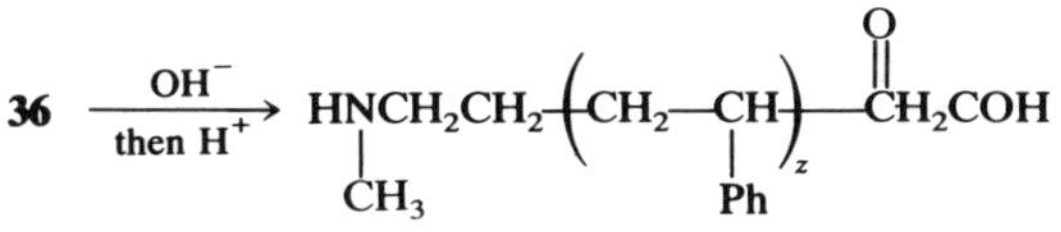

(5.22)

5.4.2. 4-Methylene-1,3-Dioxolanes

Radical polymerizations of 4-methylene-1,3-dioxolanes with various kinds of substituents on the 2-position have been studied. 4-Methylene-2-phenyl-1,3-dioxolane **37** polymerized at 65–120 °C to obtain a copolymer containing addition units, ring-opened units, and ketone units formed by elimination of benzaldehyde after ring opening [Eq. (5.23)].[39,40] The 2-phenyl-2-alkyl derivatives **38** gave a copolymer containing addition units and ketone units [Eq. (5.24)],[41] and 2-phenyl-2-methoxy-1,3-dioxolane **39** gave a copolymer containing ring-opening units and ketone units [Eq. (5.25)].[42] However, 2,2-diphenyl-4-methylene-1,3-dioxolane **40** selectively underwent ring-opening polymerization accompanying quantitative elimination of benzophenone to form polyketone [Eq. (5.26)].[43,44] Although the 1,3-dioxolane ring seems not to be strongly strained, the reaction proceeded selectively by formation of a stable diphenylmethyl radical, which does not attack the olefin but produces benzophenone. In other words, the predominant formation of benzophenone and diphenyl radical is the driving force for the polymerization of **40**.

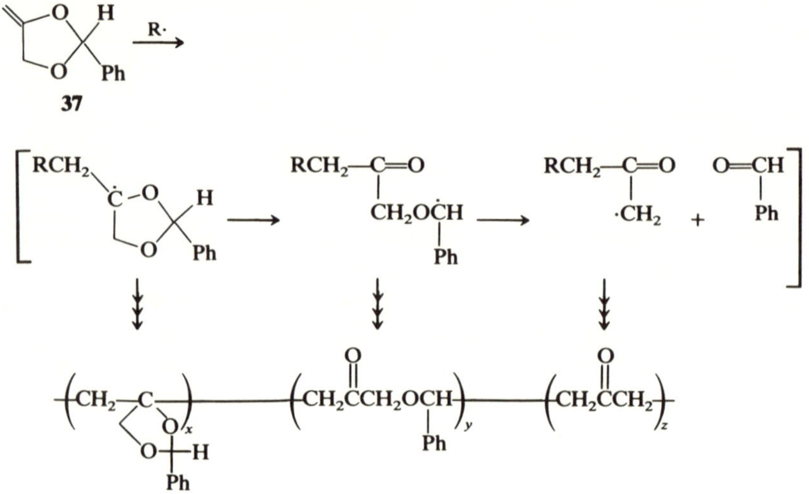

(5.23)

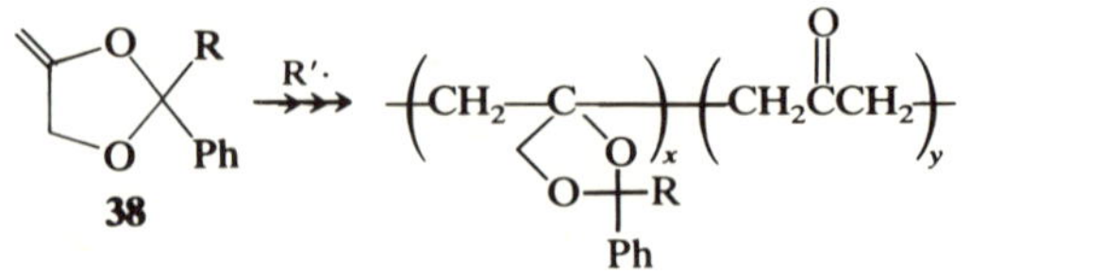

(5.24)

R = CH_3, C_2H_5, n-C_3H_5
i-C_3H_5, n-C_7H_{15}

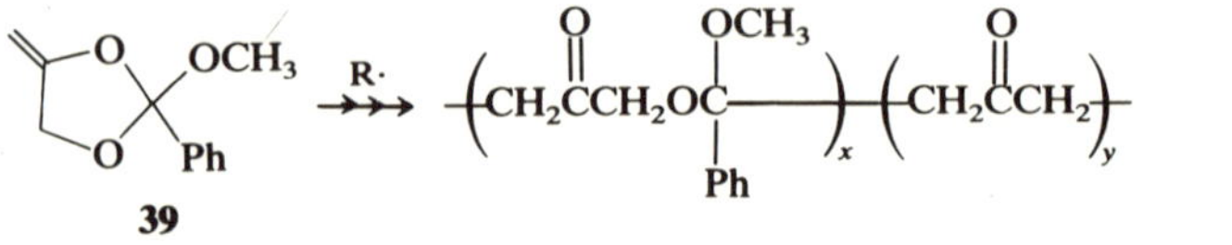

(5.25)

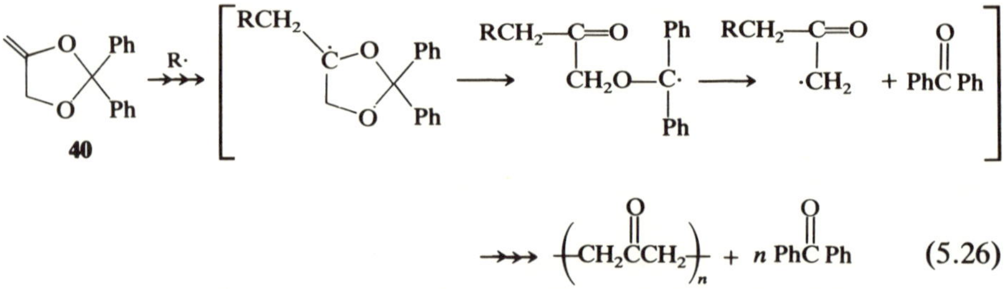

(5.26)

2,2-Diaryl-4-methylene-1,3-dioxolane **41** endowed with various kinds of *para* substituents on the phenyl groups was readily prepared and polymerized with more than 96% ring opening accompanying elimination of the corresponding 4,4′-disubstituted benzophenone [Eq. (5.27)]. It was found

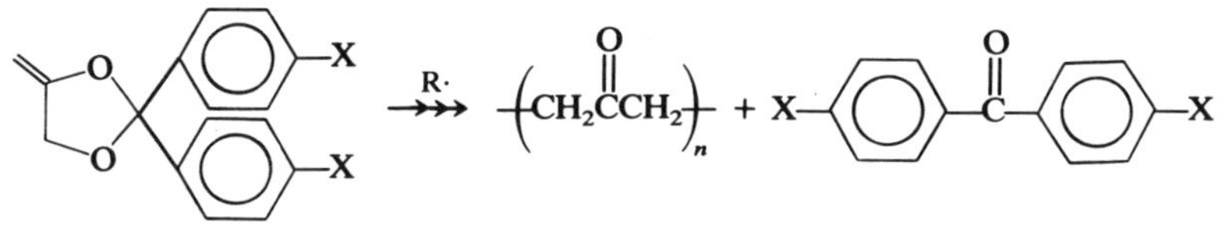

(5.27)

41

X = OMe, Me, Cl, CN

that the polymer yields decreased as the electron-withdrawing character of the substituents increased.[45] This substituent effect on the polymer yields may be explained by the theory of Imoto[46] on the activation energy of propagation $E_p^{\neq}$. Imoto derived from molecular orbital considerations that charge transfer from radical to monomer in the transition state could be promoted by electron-donating substituents, lowering $E_p^{\neq}$ and increasing polymer yield.

The phenyl groups of **40** serve not only as a radical stabilizing group, but the latter is also a bulky group which prevents **40** from undergoing vinyl polymerization. Monomer **42** having combined phenyl groups underwent only 50–63% ring opening.[47]

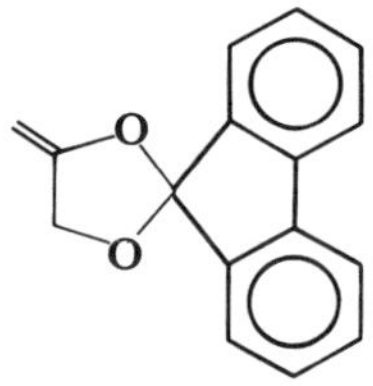

42

Although the polyketone obtained from **40** is only soluble in *m*-cresol and hot dimethylsulfoxide (DMSO), the polyketone **44** is soluble in common organic solvents [Eq. (5.28)].[48]

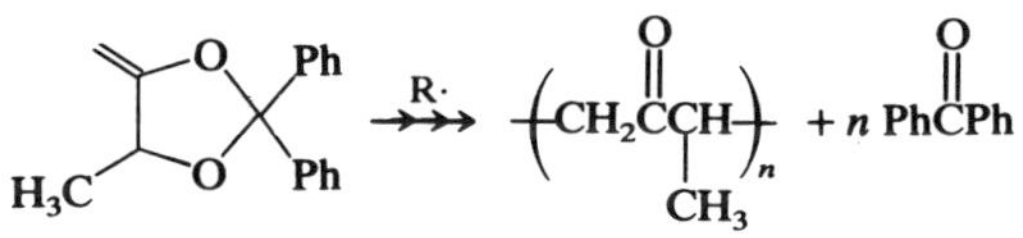

(5.28)

43 **44**

4-Methylene-1,3-dioxolane **45** carrying a carbonyl group on the 5-position underwent also quantitative ring opening to obtain a poly(α-ketoester) [Eq. (5.29)],[49] while 2-keto-4-methylene-1,3-dioxolane **46** underwent only vinyl polymerization without ring opening [Eq. (5.30)].[50]

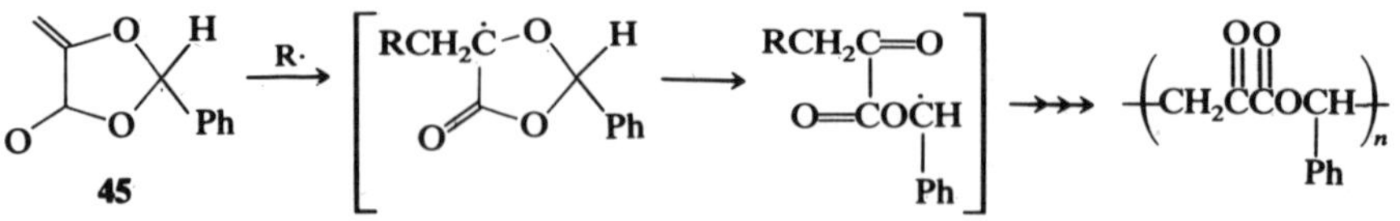

(5.29)

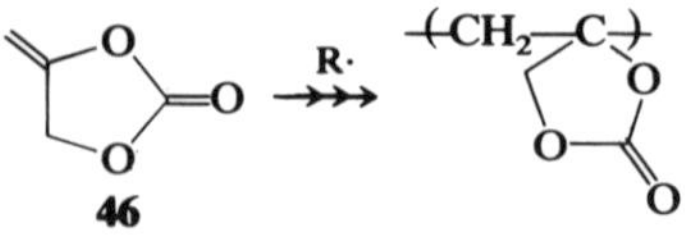

(5.30)

Copolymerization of **40** with vinyl monomers enables quantitative introduction of a keto group in the backbone of a large variety of polymers. Such polymers containing keto groups can be expected to be photodegradable. When **40** was copolymerized with St, VA, MMA, and *N*-vinyl-2-pyrrolidone in the presence of DTBP at 120 °C, the corresponding copolymers containing keto groups in the backbone, were obtained [Eq. (5.31)]. Especially in the copolymerization with St, the amount of keto groups incorporated can be controlled by changing the monomer feed.[51]

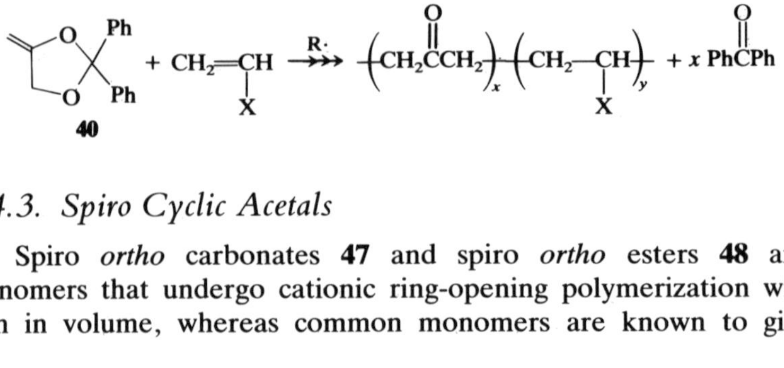

(5.31)

5.4.3. Spiro Cyclic Acetals

Spiro *ortho* carbonates **47** and spiro *ortho* esters **48** are unique monomers that undergo cationic ring-opening polymerization with expansion in volume, whereas common monomers are known to give rise to

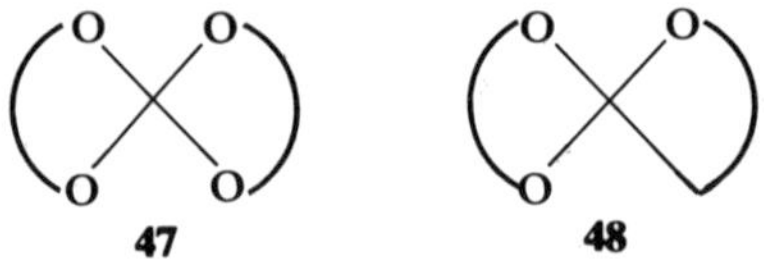

pronounced shrinkage during polymerization.[52] Introduction of unsaturation into spiro cyclic acetals **47** and **48** permits radical ring-opening polymerization; only spiro cyclic acetals containing an *exo*-methylene group in case I have been reported. Such monomers undergo double ring opening, contrary to other monomers that undergo radical ring-opening polymerization, and also show expansion or no shrinkage in radical polymerization.

The spiro *ortho* carbonate **49** bearing two *exo*-methylene groups underwent double ring-opening polymerization with DTBP at 130 °C in chlorobenzene to afford a soluble polymer having carbonate and pendant methylene moieties, although bulk polymerization gave only crosslinked polymer [Eq. (5.32)].[53] When the spiro *ortho* carbonate **50** containing an *exo*-methylene group was treated with DTBP at 120–130 °C even in bulk, a soluble polymer was obtained [Eq. (5.33)].[54]

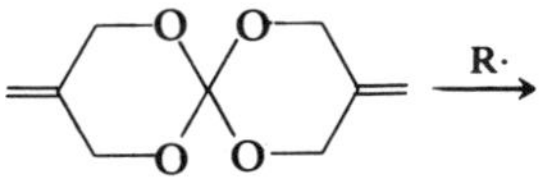

49

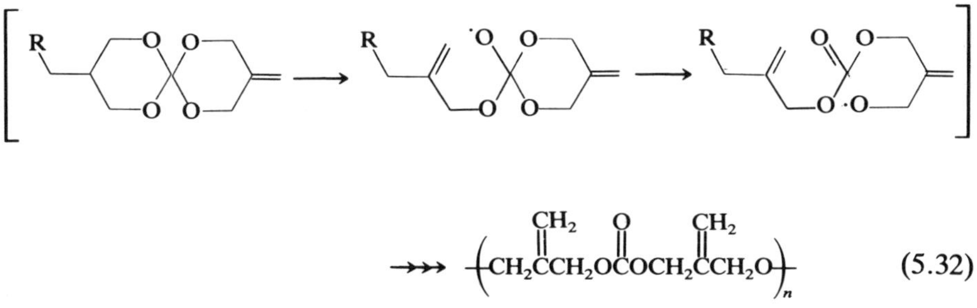

(5.32)

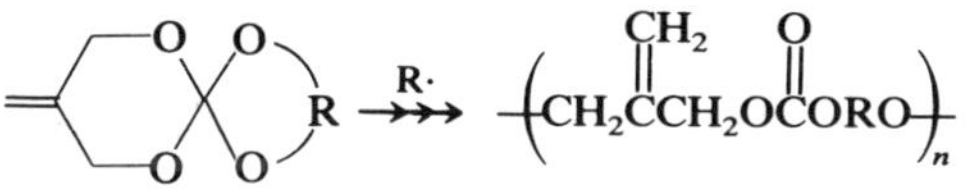

(5.33)

50

R = $+\!(CH_2)_n\!+$, n = 3, 4, 5

Investigation of this polymerization reveals two characteristic features, contrary to other radical ring-opening polymerizations. (1) When the bond C═X is formed during radical isomerization in case I, other monomers give carbonyl groups (C═O) but spiro *ortho* carbonates give carbon–carbon double bonds (C═C). (2) The propagating radical of the other monomers dealt with, except for the sulfur-containing cyclic monomers (*vida infra*), is a carbon radical, while that derived from the spiro *ortho* carbonates is a less stable carbonate group during double ring opening is suggested to be the driving force for the polymerization.

The spiro *ortho* carbonates **51**[55] and **52**[56] having the 4-methylene-1,3-dioxolane skeleton underwent only vinyl polymerization without ring opening to afford a crosslinked polymer and a soluble polymer, respectively [Eq. (5.34)].

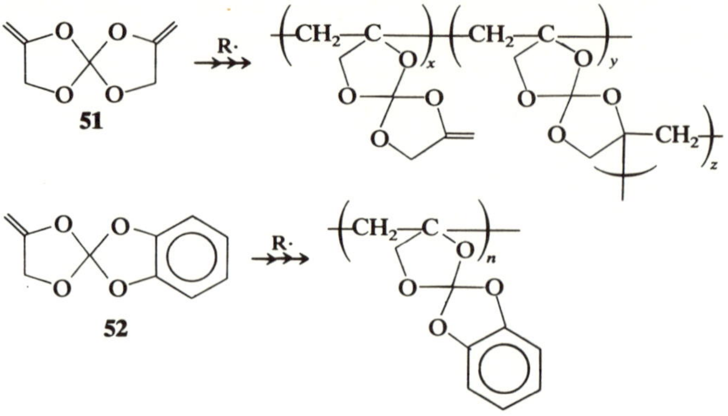

51

52

(5.34)

However, spiro *ortho* ester **53** with the 4-methylene-1,3-dioxolane skeleton polymerized with quantitative ring opening, irrespective of the ring size, to give a polymer containing successive ketone and ester groups in the backbone [Eq. (5.35)].[57] It should be noted that the radical ring-opening

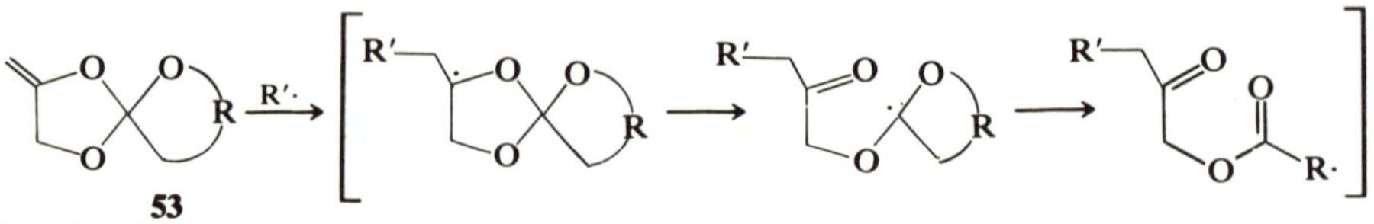

53

R = $\left(CH_2\right)_n$, $n = 3, 4, 5$

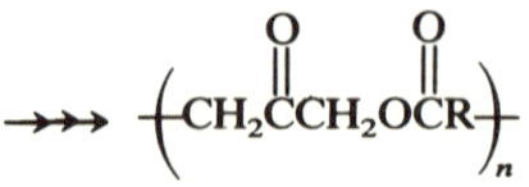

(5.35)

ability of spiro *ortho* carbonates and spiro *ortho* esters are very different even if they have similar structure elements such as the 4-methylene-1,3-dioxolane skeleton. Spiro *ortho* ester **54** bearing an *exo*-methylene group on the cyclic ether part underwent both single ring-opening and double ring-opening polymerization. No regioselectivity in opening the acetal ring was found [Eq. (5.36)].[58]

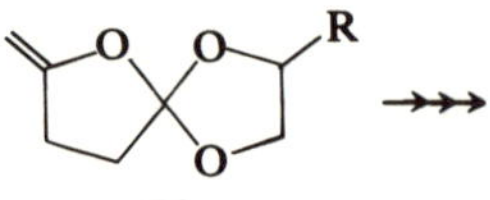

54

R = CH_3, n-Bu

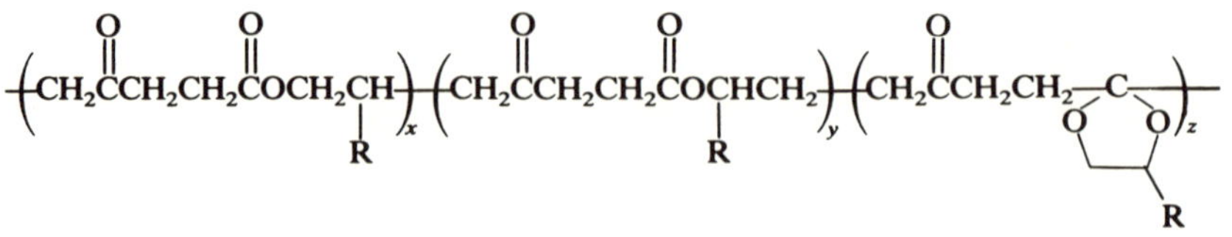

(5.36)

Copolymerization of spiro cyclic acetals with vinyl monomers makes possible the introduction of various functional groups in the backbone of common polymers. When spiro *ortho* carbonates **49** and **50** were copolymerized with St in the presence of DTBP at 120–130 °C, styrene copolymers **55** containing successive carbonate and ether functions in the backbone were obtained [Eq. (5.37)].[54] Copolymers **55** were hydrolyzed to produce a hydroxy-terminated polystyrene [Eq. (5.38)].[35]

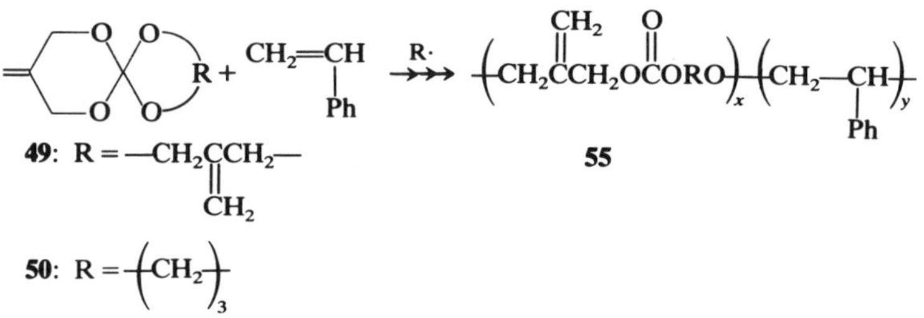
(5.37)

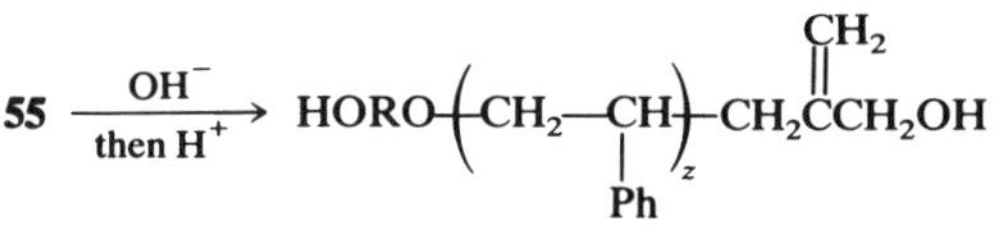
(5.38)

5.5. SULFUR-CONTAINING CYCLIC MONOMERS

Radical ring-opening polymerization of sulfur-containing cyclic monomers have been rare, but one example of the monomer in case I and four examples of the monomer in case II have been reported. Cyclic ketene dithioacetal **56** polymerized without ring opening, contrary to cyclic ketene acetals [Eq. (5.39)].[59] Apparently the dithioester group is less stable than the thioester group resulting from the cyclic *O,S*-ketene acetal **33** (see Section 5.4.1).

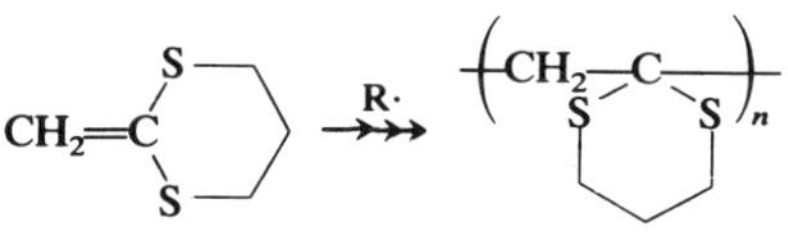
(5.39)

56

Monomers in case II that have been studied have included cyclic α-vinylsulfone **57–59** and vinylthiirane **60**. Five-membered α-vinylsulfone **57** polymerized with essentially 100% ring opening to obtain a polymer containing sulfonyl groups in the backbone [Eq. (5.40)].[60] However, in the

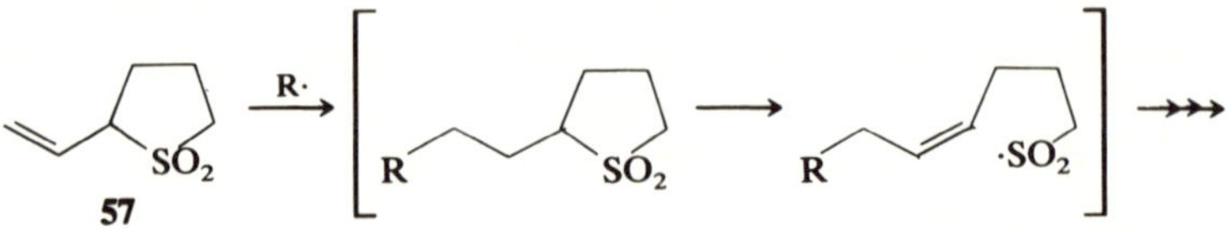

57

$$\left(CH_2CH{=}CHCH_2CH_2CH_2SO_2\right)_n \qquad (5.40)$$

polymerization of six-membered α-vinylsulfone **58**, only 43% of rings are opened, and even in the polymerization of the seven-membered **59**, which is slightly strained, only 83% of rings are opened.[61]

58 **59**

Vinylthiirane **60** underwent essentially ring opening with selective scission of the carbon-hetero bond (C—S bond) of the thiirane ring [Eq. (5.41)],[62,63] contrary to the radical ring-opening polymerization of vinyloxiranes (see Section 5.3.1).

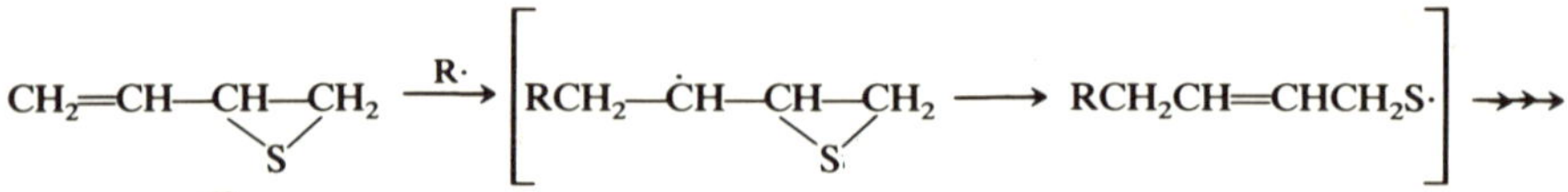

60

$$\left(CH_2CH{=}CHCH_2S\right)_n \qquad (5.41)$$

These examples imply that the driving force of radical ring-opening polymerization of sulfur-containing monomers is not formation of carbon–sulfur double bonds (case I) but generation of a radical on the sulfur atom as in case II.

Cyclic α-vinylsulfone **57** could be copolymerized with AN, VA, MAn, and ethyl acrylate in the presence of AIBN at 65 °C to afford the corresponding copolymer containing both carbon–carbon double bonds and sulfonyl groups in the backbone [Eq. (5.42)]. However, **57** could not be copolymerized with St and MMA.[60]

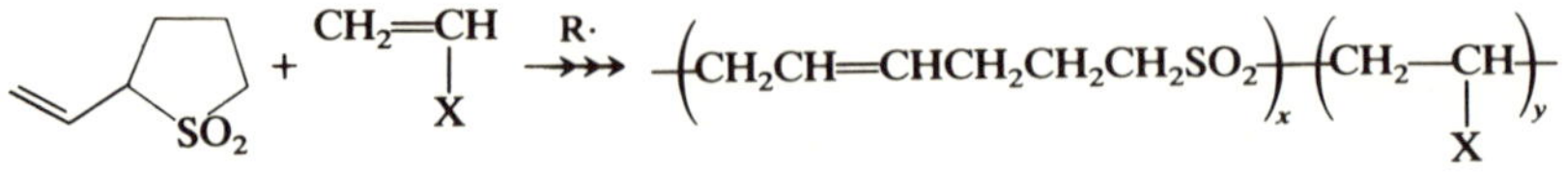

57 (5.42)

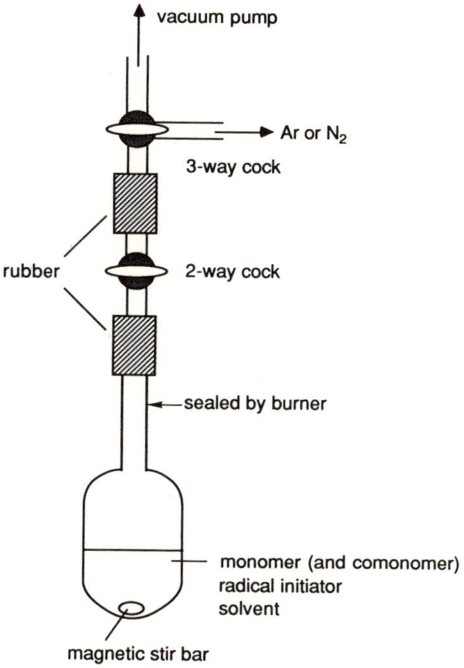

FIG. 5.1. Typical experimental apparatus of radical ring-opening polymerization.

5.6. EXPERIMENTAL PROCEDURES

A typical procedure was carried out as follows: A solution such as benzene, chlorobenzene, and DMF of the monomer to undergo radical ring-opening polymerization, a comonomer (in the copolymerization), and a radical initiator (2–5 mol%) were placed in a glass tube (Fig. 5.1). The tube was subsequently degassed three times by the standard freeze–thaw method under argon or nitrogen, sealed *in vacuo,* and then placed in a constant-temperature bath. After the reaction mixture was stirred, the tube was opened and the resulting viscous oil was dissolved in solvent, such as dichloromethane or chloroform, and precipitated by pouring the solution into a large volume of precipitating solvent, such as methanol or hexane. The precipitated polymer was collected and dried under reduced pressure.

REFERENCES

1. W. J. Bailey, P. Y. Chen, S.-C. Chen, W.-B. Chiao, T. Endo, B. Gapud, V. Kuruganti, Y.-N. Lin, Z. Ni, C.-Y. Pan, S. E. Shaffer, L. SIdney, S.-R. Wu, N. Yamamoto, N. Yamazaki, K. Yonezawa, and L. L. Zhou, *Makromol. Chem., Macromol. Symp.* **6**, 81 (1986).
2. T. Takahashi, I. Yamashita, and T. Miyakawa, *Bull. Chem. Soc. Jpn.* **37**, 131 (1964).

3. T. Takahashi and I. Yamashita, *Kogyo Kagaku Zasshi* **68**, 869 (1968).
4. T. Takahashi, *J. Polym. Sci., Part A*1 **6**, 403 (1968).
5. V. M. Zhulin, M. G. Gonkberg, A. R. Volchek, O. M. Nefedox, and A. S. Shashkov, *Vysokomol. Soyed., Ser. A* **13**, 2153 (1971).
6. T. Endo and K. Suga, *J. Polym. Sci., Polym. Chem. Ed.* **27**, 1831 (1989).
7. (a) I. S. Lishanskii, A. G. Zak, E. F. Fedorova, and A. S. Khachaturov, *Vysokomolekul. Soedin.* **7**, 966 (1965); (b) I. S. Lishanskii, A. G. Zak, Ye. I. Zherebetskaya, and A. S. Khachaturov, *Vysokomol. Soyed., Ser. A* **9**, 1895 (1967).
8. (a) I. Cho and K.-D. Ahn, *J. Polym. Sci., Polym. Lett. Ed.* **15**, 751 (1977); (b) I. Cho and K.-D. Ahn, *J. Polym. Sci., Polym. Chem. Ed.* **17**, 3169 (1979).
9. I. Cho and Ju-Y. Lee, *Makromol. Chem., Rapid Commun.* **5**, 263 (1984).
10. I. S. Lishanskii and O. S. Fomina, *Vysokomol. Soyed., Ser. A* **11**, 1398 (1969).
11. T. Endo, M. Watanabe, K. Suga, and T. Yokozawa, *J. Polym. Sci., Polym. Chem. Ed.* **25**, 3039 (1987).
12. T. Endo, M. Watanabe, K. Suga, and T. Yokozawa, *Makromol. Chem.* **190**, 1827 (1989).
13. T. Endo, M. Watanabe, K. Suga, and T. Yokozawa, *J. Polym. Sci., Polym. Chem. Ed.* **27**, 1435 (1989).
14. J. P. Kennedy, J. J. Elliott, and P. E. Butler, *J. Macromol. Sci., Chem.* **A2**, 1415 (1968).
15. T. Takahashi, *J. Polym. Sci., Part A*-1 **8**, 739 (1970).
16. Y. Hiraguri and T. Endo, *J. Polym. Sci., Polym. Lett. Ed.* **26**, 381 (1988).
17. Y. Hiraguri and T. Endo, *J. Polym. Sci., Polym. Lett. Ed.* **27**, 333 (1989).
18. L. A. Errede, *J. Polym. Sci.* **49**, 253 (1961).
19. I. Cho and J.-B. Kim, *J. Polym. Sci., Polym. Lett. Ed.* **21**, 433 (1983).
20. T. Endo and N. Kanda, *J. Polym. Sci., Polym. Chem. Ed.* **23**, 1931 (1985).
21. K. Suga and T. Endo, *Polym. Prepr. Jpn.* **36**, 140 (1987).
22. T. Endo, K. Suga, Y. Orikasa, and S. Kojima, *J. Appl. Polym. Sci.* **37**, 1815 (1989).
23. L. N. Sidney, S. E. Shaffer, and W. J. Bailey, *ACS Polym. Prepr.* **22**(2), 373 (1981).
24. W. J. Bailey, P. Y. Chen, S.-C. Chen, W.-B. Chiao, T. Endo, B. Gapud, Y.-N. Lin, Z. Ni, C.-Y. Pan, S. E. Shaffer, L. Sidney, S.-R. Wu, N. Yamamoto, N. Yamazaki, and K. Yonezawa, *J. Macromol. Sci., Chem.* **A21**, 1611 (1984).
25. M. C. Salon and A. Gandini, *Makromol. Chem., Rapid Commun.* **9**, 539 (1988).
26. (a) K. B. Wiberg, R. R. Kintner, and E. L. Motell, *J. Am. Chem. Soc.* **85**, 450 (1963); (b) Y. Izawa and Y. Ogata, *J. Org. Chem.* **35**, 3192 (1970).
27. G. F. Meijs and E. Rizzardo, *Makromol. Chem., Rapid Commun.* **9**, 547 (1988).
28. W. J. Bailey, Z. Ni, and S.-R. Wu, *J. Polym. Sci., Polym. Chem. Ed.* **20**, 3021 (1982).
29. W. J. Bailey, S.-R. Wu, and Z. Ni, *Makromol. Chem.* **183**, 1913 (1982).
30. I. Cho and M. S. Gong, *J. Polym. Sci., Polym. Lett. Ed.* **20**, 361 (1982).
31. W. J. Bailey, S.-R. Wu, and Z. Ni, *J. Macromol. Sci., Chem.* **A18**, 973 (1982).
32. T. Endo, N. Yako, K. Azuma, and K. Nate, *Makromol. Chem.* **186**, 1543 (1985).
33. T. Yokozawa, R. Hayashi, and T. Endo, *J. Polym. Sci., Polym. Chem. Ed.* **28**, 3739 (1990).
34. N. Yako and T. Endo, *Polym. Prepr. Jpn.* **34**, 154 (1985).
35. T. Endo, M. Okawara, W. J. Bailey, K. Azuma, K. Nate, and H. Yokono, *J. Polym. Sci., Polym. Lett. Ed.*, **21**, 373 (1983).
36. W. J. Bailey, Z. Ni, and S.-R. Wu, *Macromolecules* **15**, 711 (1982).
37. W. J. Bailey, A. Arfaei, P.-Y. Chen, S.-C. Chen, T. Endo, C.-Y. Pan, Z. Ni, S. Shaffer, L. Sidney, S.-R. Wu, and N. Yamazaki, *Proceedings IUPAC 28th Macromolecular Symposium,* Amherst, Massachusetts, July 12–16, (1982), p. 214.
38. W. J. Bailey, T. Endo, B. Gapud, Y.-N. Lin, Z. Ni, C.-Y. Pan, S. E. Shaffer, S.-R. Wu, N. Yamazaki, and K. Yonezawa, *J. Macromol. Sci., Chem.* **A21**, 979 (1984).
39. C.-Y. Pan, Z. Wu, and W. J. Bailey, *J. Polym. Sci., Polym. Lett. Ed.* **25**, 243 (1987).
40. C.-Y. Pan, Z. Wu, Q. R. Zhu, and W. J. Bailey, *J. Macromol. Sci., Chem.* **A25**, 27 (1988).

41. Y. Hiraguri and T. Endo, *J. Polym. Sci., Polym. Lett. Ed.* **27**, 4403 (1989).
42. M. S. Gong, S.-I. Chang, and I. Cho, *Makromol. Chem., Rapid Commun.* **10**, 201 (1989).
43. Y. Hiraguri and T. Endo, *J. Am. Chem. Soc.* **109**, 3779 (1987).
44. Y. Hiraguri and T. Endo, *J. Polym. Sci., Polym. Chem. Ed.* **27**, 2135 (989).
45. Y. Hiraguri, T. Sugizaki, and T. Endo, *Macromolecules* **23**, 1 (1990).
46. (a) M. Imoto and T. Ouchi, *Nippon Kagaku Kaishi* 585 (1986); (b) M. Imoto and T. Ouchi, *Nippon Kagaku Kaishi* 894 (1987).
47. Y. Hiraguri and T. Endo, *J. Polym. Sci., Polym. Chem. Ed.* **28**, 2881 (1990).
48. Y. Hiraguri and T. Endo, *Polym. Prepr. Jpn.*, **38**, 221 (1989).
49. W. J. Bailey and P.-Z. Feng, *ACS Polym. Prepr.* **28**(1), 154 (1987).
50. I. Cho and T.-W. Lee, *Makromol. Chem., Rapid Commun.* **10**, 453 (1989).
51. Y. Hiraguri and T. Endo, *J. Polym. Sci., Polym. Lett. Ed.* **27**, 1 (1989).
52. W. J. Bailey, R. L. Sun, H. Katsuki, T. Endo, H. Iwama, R. Tsushima, K. Saigo, and M. M. Bitritto, in *ACS Symposium No. 59, Ring-Opening Polymerization,* ed. by T. Saegusa and E. Goethals (American Chemical Society, Washington, D.C., 1977), p. 38.
53. T. Endo and W. J. Bailey, *J. Polym. Sci., Polym. Lett. Ed.* **13**, 193 (1975).
54. T. Endo and W. J. Bailey, *J. Polym. Sci., Polym. Chem. Ed.* **13**, 2525 (1975).
55. H. Tagoshi and T. Endo, *J. Polym. Sci., Polym. Chem. Ed.* **27**, 1415 (1989).
56. J. Sugiyama, T. Yokozawa, and T. Endo, *J. Polym. Sci., Polym. Chem. Ed.* **28**, 3529 (1990).
57. (a) T. Endo and W. J. Bailey, *J. Polym. Sci., Polym. Lett. Ed.* **18**, 25 (1980); (b) T. Endo, M. Okawara, N. Yamazaki, and W. J. Bailey, *J. Polym. Sci., Polym. Chem. Ed.* **19**, 1283 (1981).
58. (a) C.-Y. Pan, Y. Wang, and W.-Y. Huang, *Makromol. Chem., Rapid Commun.* **7**, 627 (1986); (b) C.-Y. Pan, Y. Wang, and W. J. Bailey, *J. Polym. Sci. Polym. Chem. Ed.* **26**, 2737 (1988).
59. S. Kobayashi, J. Kadokawa, and S. Shoda, *56th Annual Meeting of Chemical Society of Japan,* VI A 49, Tokyo (1988).
60. I. Cho, S.-K. Kim, and M. Lee, *J. Polym. Sci., Polym. Symp.* **74**, 219 (1986).
61. I. Cho and M.-H. Lee, *J. Polym. Sci., Polym. Lett. Ed.* **25**, 309 (1987).
62. F. Lautenschlaeger and H. Schnecko, *J. Polym. Sci., Part A*-1 **8**, 2579 (1970).
63. K. Suga and T. Endo, *Polym. Prepr. Jpn.* **37**, 348 (1988).

6

Phase Transfer Catalyzed Polycondensation

Sylvie Boileau

6.1. INTRODUCTION

In many situations where two species cannot react because they are separated by an interface, the lack of activity has been overcome by using the technique of "Phase Transfer Catalysis" (PTC), which allows the reaction to occur, sometimes in very selective ways. This technique, the name of which was proposed by Starks in 1971,[1] was developed nearly at the same time by two other pioneers: Makosza[2] and Brandström.[3] During the last twenty years, PTC has become a very active subject of research with deep implications especially in preparative organic, inorganic, and polymer chemistry,[4–9] because it has the advantages of being extremely simple and economical.

Reactions are conducted in a two-phase system consisting of mutually insoluble aqueous and organic layers (liquid–liquid PTC). Ionic reagents are located mainly in the aqueous phase while the substrate is in the organic phase. Alternatively, ionic reagents can be used in the solid state as a suspension in the organic phase (solid–liquid PTC). In both cases, the transport of the anionic species from the aqueous or the solid phase into the organic medium, under a highly active form, is ensured by catalytic amounts of lipophilic transfer agents such as quaternary onium salts, crown ethers, cryptands, etc. The reaction generally occurs in the organic phase. Most of

Sylvie Boileau • Collège de France, 11 place Marcelin Berthelot, 75231 Paris Cédex 05, France.

New Methods for Polymer Synthesis, edited by W. J. Mijs. Plenum Press, New York, 1992.

the simple nucleophilic substitutions in liquid–liquid systems can be represented by the following scheme [Eq. (6.1)] originally proposed by Starks:[1]

$$\begin{array}{lccc} \text{Aqueous phase} & Q^+X^- + M^+Y^- & \rightleftarrows & Q^+Y^- + M^+X^- \\ & \uparrow & & \downarrow \\ \text{Interface} & \text{- - - - - - -} & \text{- - - - - - - - - -} & \text{- - - - - - -} \\ & \uparrow & & \downarrow \\ \text{Organic phase} & [Q^+, X^-] + R\text{—}Y & \longleftarrow & [Q^+, Y^-] + R\text{—}X \end{array} \tag{6.1}$$

Quaternary onium salt $Q^+ = R_4N^+$ or R_4P^+, selected due to its high solubility in the organic phase, transfers anion Y^- into the organic phase as $[Q^+, Y^-]$, which then reacts with, for instance, an alkyl halide RX to give the substitution product R—Y. The produced Q^+X^- is rapidly reconverted into Q^+Y^- by anion exchange with nucleophile M^+Y^- from the aqueous phase. Poly(ethyleneglycol) derivatives, crown ethers, and cryptands can also be used as phase transfer catalysts preferentially for solid–liquid systems.

Phase transfer catalyzed reactions require shorter reaction times and lower temperatures than the corresponding classical reactions (thus limiting side reactions). Moreover, the process is more simple, and in the case of liquid–liquid systems the use of expensive anhydrous solvents and reagents becomes unnecessary. This is why PTC has been widely and successfully used in different fields of polymer chemistry such as chemical modification of polymers, free-radical polymerization, nucleophilic displacement polycondensation, and preparation of some monomers.

A large variety of polymers such as polyethers, polythioethers, polycarbonates, polythiocarbonates, polyesters, polysulfonates, polyphosphonates, carbon–carbon chain polymers, and so on has been prepared by phase transfer catalyzed polycondensation.[10–13]

It must be emphasized that the main criteria for obtaining linear high polymers from phase transfer catalyzed polycondensation are, first, the reaction should be clean and relatively fast, and second, the organic phase should be capable of maintaining in solution fairly high concentrations of both monomer and polymer. The most interesting features of this special type of polycondensation are as follows. It is possible to prepare polymers having high enough molecular weights, a very low polydispersity, i.e., $\bar{M}_w/\bar{M}_n \leqslant 1.3$, and well-controlled end groups. Moreover, it is not necessary to start from a stoichiometric ratio of monomers to obtain high-molecular-weight polycondensates as requested in classical polycondensations.

The influence of several parameters, such as the nature of the solvent and of the catalyst as well as the stoichiometry and concentration of the reagents, has been examined in some cases. However, a detailed knowledge of the mechanism of this type of polycondensation is needed. Selected examples of tailor-made polymers belonging to different families will be presented in this chapter together with scope and limitations of phase transfer catalyzed polycondensation.

6.2. POLYETHERS

6.2.1. Literature Survey

Among the numerous nucleophilic displacement polycondensations, the synthesis of polyethers has been most widely studied. They were prepared from dihalogeno aliphatic or aromatic compounds, XRX (or epichlorhydrin,[14]), and several bisphenols, usually with a liquid–liquid system, according to Eq. (6.2). Some selected examples are listed in Table 6.1.

$$n\text{XRX} + n\text{HO—C}_6\text{H}_4\text{—Y—C}_6\text{H}_4\text{—OH} \xrightarrow[\text{NaOH/org. solvent}]{\text{PTC}} \text{—[R—O—C}_6\text{H}_4\text{—Y—C}_6\text{H}_4\text{—O—]}_n + 2n\text{MX} \quad (6.2)$$

with X = Br, Cl; Y = $C(CH_3)_2$, SO_2, CO, etc.

The polymers are linear even when obtained from a polyhalogeno compound.[13,21] They often have well-defined halogenated end groups[11,16–20] and low polydispersities.[11,17] All the authors noted that the structure and characteristics of the polyethers are greatly influenced by several parameters such as the nature of the organic phase[11,13–15,19,20] and of the catalyst,[11,14,15,17,18,24] the structure of the bisphenol[13,18,20] and of the dihalogeno compound XRX,[11,20] and their initial molar ratio.[11,17] The influence of the reagents and catalyst concentrations,[11,17–19] of the addition of an alkaline halide to the system,[11,24] and of the stirring rate[17] was also examined as well as the influence of water in the case of a solid–liquid system.[13,21] Polyformals were also prepared from bisphenols and a large excess of dichloromethane, which also plays the role of the organic solvent.[22,23]

6.2.2. Mechanistic Aspects

The kinetics and mechanism of the polycondensation of bisphenol A (BPA) with *trans*-1,4-dichloro-2-butene (DCB) under PTC conditions have been studied in detail.[11,17] Some results obtained with the biphasic system namely, aqueous NaOH/toluene at 65 °C, using tetrabutylammonium hydrogen sulfate (TBAH) as the catalyst, are shown in Table 6.2.

In most cases, the polymers have the following structure:

$$\text{Cl—CH}_2\text{—CH=CH—CH}_2\text{—[—O—C}_6\text{H}_4\text{—C(CH}_3)_2\text{—C}_6\text{H}_4\text{—O—CH}_2\text{—CH=CH—CH}_2\text{—]}_n\text{—Cl} \quad (6.3)$$

The absence of phenolic end groups was confirmed by UV spectroscopy according to the method described by Shchori and McGrath,[25] by IR spectrophotometry as well as by ^{1}H and ^{13}C NMR. This allows measurement of $\bar{M}_n$ by different techniques, such as end group analysis (^{1}H NMR and

TABLE 6.1. Preparation of Polyethers by Phase Transfer Catalyzed Polycondensation: Selected Examples

XRX	Bisphenol	Catalyst/Base	Solvent/t(°C)	Ref.
$(p\text{-Cl}, m\text{-NO}_2C_6H_3)_2SO_2$	BPA[a]	DCHE[b]/NaOH	CH_2Cl_2/r.t.	15
$ClCH_2—CH═CH—CH_2Cl$	BPA	TBAH[c]/NaOH	Toluene/65	11,16,17
$Br(CH_2)_6Br$	$(p\text{-HOC}_6H_4)_2Y$ [Y = $C(CH_3)_2$, CH_2, S, SO_2, CO]	TBPB[d]/NaOH TBAB[d]/KOH DBC[d]	$C_6H_5NO_2$/78	18
$p\text{-ClCH}_2—C_6H_4—CH_2Cl$	BPA	DCHE/NaOH	DMSO/70	16
$p\text{-ClCH}_2—C_6H_4—CH_2Cl$	BPA	BTEAC[e]/NaOH	C_6H_6/DMSO/60	19
m- or $p\text{-XCH}_2—C_6H_4—CH_2X$ (X = Br, Cl)	$(p\text{-HOC}_6H_4)_2Y$ [Y = $C(CH_3)_2$, SO_2, CO]	Aliquat 336 18-C-6/KOH [222]	$C_6H_5NO_2$/60	20
C_6F_6	$(p\text{-HOC}_6H_4)_2Y$ [Y = $C(CH_3)_2$, S, SO_2]	18-C-6/K_2CO_3	DMAc/tol./80	13,21
CH_2Cl_2	BPA	Aliquat 336/NaOH	C_6H_5Cl/CH_2Cl_2/75	22,23

[a] BPA denotes bisphenol A.
[b] DCHE denotes dicyclohexyl-18-crown-6.
[c] TBAH denotes $Bu_4N^+SO_4H^-$.
[d] TBPB denotes $Bu_4P^+Br^-$, TBAB $Bu_4N^+Br^-$, and DBC dibenzo-18-crown-6.
[e] BTEAC denotes benzyl $Et_3N^+Cl^-$

TABLE 6.2. Preparation of Polyethers from BPA and *trans*-DCB under PTC Conditions[a]: Influence of the Ratio [BPA]/[DCB]

Run	[BPA]/[DCB]	Yield (%)	$\bar{M}_n$ (elemental analysis)	$\bar{M}_n$ NMR	$\bar{M}_n$[b] GPC	$I_p = \bar{M}_w/\bar{M}_n$
1[c]	1	0	—	—	—	—
2	1	100	2250	3100	3100	1.8
3[d]	1	84	4600	4600	3500	1.4
4[d,e]	1	71	4100	4700	3400	1.3
5	2	74	5000	5200	5300	2.5
6[f]	3	70	12500	—	[g]	—
7[f,h]	4	58	8000	6700	2600	1.3

[a] Toluene/NaOH 3 N (10 mL/10 mL); DCB, 3.07 mmol; TBAH, 0.614 mmol; 65 °C; *t*, 5 h; 500 rpm.
[b] In THF using a polystyrene calibration curve.
[c] Without any catalyst; *t*, 18 h.
[d] NaOH 6 N.
[e] *cis*-DCB.
[f] The aqueous phase contains a solid.
[g] Osmometric $\bar{M}_n$ of the soluble fraction in toluene, 52,000.
[h] Presence of phenolic end groups in the polymer.

elemental analysis) and GPC. Taking into account the fact that a polystyrene calibration curve in THF was used for GPC measurements, the agreement between the different values of $\bar{M}_n$ for a given polymer is quite satisfactory (runs 2–4 in Table 6.2).

The index of polydispersity, $I_p = \bar{M}_w/\bar{M}_n$, can be remarkably low (runs 3 and 4 in Table 6.2).

Moreover, it has been shown that $\bar{M}_n$ increases with the NaOH concentration (runs 3 and 4 in Table 6.2) as well as with the catalyst concentration. It must be also noted that polymers are obtained in good yields even by using nonstoichiometric initial monomer ratios (run 5 in Table 6.2 for which [BPA]/[DCB] is equal to 2).

All those observations allowed the authors to propose the following mechanism, which was also assumed by Jin and Chang:[18]

$$\mathrm{Na^{+-}O\!-\!-O^{-}Na^{+}_{aq} + Q^{+}X^{-}_{org} \overset{K_1}{\rightleftarrows} Na^{+-}O\!-\!-O^{-}Q^{+} + Na^{+}X^{-}_{aq}} \tag{6.4}$$

$$\underset{}{\mathrm{Na^{+-}O\!-\!-O^{-}Q^{+} + XRX_{org}}} \xrightarrow{k_1} \underset{\text{(A)}}{\mathrm{Na^{+-}O\!-\!-O\!-\!RX}} + \mathrm{Q^{+}X^{-}_{org}} \tag{6.5}$$

$$\mathrm{Na^{+-}O\!-\!-O\!-\!RX + Q^{+}X^{-}_{org}} \overset{K_2}{\rightleftarrows} \underset{\text{(A}')}{\mathrm{Q^{+-}O\!-\!-O\!-\!RX}} + \mathrm{Na^{+}X^{-}_{aq}} \tag{6.6}$$

$$\mathrm{Q^{+-}O\!-\!-O\!-\!RX + XRX} \xrightarrow{k_2} \underset{\text{(B)}}{\mathrm{XR\!-\!O\!-\!-O\!-\!RX}} + \mathrm{Q^{+}X^{-}_{org}} \tag{6.7}$$

with $^{-}O\text{--}\text{--}O^{-}$ = bisphenolate, $Q^{+}X^{-}$ = onium salt, and XRX = DCB.

Polycondensation occurs via the alkylation of a phenolate end group by DCB in the organic phase. In a first step [Eq. (6.4)], sodium bisphenolate which is located in the aqueous phase undergoes a fast cation exchange with $Q^{+}X^{-}$ to give a mixed salt, which will react with XRX through the reactive and lipophilic —$O^{-}Q^{+}$ end group leading to (A) [Eq. (6.5)]. After another cation exchange with the catalyst, (A) will give a more lipophilic species (A′) which is presumably located in the organic phase; (A′) will then react with another molecule of XRX to give (B).

It has been verified that the concentration of bisphenolate anion is equal to zero in the toluene phase, in the absence of DCB and of catalyst. No reaction between BPA and DCB occurs in the absence of TBAH (run 1 in Table 6.2). The cation of the catalyst Q^{+} transports the phenolate from the aqueous phase to the organic phase, and takes back Cl^{-} to the aqueous phase continuously. This process governs the kinetics, and it is important to note that the concentration of —O^{-}, Q^{+} is much lower than that of the chlorine end groups in the toluene phase: $[\text{—}O^{-}, Q^{+}]/[\text{TBAH}] \simeq 4\%$ (as determined in the toluene phase in the absence of DCB). This can easily explain why the polyethers have only chlorinated end groups.

Another peculiarity of this type of polycondensation can be understood, according to this mechanism. In conventional polycondensations, high degrees of polymerization require stoichiometric monomer ratios and very high conversions. In the present case, the polycondensation does not take place between stoichiometric amounts of monomers because the BPA is continuously transferred in a small amount, in a very reactive form, into the organic phase containing the DCB. Consequently there is no need for an equimolar ratio between the monomers to obtain polymers with high molecular weights.

On the basis of this mechanism, several attempts to increase the molecular weights of these polyethers were made by varying on different parameters, the main idea being to increase the concentration of active species derived from BPA in the organic phase. It has been found that $\bar{M}_n$ increases up to a plateau on increasing the stirring rate up to 500 rpm.[11]

Another way of changing $[\text{—}O^{-}, Q^{+}]$ in the toluene phase is to use various [BPA]/[DCB] ratios. The results are shown in Table 6.2. It appears that for [BPA]/[DCB] = 3, side reactions occur having undesirable effects on the kinetics and on the structure of the polymers, presumably because the volume of the aqueous phase was kept constant. Complementary experiments have to be conducted in which the BPA concentration is increased together with the volume of the aqueous phase.

Considering the proposed mechanism, it is easy to see that $[\text{—}O^{-}, Q^{+}]$ in toluene can be increased by adding an inorganic chloride to the aqueous phase. The results indicated in Table 6.3 are in good agreement with this idea. The concentration of anions was determined by UV spectroscopy.

TABLE 6.3. Determination of Anions Derived from BPA in the Organic Phase in the Absence of DCB by UV Spectroscopy[a]

Run	BPA (10^3 mol)	NaCl (10^3 mol)	[NaOH]/ 2[BPA]	$[Na^+]$/ [—OH]	$[—O^-, Q^+]:10^5$ eq in org. phase (10 mL)	$\bar{M}_n$ (elemental analysis)
8	3.07	0	4.88	4.88	0.36	2250
9	6.14	0	2.44	2.44	0.44	5000
10	9.21	0	1.22	1.22	0.44	12500
11	3.07	3.07	4.88	5.38	1.47	3300
12	3.07	6.14	4.88	5.88	0.78	1800
13	3.07[b]	0	9.76	9.76	3.37	3400

[a] Toluene/NaOH 3 N (10 mL/10 mL); TBAH, 0.614 mmol; 65 °C.
[b] *o*-Dichlorobenzene (oDCB)/NaOH 6 N.

Run 12 indicates that too much salt lowers the molecular weight of the polyether, presumably because it prevents the transfer of Q^+Cl^- back to the aqueous phase. It is obvious in run 13 that the concentration of anions in *o*-dichlorobenzene (oDCB) is much higher than in toluene.

The influence of the nature of the solvent, and more specifically of its dielectric constant, ε, has been examined, keeping in mind that the solvent used had to be a good solvent for the polymer and a bad one for BPA. The results are shown in Table 6.4. The value of $\bar{M}_n$ increases on increasing ε, except for a strongly polar solvent such as nitrobenzene. However, phenolic end groups appear in the polymers prepared in oDCB and in nitrobenzene, and the polydispersity becomes larger. In fact, the polarity of the organic phase must have an influence on both the extraction of active species $—O^-$, Q^+ and on their own reactivity.

TABLE 6.4. Preparation of Polyethers from BPA and DCB under PTC Conditions[a]: Influence of the Solvent

Run	Solvent[b]	ε (25 °C)	Yield (%)	$\bar{M}_n$ (elemental analysis)	$\bar{M}_n$ NMR	$\bar{M}_n$[c] GPC	$I_p = \bar{M}_w/\bar{M}_n$
3	toluene	2.38	84	4600	4600	3500	1.4
14	TCB	3.94	83	3300	3200	8400	1.6
15	CB	5.62	90	4900	5200	9500	2.3
16	oDCB	9.93	75	20000	—	31000	1.7
17	NB	34.82	80	11000	9700	19300	2.5

[a] Solvent/NaOH 6 N, 10 mL/10 mL; BPA, 3.07 mmol; DCB, 3.07 mmol; TBAH, 0.614 mmol; 65 °C; *t*, 5 h; 500 rpm.
[b] TCB denotes 1,2,4-trichlorobenzene, CB chlorobenzene, oDCB 1,2-dichlorobenzene, and NB nitrobenzene.
[c] In THF using a polystyrene calibration curve.

6.2.3. Scope, Limitations, and Side Reactions

Since PTC proved to be an efficient, simple way of preparing polyethers having well-defined structures, Percec[2,26–39] and several other authors[40–43] applied this technique to the synthesis of new specialty polymers. For instance, new thermotropic main chain liquid crystal polyethers have been prepared.[33–39,41] Under carefully selected reaction conditions, the polyetherification of a monofunctional oligomer bearing a phenolic end group with an α,ω-dihalogeno oligomer leads to the formation of a well-defined ABA triblock copolymer.[29] The reaction of an α,ω-dihalogeno oligomer with another α,ω-difunctional oligomer with phenolic end groups, carried out under PTC conditions, represents a new synthetic method for the obtention of perfectly alternating $(AB)_n$ block copolymers.[26,28,29,38] Segmented copolymers can be prepared by chain extension of two different α,ω-diphenolic oligomers with a dihalogeno monomer such as methylene chloride, DCB, α,ω-dibromoalkanes, and so on, through a phase transfer catalyzed etherification.[23,27] Finally, it must be mentioned that polymers with specific end groups or side groups (i.e., cyclic imino ethers) have been obtained via chemical modification involving phase transfer catalyzed Williamson reactions.[30,31]

In most of the cases which have been discussed, PTC is a very powerful method for the preparation of well-defined α,ω-difunctional polymers. However, it presents some limitations.

Cyclic oligomers have been identified, for instance, in the reaction of BPA with methylene chloride carried out under PTC conditions.[23] The initial formation of the insoluble disodium or dipotassium salt of BPA apparently creates essentially suitable conditions for a high dilution reaction which favors cyclic oligomers formation. Hay *et al.*[23] found that higher temperatures, higher concentrations of phase transfer catalyst, or the use of a more polar organic phase minimize, but do not eliminate, the formation of cyclic oligomers.

Another important side reaction which has to be avoided is the reaction of the halide groups belonging to the monomer and/or the polymer present in the organic phase with the OH^- transferred from the aqueous phase into the organic phase, particularly when the dihalogeno compound is very reactive. As an example, the chloromethyl derivative of 1,4-butane diol, $ClCH_2O—(CH_2)_4—OCH_2Cl$, undergoes an interfacial decomposition as well as a hydroxide displacement in PTC polyetherification.[47] This compound was found to polymerize through intermediate nucleophiles generated *in situ* leading to a poly(ether acetal).

The extraction constants for a series of quaternary onium species Q^+Y^- are strongly dependent on the structure of both the anion, Y^-, and cation, Q^+, and on the solvent nature. For a given Q^+ they are controlled by the hydrophobicity of Y^-.[44] For instance, it is known that the extraction constant of a salt having a phenolate as the anion is higher by a factor of about 10^5 than the extraction constant of a salt having OH^- as the anion.

The hydrolysis reaction can be minimized by lowering the polarity of the solvent, the concentration of base, and the concentration of catalyst. The influence of those parameters upon the course of polyetherifications ($\bar{M}_n$, $\bar{M}_w/\bar{M}_n$) has been correlated with changes in polymer microstructures.(11,14,18,19,22,24,40) Some examples, such as the phase transfer catalyzed polycondensations involving α,ω-dichloro or dibromop-xylene (DCPX, DBPX), illustrate the importance of the hydrolysis side reaction.(19,20,24,45,46) The polyetherification of DCPX or DBPX with BPA produces polymers which are partly insoluble in the organic phase. Depending on the nature of the solvent used, several side reactions occur which affect both the polymer microstructure and the nature of its chain ends. The α,α'-dihalogeno-p-xylenes can react with themselves according to the following equations:(45)

$$XCH_2C_6H_4CH_2X + OH^- \rightarrow XCH_2C_6H_4CH^-X \tag{6.8}$$

$$XCH_2C_6H_4CH^-X + nXCH_2C_6H_4CH_2X \xrightarrow{OH-} XCH_2\text{—}[C_6H_4CHXCH_2\text{—}]\text{—}_nC_6H_4CH_2X \tag{6.9}$$

$$XCH_2\text{—}[C_6H_4CHXCH_2\text{—}]\text{—}_nC_6H_4CH_2X \xrightarrow{OH-} XCH_2\text{—}[C_6H_4CH{=}CH]_x\text{—}[C_6H_4CHXCH_2\text{—}]_yC_6H_4CH_2X \tag{6.10}$$

The p-xylylidene units thus formed lower the solubility of the resulting polyether. The displacement of the halogenobenzylic chain ends by OH^- could occur:

$$--C_6H_4CH_2X + OH^- \rightarrow --C_6H_4CH_2OH \tag{6.11}$$

Under these conditions, low-molecular-weight polyethers with low CH_2X chain end functionality were obtained. In order to minimize these side reactions and to increase $\bar{M}_n$, Yamazaki and Imai(19) used a mixture of dipolar aprotic solvents (DMSO, DMF, etc.) and nonpolar aprotic solvents (toluene, etc.) as the organic phase. High-molecular-weight polyethers were obtained, however the resulting polymers do not contain only CH_2X chain ends. This is due to the presence of DMSO or DMF which favors the extraction of OH^-, the concentration of which cannot be neglected when that of the onium phenolates becomes very low.

Incidentally, Percec observed that when the phenolate anion is attached to a polymer chain, the extraction constant of its salt with Q^+ should be molecular weight dependent because the hydrophobicity of the phenolate anion increases with the molecular weight of the polymer.(12) This author speculates that the nucleophilicity, and consequently the reactivity of a phenol attached to a polymer chain end, is molecular weight dependent and increases with $\bar{M}_n$. Thus he could explain the efficiency of the reaction leading to segmented, ABA triblock and $(AB)_n$ alternating block copolymers.

6.2.4. Example of a Polyether Synthesis

The experimental procedure used for the preparation of a polyether from BPA and DCB (run 3 in Table 6.2) is as follows:

$$(n+1)\,ClCH_2{-}CH{=}CH{-}CH_2Cl + n\,HO{-}C_6H_4{-}C(CH_3)_2{-}C_6H_4{-}OH \rightarrow ClCH_2{-}CH{=}CH{-}CH_2{-}[O{-}C_6H_4{-}C(CH_3)_2{-}C_6H_4{-}O{-}CH_2{-}CH{=}CH{-}CH_2{-}]_n{-}Cl \quad (6.12)$$

6.2.4.1. Materials Purification

BPA (Merck) was recrystallized from toluene. *Trans*-DCB was purchased from Aldrich and distilled under nitrogen. Phase transfer catalyst TBAH (Fluka) was used as received. Pure grade commercial solvents such as toluene, chloroform, and methanol were also used without any further purification.

6.2.4.2. Polymer Synthesis

The sodium bisphenolate was prepared by mixing 3.07 mmol of BPA with 10 mL of aqueous NaOH 6 N at 65 °C. Then the catalyst TBAH (0.614 mmol), toluene (10 mL), and DCB (3.07 mmol) were successively added, and the mixture was stirred (stirring rate, 500 rpm) for 5 h at 65 °C. After cooling, the organic layer was diluted with toluene (20 mL), separated from the aqueous phase, washed with slightly acidic aqueous solution (HCl 0.1 N) and with water. The polymer was recovered by precipitation in methanol and purified by dissolution in $CHCl_3$ followed by precipitation in methanol. It was then dried under high vacuum.

It must be mentioned that the order in which the reagents are put together seems to have an influence upon the results, especially in the case of high concentrations of BPA and NaOH.

The yield of polyether, Y, is calculated by taking into account the amount of evolved HCl and the fact that n moles of BPA reacted with $n + 1$ moles of DCB, according to the structure [Eq. (6.3)]:

$$Y = (\text{weight of polymer})/\{[N_A M_A + N_B M_B \times (n-1)/n] - [2M_{HX} \times N_A \times (n-1)/n]\} \quad (6.13)$$

where N_A is the number of moles of DCB, N_B the number of moles of BPA, M_A the molecular weight of DCB, M_B the molecular weight of BPA, and M_{HX} is the molecular weight of HCl.

The polyether (white powder) was obtained in 84% yield.

6.2.4.3. Polymer Characterization

The composition of the polymer was determined by elemental analysis. The chlorine content allows one to evaluate $\bar{M}_n$. According to the structure [Eq. (6.3)], C/Cl = $(19n + 4)/2$ and $\bar{M}_n = 280n + 125$. The percents of different elements are: C % = $(19n + 4) \times 12/\bar{M}_n$; H % = $(20n + 6)/\bar{M}_n$; O % = $2n \times 16/\bar{M}_n$; Cl % = $2 \times 35.45/\bar{M}_n$. The following percents were found: C, 79.97; H, 7.27; O, 11.20; Cl, 1.54 which are in excellent agreement with the values calculated with $n = 16$: C, 80.05; H, 7.22; O, 11.19; Cl, 1.54. Thus $\bar{M}_n$ was found to be equal to 4600.

The polymer was also characterized by ^{1}H and ^{13}C NMR:[16,26] n was determined from the ratio of the surfaces of C$\underline{H}_2$Cl and C$\underline{H}_2$O protons of the end groups and main chain, respectively, and was found to be equal to 16, leading to $\bar{M}_n = 4600$.

Additionally, the molecular weight of this polyether was examined by GPC in THF at 25 °C on a Waters apparatus equipped for refractive index and UV detections, using a polystyrene calibration curve. The following values were found: $\bar{M}_n = 3500$ and $\bar{M}_w = 4900$ ($I_p = 1.4$).

6.3. POLYTHIOETHERS

6.3.1. Literature Survey

Phase transfer catalyzed polythioetherifications have been performed with aliphatic dithiols, bisthiophenols, and sodium sulfide as comonomers with dihalogeno aliphatic or aromatic compounds, XRX, according to either a liquid–liquid system or a solid–liquid system [Eqs. (6.14) and (6.15)]:

$$n\text{XRX} + n\text{HS—R'—SH} \xrightarrow[\text{NaOH/org. solvent}]{\text{PTC}} \text{—[—R—S—R'—S—]}_n + 2n\,\text{NaX} \quad (6.14)$$

$$n\text{XRX} + n\text{Na}_2\text{S} \xrightarrow[\text{org. solvent}]{\text{PTC}} \text{—[—R—S—]}_n + 2n\,\text{NaX} \quad (6.15)$$

with X = Br, Cl.

Some selected examples are listed in Table 6.5. High yields of polythioethers are generally observed. In most cases, their molecular weights have been estimated by viscometric measurements only, presumably

TABLE 6.5. Preparation of Polythioethers by Phase Transfer Catalyzed Polycondensation: Selected Examples

XRX	Disulfide	Catalyst/Base	Solvent/t (°C)	Ref.
$ClCH_2$—CH_2Cl	*aq. NaS_x*	$Bu_4N^+OH^-$	DCE[a]/reflux	44,45
$Br(CH_2)_4Br$	$HS(CH_2)_6SH$	DCHE[b]/KOH	none/80	10,46
$ClCH_2$—CO—C_6H_4—CO—CH_2Cl	$(p\text{-}HSC_6H_4)_2O$	CTMAC[c]/KOH	DCE[a]/30	47
C_6F_6	$(p\text{-}HSC_6H_4)_2O$	18-C-6/K_2CO_3	DMAC[d]/80	13
$Br(CH_2)_mBr$ ($m = 6,8$)	aq. Na_2S	CTMAC[c]	none/100	10,47–49
$Br(CH_2)_mBr$ ($m = 2$–12)	$(p\text{-}HSC_6H_4)_2$	TBAH[e]/NaOH	oDCB[f]/80	50
2,6–DBPy[g]	dry Na_2S	15-C-5	DMF/120	51

[a] DCE denotes 1,2-dichloroethane.
[b] DCHE denotes dicyclohexyl-18-crown-6.
[c] CTMAC denotes cetyltrimethylammonium chloride.
[d] DMAc denotes dimethylacetamide.
[e] TBAH denotes tetrabutylammonium hydrogen sulfate.
[f] oDCB denotes *o*-dichlorobenzene.
[g] 2,6-DBPy denotes 2,6-dibromopyridine.

because these polymers are less soluble in common organic solvents than the corresponding polyethers. Detailed studies of their microstructure have not been made except for polythioethers prepared from 4,4′-dithiobiphenyl and α,ω-dibromoalkanes[50] and for those derived from 2,6-dibromopyridine and Na_2S.[51] All the authors noted that the molecular weights of the polythioethers depend on several factors, such as the nature of the catalyst,[10,46–48,51–54] its concentration,[47,51] as well as the nature of the organic phase.[10,13,46,47,51–54]

6.3.2. Mechanistic Aspects

Most of the systems described in the literature are of the liquid–liquid type even when Na_2S was used,[10,44,45,49] except in the case of the polycondensation of 2,6-dibromopyridine (DBPy) with Na_2S.[51] This example of a solid–liquid phase transfer catalyzed process has been studied in detail, and some typical results are listed in Table 6.6.

It has been shown that, in most cases, the polymers have the following structure:

$$(n + 1)\text{Br—Py—Br} + n\text{Na}_2\text{S} \xrightarrow[\text{org. solvent/120 °C}]{\text{15-C-5}} \text{Br—Py—[—S—Py—]—}_n\text{Br} + 2n\,\text{NaBr} \tag{6.16}$$

There is a good agreement between the values of $\bar{M}_n$ of these polythioethers determined by different techniques, such as end groups analysis (^{13}C NMR and elemental analysis) and GPC. For instance, the results obtained for run 2 (Table 6.6) are as follows: $\bar{M}_n$ (NMR) = 750; $\bar{M}_n$ (elemental analysis) =

TABLE 6.6. Preparation of Polythioethers from 2,6-DBPy and Na_2S under PTC Conditions[a]: Influence of the Solvent

Run	Solvent	Yield (%)	MC[b] (%)	$\bar{M}_n$ (elemental analysis)	$\bar{M}_n$[c] (GPC)
1[d]	oDCB	traces	—	—	—
2	oDCB	46	0	850	700
3	NMP[e]	73	4	1250	1050
4	DMF	75	3	1250	1300
5	DMSO	78	3	1850	1350

[a] DBPy, 21.1 mmol; [DBPy]/[Na_2S], 1; 15-C-5, 2.11 mmol; solvent, 4 ml; 120 °C; *t*, 6 h; 600 rpm.
[b] Percent of cyclic trimer determined by GPC.
[c] In benzyl alcohol at 130 °C using a self-calibration curve.
[d] Without any catalyst.
[e] NMP denotes *N*-methylpyrrolidone.

850, and $\bar{M}_n$ (GPC) = 700. In all cases, the index of polydispersity, $I_p = \bar{M}_w/\bar{M}_n$, was found to be lower than 1.1.

In some experiments, the presence of a cyclic trimer containing three thiopyridyl units has been detected in the GPC diagram. This compound has been isolated and fully identified by mass spectroscopy, ^{1}H and ^{13}C NMR, as well as by elemental analysis. It can be seen that the percent of this macrocyclic compound is very low in the experiments described in Table 6.6. The mechanism of formation of this cyclic trimer is not yet elucidated in this solid–liquid system while a mechanism for cyclization has been recently proposed in the case of liquid–liquid phase transfer catalyzed polythioetherifications(49) (see Section 6.3.3 below).

From the results of Table 6.6, it can be seen that an increase in the dielectric constant of the organic phase leads to an increase in the polymer yield and in its molecular weight with a negligible amount of side reactions.

Several catalysts have been tried, such as quaternary ammonium salts, dicyclohexyl-18-crown-6, and the macrobicyclic ligand [221]. The best results were obtained with the 15-crown-5 polyether while the ammonium salts were found to be inefficient, as expected for a solid–liquid system. It must be mentioned that the reverse situation was observed for polythioetherifications performed with an aqueous Na_2S solution.(10) Since the S^{2-} anion is strongly hydrated, its transfer from the aqueous phase to the organic phase is difficult and requires highly lipophilic catalysts. Cetyltrimethylammonium chloride or bromide, and cetyltributylphosphonium bromide were found to be particularly effective for the preparation of high-molecular-weight polythioethers —[—(CH_2)—S—]—$_n$, while tetrabutylammonium chloride was ineffective.

6.3.3. *Scope, Limitations, and Side Reactions*

Several authors applied PTC to the synthesis of polythioethers having special properties. For instance, new thermotropic polythioethers and copolythioethers based on 4,4′-dithiobiphenyl have been prepared.(50) This technique has also been used for the synthesis of a new polysulfide by terpolymerization of malononitrile, carbon disulfide, and α,α'-dichloro-*p*-xylene.(55)

The well-defined α,ω-dibromothiopyridyl oligomers can be used as precursors for the elaboration of more complicated macromolecular structures such as ABA triblock copolymers, perfectly alternating $(AB)_n$ block copolymers, as well as well-defined networks with specific properties. Some work in this field is in progress in our laboratory.

Except for these last studies which present very promising developments, there are a number of limitations which prevent an extensive use of PTC for the synthesis of polythioethers.

In most cases, insoluble polymers are formed in the organic phase affording a limitation of the molecular weights. For instance, this is the case for polythioethers prepared from 2,5-disubstituted-1,4-bis(chloromethyl)-benzenes and Na_2S.[54] None of the polymers obtained by phase transfer catalyzed polycondensation of hexafluorobenzene with several bisthiophenols is soluble.[13] This could be explained by the nature itself of the polythioethers and/or by the reaction of more than two fluorines of C_6F_6.

Side reactions caused by the presence of hydroxyl anions have been shown by Shaffer and Percec in the case of the synthesis of polythioethers and copolythioethers based on 4,4'-dithiobiphenyl and α,ω-dibromoalkanes.[50] The presence of alkyl–alkyl ether units —CH_2—O—CH_2 in the polymers was clearly demonstrated by ^1H NMR analysis of the polythioethers. The mechanism in Eq. (6.17) was proposed. Once a

$$
\begin{array}{c}
\cdots C_6H_4{-}S{-}(CH_2)_n{-}Br \rightleftharpoons \cdots C_6H_4{-}S^+(CH_2)_nBr^- \\
\downarrow \text{aq. phase, } OH^- \\
Br{-}(CH_2)_nBr + \cdots C_6H_4-S-(CH_2)_n{-}OH \\
\downarrow \\
\cdots C_6H_4{-}S{-}(CH_2)_n{-}O{-}(CH_2)_n{-}Br
\end{array}
\qquad (6.17)
$$

bromoalkyl sulfide is formed, one can assume the formation of a sulfonium species which will enhance the reactivity of the bromine of the corresponding end group relative to that of the halogens of the α,ω-dibromoalkane. Moreover, this species would allow transfer to the aqueous phase where the concentration of OH^- is high, thus leading to an alcohol chain end which can subsequently form an ether linkage. The presence of such defects in the structures of these polythioethers has an influence upon their mesomorphic properties.

Another type of side reaction encountered in phase transfer catalyzed polythioetherifications is the cyclic oligomers formation, which has been studied in some detail in the case of polycondensation of α,ω-dibromoalkanes with Na_2S.[49] In liquid–liquid systems in which the monomer is used as the organic phase, dibromoalkanes containing fewer than six methylene units lead to large, and sometimes quantitative, amounts of cyclic thioethers. If the number of methylene units of the dibromoalkanes is higher than six, linear polymers could be formed quantitatively. A mechanism has been proposed which implies the participation of cyclic thioethers to either facilitate cyclic formation or enhance polymer reactivity as controlled by the length of the dibromoalkane. As an interesting consequence, very high molecular weight polymers ($\bar{M}_n \simeq 700{,}000$) with a narrow polydispersity ($I_p = 1.24$) have been obtained from 1,8-dibromo octane by performing the polycondensation with Na_2S in the presence of tetrahydrothiophene.

6.3.4. Example of a Polythioether Synthesis

The experimental procedure used for the preparation of a polythioether from 2,6-DBPy and Na_2S, according to Eq. (6.16), is as follows (run 4 in Table 6.6).

6.3.4.1. Materials Purification

Na_2S, $9H_2O$ (Aldrich) was placed in a dryer containing P_2O_5 under vacuum for 15 days. It was then dried by heating at 180 °C for 48 h under high vacuum (10^{-5} mm). The phase transfer catalyst 15-C-5 and 2,6-DBPy were purchased from Aldrich and were used as received. Pure grade commercial DMF (Prolabo) was distilled before use while the other solvents (chloroform, n-hexanes) were used without any further purification.

6.3.4.2. Polymer Synthesis

In a 100-mL three-necked flask equipped with a mechanical stirrer, a nitrogen inlet, and a reflux condenser, were placed 42.2 mmol of Na_2S, 8 mL of DMF, 42.2 mmol of 2,6-DBPy, and 4.22 mmol of 15-C-5. The mixture was heated at 120 °C and stirred at 600 rpm for 6 h. After cooling, 70 mL of chloroform were added. The organic phase was separated from the inorganic salts, and washed several times with distilled water, twice with a slightly acidic aqueous solution (H_2SO_4, pH 3.5), then with water. After drying on Na_2SO_4, the major part of the solvent was evaporated and the polymer was recovered by precipitation in n-hexanes (*ca* 2 L). It was then dried under high vacuum.

The yield in polythioether, Y, is calculated by taking into account the amount of formed NaBr and the fact that n moles of Na_2S reacted with $n + 1$ moles of 2,6-DBPy, according to Eq. (6.16) (see Section 6.2.4.2). A 75% yield of beige powder was obtained.

6.3.4.3. Polymer Characterization

The molecular weight of the polymer was determined by GPC in benzyl alcohol at 130 °C. A self-calibration curve established from 2,6-DBPy and several models which have been synthesized and characterized separately allowed the determination of the percent of cyclic oligomer. In the present case, it was found to be equal to 3%, and $\bar{M}_n$ was equal to 1300 with $I_p = 1.1$.

Composition of the polymer was determined by elemental analysis. The bromine content allows an evaluation of $\bar{M}_n$, by assuming that the structure [Eq. (6.16)] is valid and by taking into account the amount of macrocyclic compound which is known from the GPC diagram (3%). The values found for the different elements are in excellent agreement with the proposed structure [Eq. (6.16)]. The value of $\bar{M}_n$ was found to be equal to 1250. The polymer was also characterized by ^{13}C NMR. By adding chromium acetylacetonate (0.025 mol L^{-1}) to the polymer solution in $CDCl_3$, it was possible to deduce n from the ratio of the surfaces corresponding to the C̱-Br and C̱-S peaks ($\bar{M}_n = 1350$). The agreement between the values of $\bar{M}_n$ determined by different methods is excellent.

6.4. POLYCARBONATES AND POLYTHIOCARBONATES

6.4.1. Literature Survey

Polycarbonates as well as polythiocarbonates can be prepared by using either a solid–liquid or a liquid–liquid system according to different types of reaction. Some selected examples are listed in Table 6.7.

The first synthetic route to be studied was the reaction between dibromo-*p*-xylene (DBPX) and potassium carbonate in the presence of 18-crown-6 ether as the catalyst:[56]

$$(n+1)\ BrCH_2—C_6H_4—CH_2Br + nK_2CO_3 \xrightarrow{18—C—6} BrCH_2—C_6H_4—CH_2—[OCOO—CH_2—C_6H_4—CH_2]_n—Br + 2n\ KBr \quad (6.18)$$

The system was improved by using dibromo compounds, potassium salts of diols or diphenols, and carbon dioxide:[57,67]

$$nBr—R_1—Br + nKO—R_2—OK + 2n\ CO_2 \xrightarrow{18—C—6} [—OCOO—R_1—OCOO—R_2]—_n + 2n\ KBr \quad (6.19)$$

Polycarbonates were also obtained by replacing carbon dioxide by ethylene or propylene carbonate in Eq. (6.19).[68] The influence of the nature of the solvent, of the amount of catalyst and reagents, of the reaction time, and of temperature was investigated.

Another way of preparing polycarbonates is the transcarbonatation reaction studied by Kuran *et al.*[69] PTC is used in the first step to prepare a biscarbonate such as 1,4-bis-(cyclohexyloxycarbonyloxymethyl) benzene, which can then react with a diol in the presence of a convenient catalyst [$Ti(OR)_4$] to give polymers having quite high molecular weights. However, no indication relative to the molecular-weight distribution and the nature of the end groups of the polymers is given.

TABLE 6.7. Preparation of Polycarbonates and Polythiocarbonates by Phase Transfer Catalyzed Polycondensation: Selected Examples

Reaction	Catalyst/Base	Solvent/*t* °(C)	Ref.
p-$BrCH_2—C_6H_4—CH_2Br + CO_3K_2$	18-C-6	Diglyme/105	56
p-$BrCH_2—C_6H_4—CH_2Br + KO—C_6H_{10}—OK + CO_2$	18-C-6	DOX[a]/100	57
BPA + $ClCOO—CH_2—CH_2—O—CH_2—CH_2—OCOCl$	TBAH[b]/NaOH	CH_2Cl_2/20	11,58–60
(*p*-$HOC_6H_4)_2Y + COCl_2$ [$Y = C(C_6H_5)_2$—]	$Me_4N^+Cl^-$/NaOH	CH_2Cl_2/20	61,62
p-$BrCH_2—C_6H_4—CH_2Br$ + (*p*-$KS—C_6H_4)_2O + CS_2$	18-C-6	DOX[a]/50	63
p-$BrCH_2—C_6H_4—CH_2Br + KO—C_6H_{10}—OK + CS_2$	18-C-6	CS_2/−78	64
BPA + $CSCl_2$	TBAB[c]/NaOH	CH_2Cl_2/20	65,66

[a] DOX denotes dioxane.
[b] TBAH denotes $Bu_4N^+SO_4H^-$.
[c] TBAB denotes $Bu_4N^+Br^-$.

More recently, several liquid–liquid systems were examined. Among them, an interesting route consists in reaching a diol with a bis chloroformate compound:[11,58–60,70]

$$n\text{ClCOO—R—OCOCl} + n\text{HO—R'—OH} \xrightarrow[\text{NaOH/org. solvent}]{\text{PTC}} \text{—[COO—R—OCOO—R'—O]}_n\text{—} + 2n\,\text{NaCl} + 2n\,\text{H}_2\text{O} \quad (6.20)$$

Finally, phosgenation of several diphenols was carried out under PTC conditions leading to polycarbonates with high molecular weights.[61,62]

Similar processes have been used for the synthesis of polythiocarbonates. The reaction of dibromo-*p*-xylene (DBPX) with carbon disulfide and potassium salts of dithiols or of diols or diphenols carried out in the presence of crown ethers leads to polymers containing (—SCSS—) and (—OCSS—) units, respectively:[63,64]

$$n\text{BrCH}_2\text{—C}_6\text{H}_4\text{—CH}_2\text{Br} + \text{nKZ—R—ZK} + n\text{CS}_2 \xrightarrow{\text{18—C—6}} \text{[—CH}_2\text{—C}_6\text{H}_4\text{—CH}_2\text{—SSCZ—R—ZCSS]}_n + 2n\,\text{KBr} \quad (6.21)$$

However, low-molecular-weight polymers are obtained, the structure of which is rather complex due to side reactions (see Section 6.4.3 below). Thiophosgenation of diphenols under PTC conditions was studied in some detail.[65,66,71–73] The effect of the nature of the catalyst and the solvent, and of the way of thiophosgene addition was examined. Finally, copoly(carbonate-thiocarbonate)s were synthesized from BPA and mixtures of phosgene and thiophosgene, in a liquid–liquid system ($NaOH/CH_2Cl_2$) using quaternary ammonium salts as catalysts.[74]

6.4.2. Mechanistic Aspects

The following mechanism has been proposed for the solid–liquid reaction between DBPX and potassium carbonate with 18-crown-6 as the catalyst [Eq. (6.18)].[56] It is assumed that the crown-ether strongly adsorbs on the surface of K_2CO_3 prior to forming the complex with K^+:

$$\text{K}_2\text{CO}_3\,(\text{massive}) + \text{CE} \xrightleftharpoons{\text{adsorption}} \text{CE}\cdot\text{K}_2\text{CO}_3\,(\text{massive}) \quad (6.22)$$

$$\text{CE}\cdot\text{K}_2\text{CO}_3\,(\text{massive}) \rightleftharpoons \text{KCO}_3^-\,\text{(K}^+\text{)} + \text{K}_2\text{CO}_3\,(\text{massive}) \quad (6.23)$$

where KCO_3^- (K+) is the complexed salt soluble in the organic phase which reacts very rapidly with DBPX:

$$BrCH_2—C_6H_4—CH_2Br + KCO_3^-\text{(K}^+\text{)} \rightarrow BrCH_2—C_6H_4—CH_2OCOO^-\text{(K}^+\text{)} + KBr \quad (6.24)$$

$$BrCH_2—C_6H_4—CH_2OCOO^-\text{(K}^+\text{)} + DBPX$$
$$\rightarrow BrCH_2—C_6H_4—CH_2OCOOCH_2—C_6H_4—CH_2Br + \text{(K}^+\text{)}\ Br^- \quad (6.25)$$

The polycarbonate is formed by repeating the reactions shown in Eqs. (6.23) and (6.24). It must be mentioned that all the end groups of the polymer have the structure of benzyl bromide as in the case of polyethers (see Section 6.2.2).

A simple kinetic treatment of these equations can explain that the polymer yield is directly proportional to the amount of K_2CO_3, and that it increases at first and then reaches a constant value with an increase in the concentration of 18-crown-6.

In the case of the reaction of CO_2 with dibromo compounds and potassium salts or diols or diphenols [Eq. (6.19)], the following mechanism has been suggested:(57,67)

$$n KO—R—OK + 2n\ CO_2 \xrightarrow[\text{fast}]{\text{CE}} n\ \text{(K}^+\text{)}\ ^-OCOO—R—OCOO^-\text{(K}^+\text{)} \quad (6.26)$$

$$n\ \text{(K}^+\text{)}\ ^-OCOO—R—OCOO^-\ \text{(K}^+\text{)} + n DBPX$$
$$\xrightarrow[\text{slow}]{} —[OCOO—R—OCOO—CH_2—C_6H_4—CH_2—]_n + 2n\ \text{(K}^+\text{)}Br^- \quad (6.27)$$

The formation of the complexed carbonate [Eq. (6.26)] is relatively fast because of the greater stability of the carbonate anion in comparison with the alkoxide anion. The reaction shown in Eq. (6.27) is slow and thus it is the rate-determining step in the overall process. It must be noted that no information relative to the nature of the end groups of the polymers and to their molecular weight distributions is given.

Concerning the liquid–liquid systems used for the preparation of polycarbonates, only the case of the reaction of bischloroformates with bisphenols has been studied in detail.(11,58–60) Some typical results obtained with diethylene glycol bischloroformate ($DEGCl_2$) and BPA or BPAF (the corresponding perfluoro bisphenol) are listed in Table 6.8. Several parameters proved to affect the results, the most important ones being the amount of sodium hydroxide and the reaction time as will be discussed in the next section. However, by choosing carefully those factors as well as the solvent, it has been shown that polymers having the following well-defined structure:

$$HOCH_2CH_2OCH_2CH_2—[OCOO—R—OCOO—CH_2CH_2OCH_2CH_2—]_nOH \quad (6.28)$$

TABLE 6.8. Preparation of Polycarbonates from BPA or BPAF and $DEGCl_2$ under PTC Conditions[a]: Influence of NaOH Concentration

Run	Diphenol	[NaOH]	Yield (%)	$\bar{M}_n$[b] (GPC)	$I_p = \bar{M}_w/\bar{M}_n$
1	BPA	1 N	28	3300	1.7
2	BPA	3 N	53	4700	1.5
3[c]	BPA	3 N	70	40000	2.0
4	BPA	6 N	0	—	—
5	BPAF	1 N	71	15000	2.5
6	BPAF	3 N	66	39000	1.8
7[c]	BPAF	3 N	25	12000	1.5
8	BPAF	6 N	56	15500	1.6

[a] oDCB/aq. NaOH (20 mL/20 mL); diphenol, 5 mmol; $DEGCl_2$, 5 mmol; TBAH, 1 mmol; 20 °C; $t = 3$ h; 500 rpm.
[b] In THF using a polystyrene calibration curve.
[c] Toluene/aq. NaOH (20 mL/20 mL).

with hydroxyl end groups resulting from the hydrolysis of the chloroformate groups could be prepared. Their molecular weights are much higher than those of the polycarbonates prepared by other methods using PTC. Finally, they have quite a narrow molecular-weight distribution.

6.4.3. *Scope, Limitations, and Side Reactions*

The synthesis of copolycarbonates containing a tertiary diol structure has been achieved for the first time, by using a solid–liquid phase transfer catalyzed polycondensation of the bis(carbonylimidazolide) of 2,5-dimethyl-2,5-hexane diol with various other diols in the presence of powdered potassium carbonate.[(75)] Allylic and benzylic polycarbonates were prepared in a similar way.[(76–78)] These classes of polycarbonates undergo multiple, simultaneous, main-chain cleavage reactions with the formation of volatile products only. These materials may find applications in thin-film technology due to their unique thermolytic behavior.

On the other hand, a liquid–liquid phase transfer catalyzed synthesis of polycarbonate/polysiloxane block-copolymers has been investigated.[(79)] It involves the reaction of a preformed carboxypropyl-terminated polydimethylsiloxane oligomer of various molecular weights with BPA, base, phosgene, and the phase transfer catalyst. These copolymers could be of interest as thermoplastic elastomers, elastoplastics, or biomaterials. Finally, the synthesis of high-molecular-weight polyester-carbonate copolymers has been carried out under PTC conditions starting from BPA, phosgene, and different dicarboxylic aromatic acids.[(80)]

Several side reactions can occur during the synthesis of polycarbonates and polythiocarbonates by using the PTC methods, which are similar to those described previously for polyethers and polythioethers.

Cyclic carbonates were formed from the reaction of potassium carbonate with 1,2-dibromoethane and 1,3-dibromopropane in the presence of 18-crown-6 ether while linear polycarbonates were obtained with longer aliphatic dibromides.[57]

The presence of reactive basic species in these two phase systems is responsible for side reactions which can lead to hydrolysis of the carbonate functions as well as to several undesirable structures of the polymers.

It was noted that when the reaction between DBPX and CO_3K_2 with 18-crown-6 as the catalyst was performed at elevated temperatures (160 °C), a yellow-colored insoluble polymer was obtained which contains both ether and xylylidene structures.[56]

The situation is more complex in the case of the reaction of CO_2 with potassium salts of diols or diphenols and dibromo compounds due to the higher reactivity of the alkoxide species compared with that of the carbonate species toward the dibromide and/or CO_2. Under certain conditions, the reaction products of potassium cyclohexanolate, CO_2 and DBPX contain *p*-xylylidene chains and ether bonds.[69] This can be avoided by preparing first the potassium carbonate and then adding the dibromo compound. The carbonate linkages content in polymers obtained from diphenolates, dibromo compounds, and cyclic carbonates was found to depend on the kind of diphenolate and of dihalogeno compound as well as on the reaction temperature.[69]

Concerning the liquid–liquid systems, hydrolysis of the polycarbonates prepared by reaction of phosgene with diphenols has been studied in relation to the nature of diphenols and of the catalyst.[62] The effect of NaOH concentration was studied in oDCB for polycarbonates obtained from BPA and from BPAF with $DEGCl_2$.[11,58–60] From the results shown in Table 6.8, it appears that $\bar{M}_n$ increases and then decreases on increasing the concentration of NaOH; in run 4 no polymer was recovered after 3 h reaction. The carbonate linkages of the growing chains are attacked by OH^- transferred from the aqueous phase to the organic phase and/or by the phenolate anions. A competition between the chain growth and its degradation occurs when the transfer of bisphenolate and OH^- anions in the organic phase is important. Hydrolysis reactions can be minimized by lowering the concentration of NaOH, by using a low polar solvent such as toluene, and by stopping the reaction after a very short time. With an adequate adjustment of these parameters, it has been possible to prepare well-defined polycarbonates from BPA and $DEGCl_2$ having a narrow molecular-weight distribution ($\bar{M}_w/\bar{M}_n \simeq 1.4$) with hydroxyl end groups [Eq. (6.28)].[58–60] A physicochemical study of end-to-end cyclization dynamics in solution was performed on this polymer after a quantitative functionalization of both ends with a fluorescent probe such as 4-(1-pyrenyl)butyroyl chloride. The rate of formation of the resulting intramolecular excimer in highly dilute solutions was measured by using fluorescence techniques. This type of measurement could be done for the first time on a polycondensate because PTC is a unique method to obtain a polymer sample quantitatively labelled on each end, with a narrow molecular-weight distribution.

Side reactions are more pronounced in the case of polythiocarbonates. A quantitative incorporation of CS_2 units in polymers synthesized from carbon disulfide, potassium salts of diols, and dibromo-*p*-xylene was found to occur only on performing the experiments in an excess of CS_2 at −78 °C, with 18-crown-6 as the catalyst.(64) The molar fraction of ether linkages can reach 50% on using solvents such as dioxane, benzene, etc., at 80 °C. On the other hand, the influence of the phase transfer catalyst on the hydrolysis of polythiocarbonates obtained by reaction of diphenols with thiophosgene was found to play an important role.(72,73) Finally, for copoly(carbonate-thiocarbonate)s prepared from BPA and a mixture of phosgene/thiophosgene, it has been found that the reactivities of both acid dichlorides are similar, but the resistances to hydrolysis are different, the carbonate groups being more stable than the thiocarbonate groups.(74)

In conclusion, the different reactions investigated for the preparation of polycarbonates and polythiocarbonates seem to be more delicate and sensitive to different parameters than that involved for the preparation of polyethers. Among them, only two lead to polymers having a well-defined structure with end groups of the same nature, i.e., the reaction of potassium carbonate with dibromo compounds(56) and the reaction of bischloroformates with diphenols.(11,58–60,70) It is noteworthy that only low-molecular-weight polymers are obtained according to the first method while the second one leads to polycarbonates with high values of $\bar{M}_n$ (up to 40,000).

6.4.4. *Example of a Polycarbonate Synthesis*

The experimental procedure used for the preparation of a polycarbonate from BPA and $DEGCl_2$, according to Eqs. (6.20) and (6.28), is as follows (run 2 in Table 6.8). It is quite similar to that described for the synthesis of polyethers (Section 6.2.4).

6.4.4.1. *Materials Purification*

BPA (Merck) was recrystallized from toluene. $DEGCl_2$ kindly provided by the SNPE was stored in the dark at −30 °C and used as received. The phase transfer catalyst TBAH (Fluka) was used as received and the pure grade commercial solvents were also used without any further purification.

6.4.4.2. Polymer Synthesis

The sodium bisphenolate was first prepared by stirring 5 mmol of BPA with 20 mL of aqueous NaOH 3 N at 20 °C. Then the catalyst TBAH (1 mmol), oDCB (10 mL), and a solution of 5 mmol of $DEGCl_2$ in 10 mL of oDCB were successively added under nitrogen. The mixture was stirred (stirring rate 500 rpm) for 3 h at 20 °C. The organic layer was diluted with methylene chloride, separated from the aqueous phase, washed with slightly acidic aqueous solution (HCl 0.1 N) and with water. The polymer was recovered by precipitation in methanol and dried under high vacuum.

The yield of polycarbonate, Y, has been defined by the ratio m/m_0 with m = the amount of recovered polymer and $m_0 = N_0(M_A + M_B - 2M_{HCl})$, N_0 being the initial number of moles of each reagent, M_A and M_B the molecular weights of BPA and $DEGCl_2$. A 53% yield of a white powder was obtained.

6.4.4.3. Polymer Characterization

It was impossible to evaluate $\bar{M}_n$ from the elemental analysis. IR and ^{1}H NMR spectra were consistent with the hydroxy-terminated structure for this sample [Eq. (6.28)]. The molecular weight of this polymer was measured by GPC in THF at 25 °C, and the following values were found: $\bar{M}_n = 4700$, $\bar{M}_w = 6900$, and $I_p = 1.5$, using a polystyrene calibration curve. After labeling with 4-(1-pyrenyl)butyroyl chloride according to a standard procedure,[60] the polymer was purified by four precipitations from methylene chloride solution into methanol to remove all unattached chromophore. Before fractionation on a preparative column for the fluorescence studies, the following values were found: $\bar{M}_n = 5100$, $\bar{M}_w = 6900$, and $I_p = 1.4$. It was possible to obtain three fractions having I_p equal to 1.1.

6.5. POLYESTERS

6.5.1. Literature Survey

The liquid–liquid process involving a diol or a bisphenol and a dichlorocarbonyl compound was successfully applied to the synthesis of polyesters. It is now widely used by numerous authors to prepare well-defined polymers, and especially liquid crystal polyesters, thermally stable polymers which can be easily processed, block-copolymers, etc. Due to the large amount of papers which were published in this domain, only a few selected references are given.[14,80–94]

On the other hand, the preparation of polyesters is also a very nice example of application of the solid–liquid process. Most of the time, the starting materials are alkaline dicarboxylates and dihalogeno compounds,[11,95–97] according to Eq. (6.29):

$$(n + 1)\ X\text{—}R\text{—}X + nM_t\text{—}OOC\text{—}R'\text{—}COOM_t \xrightarrow{PTC} X\text{—}[R\text{—}OOC\text{—}R'\text{—}COO]\text{—}_nR\text{—}X + 2n\,M_tX \quad (6.29)$$

with X = Br, Cl and M_t = K, Na.

The self-condensation of salts of 4-(*p*-bromoacetylphenyl) butanoic acid and related compounds was also investigated.[98]

Some selected examples of solid–liquid phase transfer catalyzed polyesterification are listed in Table 6.9. In most cases, well-defined polymers were obtained, the molecular weight of which is generally lower than 10,000. Studies of their microstructure have been made in the case of polyesters derived from fumaric and maleic acid salts,[96] and for those prepared from 2,6-naphthalene dicarboxylates.[11,97] All the authors observed that the yields and the molecular weights of the polyesters depend on several factors, such as the solvent, the temperature, and the catalyst. Attempts to perform experiments under liquid–liquid PTC conditions were not successful.

6.5.2. Mechanistic Aspects

Most of the systems described in the literature are of the liquid–liquid type except those relative to the polycondensation of alkaline dicarboxylates with dihalogeno compounds. The reaction of potassium 2,6-naphthalene dicarboxylate with various α,ω-dibromo alkanes and 1,4-dichloro-2-butene

TABLE 6.9. Preparation of Polyesters by Solid–Liquid Phase Transfer Catalyzed Polycondensation: Selected Examples

Reaction	Catalyst	Solvent/t (°C)	Ref.
m-$BrCH_2$—C_6H_4—CH_2Br + $KO_2C(CH_2)_8CO_2K$	Aliquat 336	ACN[a]:C_6H_6 (1:1)/80	95
p-$BrCH_2$—C_6H_4—CH_2Br + *cis* or *trans* KO_2C—CH=CH—CO_2K	TEBAC[b]	ACN[a]/80	96
$Br(CH_2)_xBr$ + KO_2C Naph CO_2X[c] (x = 4, 5, 6, 7, 8)	DCHE[d]	NMP[e]/110	11,97
m-$BrCH_2$—C_6H_4—$(CH_2)_xCO_2K$ (x = 0, 2, 3)	18-C-6	Diglyme/100	98

[a] ACN denotes acetonitrile.
[b] TEBAC denotes triethylbenzylammonium chloride.
[c] Potassium 2,6-naphthalene dicarboxylate.
[d] DCHE denotes dicyclohexyl-18-crown-6.
[e] NMP denotes *N*-methyl-2-pyrrolidone.

TABLE 6.10. Preparation of Polyesters from the Potassium Salt of 2,6-Naphthalene Dicarboxylic Acid and Dihalogeno Compoundx, XRX, under PTC conditions[a]

Run	$Br(CH_2)_xBr$	Yield (%)	$\bar{M}_n$ (elemental analysis)	$\bar{M}_n$ NMR	$\bar{M}_n$[b] GPC	$I_p = \bar{M}_w/\bar{M}_n$
1	DCB[c]	84	3000	—	3800	2.3
2	$x = 4$	97	1800	1300	3800	1.4
3	$x = 5$	94	4200	—	3600	2.3
4	$x = 6$	87	2600	2300	5300	1.5
5[d]	$x = 6$	86	2800	2600	6400	1.5
6[e]	$x = 6$	74	1700	1700	3200	1.3
7[f]	$x = 6$	20	1050	1050	1800	1.4
8	$x = 7$	87	26000	—	11000	—
9	$x = 8$	90	33000	—	11000	—
10	$x = 9$	87	28000	—	—	—

[a] XRX, 5 mmol; $[XRX]/[K^+]$, 1.9; DCHE, 1.05 mmol; solvent, NMP (4 ml); 110 °C; *t*, 3 h; 700 rpm.
[b] In *m*-cresol or in a mixture of phenol and 1,2,4-trichlorobenzene at 110° C.
[c] DCB denotes 1,4-dichloro-2-butene.
[d] Catalyst, [222].
[e] $[XRX]/[K^+]$, 2.4.
[f] XRX denotes $Cl(CH_2)_6Cl$.

(DCB) has been studied in some detail. Typical results are listed in Table 6.10.

It has been shown that in most cases, the polymers have the structure indicated in Eq. (6.29), which has been confirmed by ^1H and ^{13}C NMR on polyesters of low molecular weights. There is good agreement between the values of $\bar{M}_n$ determined by elemental analysis and by ^1H NMR while the corresponding values determined by GPC are higher. This might come from the fact that a calibration curve based on polystyrene was used.

The influence of different factors, such as the nature of the catalyst or the stoichiometry of the reagents, was examined. It turned out that the former has no appreciable effect as long as a crown ether or a cryptand is used as the catalyst (runs 4 and 5) while the latter is more important (runs 4 and 6). Better results are obtained with a slight excess of dipotassium salt. Attempts were also made to study the influence of the nature of the halogen X: 1,6-dichlorohexane leads to poor results compared to 1,6-dibromohexane (runs 4 and 7). This illustrates the fact that Br^- is a better nucleofuge than Cl^- in a polar aprotic solvent such as NMP.

6.5.3. Scope, Limitations, and Side Reactions

PTC is more and more used for the reaction of bisphenols or diols with dichlorocarbonyl compounds, in liquid–liquid systems, leading to polyesters having special properties.[14,80–91]

The solid–liquid process gives oligomers having well-defined chloromethyl or bromomethyl end groups. Polyesters containing 2,6-naphthalene dicarboxylic moieties linked by 4 to 8 methylene groups were prepared with excellent yields by this technique. These polymers, although not thermotropic, show very interesting thermal properties and it seems worthwhile studying similar systems.[11,97]

In the case of polyesters formed from α,ω-dihalogeno compounds and alkali metal salts of maleic or fumaric acids, it must be mentioned that no isomerization was observed during the reaction, contrary to what is observed when other techniques of polyesterification are used.[96] However, some limitations are observed which prevent an extensive use of this technique for the synthesis of these polyesters. Yields were found to be far from quantitative and molecular weights were relatively low.[95,96] This was overcome by using the cesium salts of the diacids in NMP, without any phase transfer catalyst.[99] When cesium maleate or fumarate was reacted with 1,4-dibromobutane in NMP at 110 °C, no isomerization was observed during the reaction but a substantial amount of cyclics was formed besides high-molecular-weight polyesters.[100]

6.5.4. Example of Polyester Synthesis

The experimental procedure used for the preparation of a polyester from potassium 2,6-naphthalene dicarboxylate and 1,6-dibromohexane, according to Eq. (6.29), is as follows (run 4 in Table 6.10):

6.5.4.1. Materials Purification

Potassium 2,6-naphthalene dicarboxylate, 1,6-dibromohexane, and the phase transfer catalyst dicyclohexyl-18-crown-6 (DCHE) were purchased from Aldrich and used as received. Pure grade commercial NMP (Merck) was distilled under nitrogen before use, while the other solvents (acetone, methanol, 1,1,2,2-tetrachloroethane, n-hexane) were used without further purification.

6.5.4.2. Polymer Synthesis

Typically, 5.25 mmol of the potassium salt, 5 mmol of 1,6-dibromohexane, 0.525 mmol of DCHE, and 4 mL of NMP were mechanically stirred (700 rpm) for 3 hours at 110 °C. After cooling, water was added to dissolve inorganic salts and precipitate the polymer. The solids was filtered, washed with water, acetone, slightly HCl solutions, and methanol. It was then dissolved in 1,1,2,2-tetrachloroethane and precipitated into n-hexane. After filtration, it was dried under high vacuum.

The yield in polyester, Y, is calculated by taking into account the amount of formed KBr and the fact that n moles of potassium salt reacted with $n + 1$ moles of 1,6-dibromohexane, according to Eq. (6.29) (see Section 6.2.4.2). An 87% yield of a beige powder was obtained.

6.5.4.3. *Polymer Characterization*

The molecular weight of the polymer was determined by GPC in phenol/1,2,4-trichlorobenzene (50/50) at 110 °C, using a polystyrene calibration curve. In the present case, $\bar{M}_n$ was found to be equal to 5300 with $I_p = 1.5$.

Composition of the polymer was determined by elemental analysis. The bromine content allows an evaluation of $\bar{M}_n$, by assuming that the structure [Eq. (6.29)] is valid. The values found for the different elements are in excellent agreement with the proposed structure. The value of $\bar{M}_n$ was found to be equal to 2600. The polymer was also characterized by ^{1}H NMR an it was possible to deduce n from the ratio of the surfaces of $C\underline{H}_2Br$ and of $COO{-}C\underline{H}_2$ protons of the end groups and of the main chain, respectively ($\bar{M}_n = 2300$).

6.6. CONCLUSION

PTC has been successfully applied to the synthesis of other polymers than those described in this paper such as, for instance, polythioesters,[101] polysulfonates,[10,102,103] polysulfones,[104–106] polysulfonamides,[107] polyphosphonates,[10,108,109] polyphosphoanhydrides,[7] carbon–carbon chain polymers,[10,110–112] Sb(V) polyamines,[7] polyarenyl platinum ethers,[113] as well as polysilanes.[114]

Detailed studies have been made in the case of polyethers, polythioethers, polycarbonates, and polyesters. Mechanisms have been proposed for this new type of polycondensation as well as for some side reactions which are mainly the formation of cyclics and irregularities in the structure of the polymers coming from hydrolysis reactions. Those side reactions can be minimized by playing with different factors, such as the catalyst (nature and concentration), NaOH concentration (in the case of a liquid–liquid process), the initial molar ratio of the reagents, and the solvent.

As a general observation, the main criteria for obtaining linear high polymers from phase transfer catalyzed polycondensation are, first, the reaction should be clean and relatively fast, and second, the organic phase should be capable of maintaining in solution fairly high concentrations of both monomer and polymer.[13,20,40]

The major advantages presented by phase transfer catalyzed polycondensations are the simplicity of the process and the possibilities of preparation of polymer having well-defined end groups and remarkably narrow molecular-weight distributions without requirement for an accurate control of the stoichiometry of the reagents. Especially, this very promising and fruitful technique allows the synthesis of α,ω-difunctional oligomers, which can be further used to prepare new specialty polymers. For instance, networks as well as ABA triblock copolymers, alternating copolymers $(AB)_n$, and segmented copolymers have been synthesized and can find applications in various domains such as thermoplastic elastomers, liquid crystal, thermally stable or complexing materials.

In conclusion, phase transfer catalyzed polycondensation represents a very promising field for both fundamental and applied research.

ACKNOWLEDGMENTS

I would like to thank Drs. H. Ben Romdhane, R. Dorigo, O. Mahamat, M. Majdoub, F. Méchin, T. D. N'Guyen, H. Sleiman, Mrs. N. d'Haeyer, and A. de Leuze, who participated in the laboratory investigations in this field. I would like to thank also Drs. M. Bartholin and B. Sillion from the CEMOTA for many stimulating discussions.

REFERENCES

1. C. M. Starks, *J. Am. Chem. Soc.* **93**, 195–199 (1971).
2. M. Makosza and B. Serafin, *Rocz. Chem.* **39**, 1223–1230 (1965).
3. A. Brandström and K. Gustavii, *Acta Chem. Scand.* **23**, 1215–1218 (1969).
4. W. P. Weber and G. W. Gokel, *Phase Transfer Catalysis in Organic Chemistry* (Springer-Verlag, Berlin, Heidelberg, New York, 1977).
5. C. M. Starks and C. Liotta, *Phase Transfer Catalysis: Principles and Techniques* (Academic Press, New York, 1978).
6. E. V. Dehmlow and S. S. Dehmlow, *Phase Transfer Catalysis,* 2nd rev. ed. (Verlag Chemie, Weinheim, Deerfield Beach Florida, Basel, 1983).
7. *Crown Ethers and Phase Transfer Catalysis in Polymer Science,* ed. by L. J. Mathias and C. E. Carraher Jr (Plenum Press, New York, London, 1984).
8. *Phase Transfer Catalysis New Chemistry, Catalysts, and Applications,* ed. by C. M. Starks, *ACS Symp. Ser.* **326** (Am. Chem. Soc., Washington, 1987).
9. F. Montanari, D. Landini, and F. Rolla, in *Topics in Current Chemistry,* **101** (Springer-Verlag, Berlin, Heidelberg, New York, 1982), pp. 147–200.
10. Y. Imai, *J. Macromol. Sci., Chem.* **A15**, 833–852 (1981).
11. O. Mahamat, M. Majdoub, F. Méchin, H. Sleiman, and S. Boileau, in *Recent Advances in Mechanistic and Synthetic Aspects of Polymerization,* ed. by M. Fontanille and A. Guyot (D. Reidel, Dordrecht, 1987), pp. 191–206.
12. V. Percec, in *Phase Transfer Catalysis New Chemistry Catalysts, and Applications,* ed. by C. M. Starks, *ACS Symp. Ser.* **326** (Am. Chem. Soc., Washington, 1987), pp. 96–115.

13. R. Kellman, R. F. Williams, G. Dimotsis, D. J. Gerbi, and J. C. Williams, in *Phase Transfer Catalysis New Chemistry, Catalysts, and Applications,* ed. by C. M. Starks, *ACS Symp. Ser.* **326** (Am. Chem. Soc., Washington, 1987), pp. 128–142.
14. A. K. Banthia, D. Lundeford, D. C. Webster, and J. E. McGrath, *J. Macromol. Sci., Chem.* **A15**, 943–966 (1981).
15. Y. Imai, M. Ueda, and M. Ii, *J. Polym. Sci., Polym. Lett. Ed.* **17**, 85–89 (1979).
16. T. D. N'Guyen and S. Boileau, *Polym. Bull.* **1**, 817–823 (1979).
17. T. D. N'Guyen and S. Boileau, in *Crown Ethers and Phase Transfer Catalysis in Polymer Science,* ed. by L. J. Mathias and C. E. Carraher Jr. (Plenum Press, New York, 1984), pp. 59–67.
18. J. L. Jin and J. H. Chang, in *Crown Ethers and Phase Transfer Catalysis in Polymer Science,* ed. by L. J. Mathias and C. E. Carraher Jr. (Plenum Press, New York 1984), pp. 91–104.
19. N. Yamazaki and Y. Imai, *Polym. J.* **15**, 603–608 (1983).
20. G. G. Cameron and K. S. Law, *Makromol. Chem. Rapid Commun.* **3**, 99–102 (1982).
21. D. J. Gerbi, G. Dimotsis, J. L. Morgan, R. F. Williams, and R. Kellman, *J. Polym. Sci., Polym. Lett. Ed.* **23**, 551–556 (1985).
22. A. S. Hay, F. J. Williams, H. M. Relles, B. M. Boulette, P. E. Donahue, and D. S. Johnson, *J. Polym. Sci., Polym. Lett. Ed.* **21**, 449–457 (1983).
23. A. S. Hay, F. J. Williams, H. M. Relles, and B. M. Boulette, *J. Macromol. Sci., Chem.* **A21**, 1065–1079 (1984).
24. N. Yamazaki and Y. Imai, *Polym. J.* **17**, 377–384 (1985).
25. E. Shchori and J. E. McGrath, *J. Appl. Polym. Sci., Appl. Polym. Symp.* **34**, 103–117 (1978).
26. V. Percec and B. C. Auman, *Makromol. Chem.* **185**, 617–627 (1984).
27. V. Percec and B. C. Auman, *Polym. Bull.* **10**, 385–390 (1983).
28. V. Percec, B. C. Auman, and P. L. Rinaldi, *Polym. Bull.* **10**, 391–396 (1983).
29. V. Percec and H. Nava, *Makromol. Chem. Rapid Commun.* **5**, 319–326 (1984).
30. V. Percec, H. Nava, and J. M. Rodriguez-Parada, *J. Polym. Sci., Polym. Lett. Ed.* **22**, 523–532 (1984).
31. V. Percec, H. Nava, and J. M. Rodriguez-Parada, *Polym. Bull.* **12**, 261–268 (1984).
32. V. Percec and B. C. Auman, *Polym. Bull.* **12**, 253–260 (1984).
33. V. Percec, T. D. Shaffer, and H. Nava, *J. Polym. Sci., Polym. Lett. Ed.* **22**, 637–647 (1984).
34. T. D. Shaffer and H. Nava, *J. Polym. Sci., Polym. Lett. Ed.* **23**, 185–194 (1985).
35. T. D. Shaffer and V. Percec, *Makromol. Chem. Rapid Commun.* **6**, 97–104 (1985).
36. T. D. Shaffer, M. Jamaludin, and V. Percec, *J. Polym. Sci., Polym. Chem. Ed.* **23**, 2913–2930 (1985).
37. T. D. Shaffer, M. Jamaludin, and V. Percec, *J. Polym. Sci., Polym. Chem. Ed.* **24**, 15–27 (1986).
38. T. D. Shaffer and V. Percec, *Makromol. Chem.* **187**, 111–123 (1986).
39. T. D. Shaffer and V. Percec, *Makromol. Chem.* **187**, 1431–1439 (1986).
40. L. H. Tagle, F. R. Diaz, and A. M. Munoz, *Polym. Bull.* **11**, 493–496 (1984).
41. P. Keller, *Makromol. Chem. Rapid Commun.* **6**, 255–259 (1985).
42. P. H. Parsania, P. P. Shah, K. C. Patel, and R. D. Patel, *J. Macromol. Sci., Chem.* **A22**, 1495–1508 (1985).
43. M. A. Shimizu, M. A. Kakimoto, and Y. Imai, *J. Polym. Sci., Polym. Chem. Ed.* **25**, 2385–2393 (1987).
44. G. R. Pettit and G. R. Pettit III, *J. Chem. Educ.* **55**, 472 (1978).
45. G. R. Pettit, *J. Polym. Sci., Polym. Chem. Ed.* **18**, 345–347 (1980).
46. Y. Imai, A. Kato, M. Ii, and M. Ueda, *J. Polym. Sci., Polym. Lett. Ed.* **17**, 579–583 (1979).
47. Y. Imai and M. Ueda, in *Crown Ethers and Phase Transfer Catalysis in Polymer Science,*

ed. by L. J. Mathias, and C. E. Carraher Jr. (Plenum Press, New York, 1984), pp. 121–137.
48. M. Ueda, Y. Oishi, N. Sakai, and Y. Imai, *Macromolecules* **15**, 248–251 (1982).
49. T. D. Shaffer and M. C. Kramer, *Makromol. Chem.* **191**, 71–79 (1990).
50. T. D. Shaffer and V. Percec, *J. Polym. Sci., Polym. Chem. Ed.* **24**, 451–467 (1986).
51. H. Ben Romdhane, Thèse (Université Pierre et Marie Curie, Paris, 1989).
52. Y. Imai, M. Ueda, M. Komatsu, and H. Urushibata, *Makromol. Chem. Rapid. Commun.* **1**, 681–685 (1980).
53. M. Ueda, N. Sakai, M. Komatsu, and Y. Imai, *Makromol. Chem.* **183**, 65–72 (1982).
54. L. H. Tagle, F. R. Diaz, M. P. de la Maza, and J. C. Vega, *J. Polym. Sci., Polym. Chem. Ed.* **24**, 495–501 (1986).
55. D. H. Suh, J. C. Won, D. K. Kim, and J. C. Jung, *J. Polym. Sci., Polym. Lett. Ed.* **26**, 83–87 (1988).
56. K. Soga, S. Hosoda, and S. Ikeda, *J. Polym. Sci., Polym. Lett. Ed.* **15**, 611–614 (1977); *J. Polym. Sci., Polym. Chem. Ed.* **17**, 517–529 (1979).
57. K. Soga, Y. Toshida, S. Hosoda, and S. Ikeda, *Makromol. Chem. Rapid Commun.* **178**, 2747–2751 (1977); *Makromol. Chem.* **179**, 2379–2386 (1979).
58. M. Majdoub, Thèse Doctorat 3° cycle (Paris, 1985).
59. S. Boileau, F. Méchin, K. Sienicki, and M. A. Winnik, *Eur. Polym. J.* **24**, 307–311 (1988).
60. S. Boileau, F. Méchin, J. M. G. Martinho, and M. A. Winnik, *Macromolecules* **22**, 215–220 (1989).
61. J. A. Mikroyannidis, *Eur. Polym. J.* **21**, 895–898, 1031–1034 (1985).
62. L. H. Tagle and F. R. Diaz, *Eur. Polym. J.* **23**, 109–112 (1987).
63. G. Rokicki, W. Kuran, and J. Kielkiewicz, *Makromol. Chem. Rapid Commun.* **180**, 2779–2782 (1979).
64. K. Soga, Y. Toshida, I. Hattori, K. Nagata, and S. Ikeda, *Makromol. Chem.* **181**, 979–984 (1980).
65. L. H. Tagle, F. R. Diaz, J. C. Vega, and P. F. Alquinta, *Makromol. Chem.* **186**, 915–921 (1985).
66. L. H. Tagle, F. R. Diaz, and P. E. Riveros, *Polym. J.* **18**, 501–504 (1986).
67. G. Rokicki, W. Kuran, and J. Kielkiewicz, *J. Polym. Sci., Polym. Chem. Ed.* **20**, 967–976 (1982).
68. G. Rokicki, J. Pawlicki, and W. Kuran, *Polym. J.* **17**, 509–516 (1985).
69. G. Rokicki, B. Pogorzelska-Marciniak, and W. Kuran, *Polym. J.* **14**, 847–854 (1982).
70. F. Méchin, Thèse (Université Pierre et Marie Curie, Paris, 1989).
71. L. H. Tagle, F. R. Diaz, and N. Valdebenito, *Polym. Bull.* **18**, 479–483 (1987).
72. L. H. Tagle, F. R. Diaz, and M. P. de la Maza, *Polym. Bull.* **18**, 485–486 (1987).
73. L. H. Tagle, F. R. Diaz, and P. Salas, *J. Macromol. Sci., Chem.* **A26**, 1321–1334 (1989).
74. L. H. Tagle, F. R. Diaz, and P. Salas, *Eur. Polym. J.* **25**, 549–551 (1989).
75. F. M. Houlihan, F. Bouchard, J. M. F. Fréchet, and C. G. Willson, *Macromolecules* **19**, 13–19 (1986).
76. J. M. J. Fréchet, F. M. Houlihan, F. Bouchard, B. Kryczka, and C. G. Willson, *J. Chem. soc., Chem. Commun.* 1514–1516 (1985).
77. J. M. J. Fréchet, F. Bouchard, F. M. Houlihan, E. Eichler, B. Kryczka, and C. G. Willson, *Makromol. Chem. Rapid Commun.* **7**, 121–126 (1986).
78. J. M. J. Fréchet, F. Bouchard, E. Eichler, F. M. Houlihan, T. Iizawa, B. Kryczka, and C. G. Willson, *Polym. J.* **19**, 31–49 (1987).
79. J. S. Riffle, R. G. Freelin, A. K. Banthia, and J. E. McGrath, *J. Macromol. Sci., Chem.* **A15**, 967–998 (1981).
80. F. L. Keshan, R. G. Freelin, J. S. Riffle, I. Yilgor, and J. E. McGrath, *J. Polym. Sci., Polym. Chem. Ed.* **22**, 679–704 (1984).

81. Z. K. Brzozowski, J. Dubczynski, and J. Petrus, *J. Macromol. Chem.* **A13**, 875–886 (1979).
82. Z. K. Brzozowski, J. Petrus, and J. Dubczynski, *J. Makromol. Chem.* **A13**, 887–897 (1979).
83. A. K. Banthia, D. C. Webster, and J. E. McGrath, *Org. Coatings and Plastics Chem.* **42**, 127–133 (1980).
84. Y. Imai and S. Tassavori, *J. Polym. Sci., Polym. Chem. Ed.* **22**, 1319–1325 (1984).
85. C. P. Yang, Y. Oishi, M. A. Kakimoto, and Y. Imai, *J. Polym. Sci., Polym. Chem. Ed.* **28**, 1353–1359 (1990) and references cited therein.
86. H. B. Tsai and Y. D. Lee, *J. Polym. Sci., Polym. Chem. Ed.* **25**, 3405–3412 (1987) and references cited therein.
87. L. Makaruk, H. Polanska, E. Ksiezakowska, and B. Wazynska, *Polym. J.* **17**, 1055–1064 (1985).
88. S. J. Havens and V. L. Bell, *J. Polym. Sci., Polym. Chem. Ed.* **24**, 901–910 (1986).
89. Y. Ozcayir and A. Blumstein, *J. Polym. Sci., Polym. Chem. Ed.* **24**, 1217–1225 (1986).
90. J. E. Mulvaney, F. Figueroa, and S. J. Wu, *J. Polym. Sci., Polym. Chem. Ed.* **24**, 613–620 (1986).
91. F. Akutsu, S. Yoshioka, Y. Suzuki, K. Naruchi, M. Miura, and K. Nagakubo, *Makromol. Chem. Rapid Commun.* **8**, 113–117 (1987).
92. J. I. Jin, E. J. Choi, and B. W. Jo, *Macromolecules* **20**, 934–939 (1987).
93. H. R. Kricheldorf and G. Schwarz, *Makromol. Chem.* **188**, 1281–1294 (1987).
94. W. Podkoscielny and D. Wdowicka, *J. Appl. Polym. Sci.* **35**, 1779–1789 (1988).
95. G. G. Cameron and K. S. Law, *Polymer* **22**, 272–273 (1981).
96. J. Kielkiewicz, W. Kuran, and B. Pogorzelska, *Makromol. Chem. Rapid Commun.* **2**, 255–260 (1981).
97. H. Sleiman, Thèse Doctorat 3° cycle (Paris, 1985).
98. G. G. Cameron, G. M. Buchan, and K. S. Law, *Polymer* **22**, 558–560 (1981).
99. G. C. East and M. Morshed, *Polymer* **23**, 168–170, 1555–1557 (1982).
100. N. Lacoudre, A. Leborgne, M. Sépulchre, N. Spassky, J. Djonlagic, and M. S. Jacovic, *Makromol. Chem.* **187**, 341–350 (1986).
101. W. Podkoscielny and S. Szubinska, *J. Appl. Polym. Sci.* **35**, 1853–1861 (1988).
102. Y. Imai, M. Ueda, and M. Ii, *Makromol. Chem.* **179**, 2085–2087 (1978).
103. M. Podgorski and W. Podkoscielny, *J. Appl. Polym. Sci.* **34**, 639–650 (1987).
104. M. Sato, H. Kondo, and M. Yokoyama, *Makromol. Chem. Rapid Commun.* **3**, 821–824 (1982).
105. M. Sato and M. Yokoyama, *Makromol. Chem.* **185**, 629–634 (1984).
106. M. Sato, *Makromol. Chem. Rapid Commun.* **5**, 151–155 (1984).
107. Y. Imai and H. Kamata, *Kobunshi Ronbunshu,* **40**, 165–168 (1983).
108. Y. Imai, N. Sato, and M. Ueda, *Makromol. Chem. Rapid Commun.* **1**, 419–422 (1980).
109. Y. Imai, H. Kamata, and M. Kakimoto, *J. Polym. Sci., Polym. Chem. Ed.* **22**, 1259–1265 (1984).
110. Y. Imai, T. Q. N'Guyen, and M. Ueda, *J. Polym. Sci., Polym. Lett. Ed.* **19**, 205–210 (1981).
111. Y. Imai, A. Kameyana, T. Q. N'Guyen, and M. Ueda, *J. Polym. Sci., Polym. Chem. Ed.* **19**, 2997–3003 (1981).
112. M. Ueda, K. Sato, and M. Sato, *Makromol. Chem. Rapid Commun.* **4**, 613–615 (1983).
113. I. K. Ahmed, J. K. Jawad, and M. A. M. Rashied, *Eur. Polym. J.* **23**, 163–166 (1987).
114. S. Gauthier and D. J. Worsfold, *Macromolecules* **22**, 2213–2218 (1989).

7

Activated Polycondensation

Fukuji Higashi

7.1. INTRODUCTION

Condensation polymers such as polyamides and polyesters have usually been prepared by reactions between diacid chlorides, diamines, and diols with the aid of low-temperature solution or interfacial methods, and by ester exchange reaction of diol acetates with diacids at high temperatures under reduced pressures. In these reactions activated monomers such as diacid chlorides should be prepared and purified before use. Unmodified monomers can be used for 6,6-nylon, which is obtained by melt polymerization of the nylon salt from the diacid and the diamine. However, this is not applicable to the preparation of the polymers which have melting points higher than the decomposition temperatures.

We shall describe polymerizations in which unmodified monomers can be activated *in situ* during polymerization by condensing agents derived from phosphorus and sulfur compounds. The polycondensation is called "direct polymerization," because unmodified monomers can be used in the reaction. The reaction easily produces high-molecular-weight polymers simply by mixing monomers with the activating agents followed by reaction at moderate temperature for several hours (see Section 7.2).

In this way polyamides, polyureas, and polyesters can be prepared under mild conditions (see Section 7.3).

Fukuji Higashi • Department of Polymer Engineering, Tokyo University of Agriculture and Technology, Koganei-shi, Tokyo 184, Japan.

New Methods for Polymer Synthesis, edited by W. J. Mijs. Plenum Press, New York, 1992.

7.2. ACTIVATING AGENTS AND MECHANISM OF ACTIVATION

High-energy phosphate compounds are involved very effectively in many chemical reactions in living cells. Some of them are of biochemical interest and can be divided into three types, according to their active structure element as shown in Table 7.1. They are used in energy-requiring reactions, transferring a phosphoryl group to substrates, and activating them for subsequent reactions. The reactions promoted by the phosphates proceed smoothly, selectively, and completely under mild conditions.

If we could build up an organic reaction system similar to the biological ones mentioned above, we could develop a widely applicable direct process that could give polymers readily and effectively from unmodified monomers as seen in protein and polysaccharide syntheses from amino acids and glucose, respectively.

There have been attempts to develop effective condensing agents like ATP, and most of them that have been found useful for the direct process can be represented by three active intermediates of the P—O—P type, the N—P type, and the C—O—P type structures, according to those in Table 7.1. They were usually formed by effective reactions such as oxidation, cleavages of anhydride bonds in P_2O_5, of ester bounds in phosphites, and phosphorus–halogen bonds in phosphorus halides, resulting in the three structure elements mentioned above.

Similarly, some sulfur compounds have also been used as condensing agent equally effective as phosphorus compounds (N—S type and C—O—S type intermediates).

TABLE 7.1. Some of the Phosphate Compounds of Biological Interest

Type of intermediate	Active structure	Compound
P—O—P	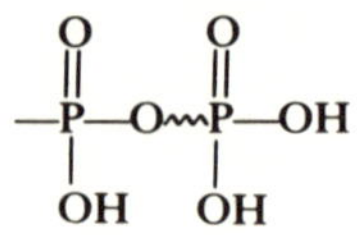	ATP
N—P	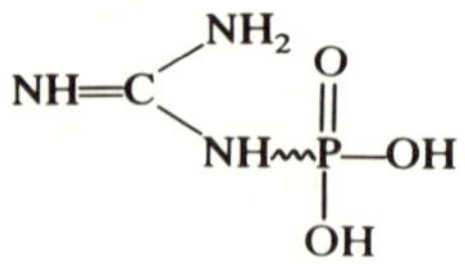	Phosphocreative Phosphoarginine
C—O—P	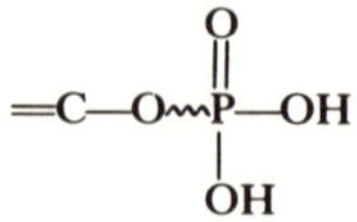	Phosphoenolpyruvate

7.2.1. *Phosphorus Compounds*

7.2.1.1. *The P—O—P Type Intermediate*

A most typical condensing agent of this type is polyphosphoric acid (PPA). PPA is prepared by the reaction of phosphorus pentoxide (P_2O_5) with phosphoric acid, and is commercially available. As it is also a good solvent for condensation polymers, the reactions are usually carried out in PPA. In the reaction,[1] PPA was found to react like ATP with carboxyl groups via P—O—P fission to form the acylphosphate **1** followed by subsequent reactions such as aminolysis to give an amide [Eq. (7.1)].

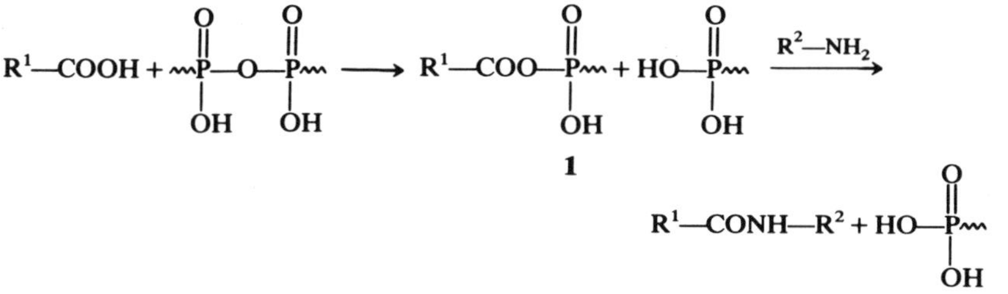

(7.1)

A PPA derivative, polyphosphoric ester (PPE), which is prepared by the reaction of P_2O_5 with ether, has also been used as a condensing agent as well as a solvent in successful polymerizations of nucleotides, peptides, and saccharides.[2] A reaction mixture of P_2O_5 and methanesulfonic acid was found very effective as a condensing agent and as a solvent similar to PPA, and used for amide formation [Eq. (7.2)].[3] In addition, the mixture activated a carboxylic acid highly enough to acylate aromatic compounds to give ketones [Eq. (7.3)]. The key reaction intermediate in these reactions had been proposed to be a mixed anhydride **2** (PMMA) of the phosphoric acid and the sulfonic acid, which reacts with a carboxylic acid to give a carboxylic sulfonic anhydride **3** with a high acylating power.

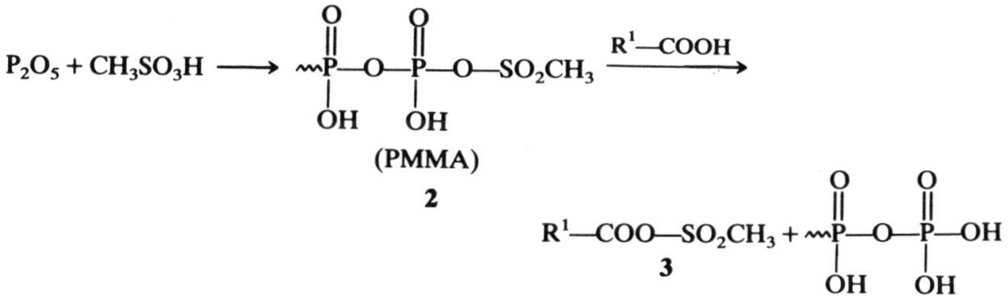

$$\mathbf{3} + R^2\text{—}NH_2 \longrightarrow R^1\text{—}CONH\text{—}R^2 + CH_3SO_3H \qquad (7.2)$$

$$\mathbf{3} + R^3\text{—}H \longrightarrow R^1\text{—}CO\text{—}R^3 + CH_3SO_3H \qquad (7.3)$$

7.2.1.2. *The N—P Type Intermediate*

When triphenylphosphine is oxidized with polyhalogenides such as hexachloroethane in the presence of tertiary amines like pyridine, triphenylphosphine dichloride **4** is initially formed. Then with pyridine the P—N type intermediate, pyridine-*N*-phosphonium salt **5**, is formed. The *N*-phosphonium salt reacts with a carboxylic acid to give **6**, which could yield an amide by aminolysis [Eq. (7.4)].[4]

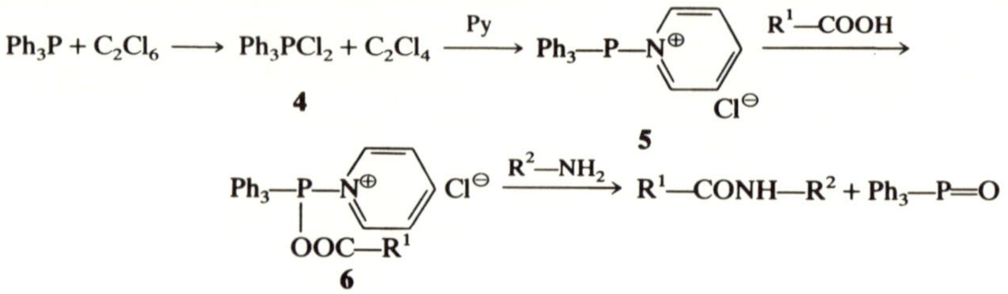

(7.4)

The resulting phosphine oxide can regenerate **4** by the reaction with oxalyl chloride [Eq. (7.5)], and the reaction was applied to the ester synthesis.[5]

$$Ph_3P{=}O + (COCl)_2 \rightarrow \mathbf{4} + CO + CO_2 \qquad (7.5)$$

Another N—P type condensing agent could be prepared by an acid-promoted cleavage of P—O—C bonds in aryl phosphites [$P(OAr)_3$, HO—$P(OAr)_2$] and diaryl alkylphosphonates [R—P(O)$(OAr)_2$].[6] For example, triphenyl phosphite reacted with a carboxylic acid in the presence of tertiary amines like pyridine to form the *N*-phosphonium salt **7**, which gave the corresponding amide and ester by aminolysis and alcoholysis, respectively [Eq. (7.6)]. However, phenyl esters can be formed from **7** in an intramolecular esterification in the absence of nucleophile [Eq. (7.7)]. This side reaction might accompany the amidation and esterification to some

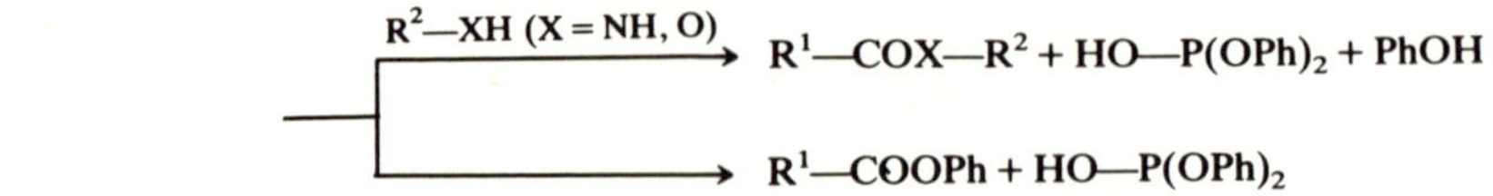

(7.6)

(7.7)

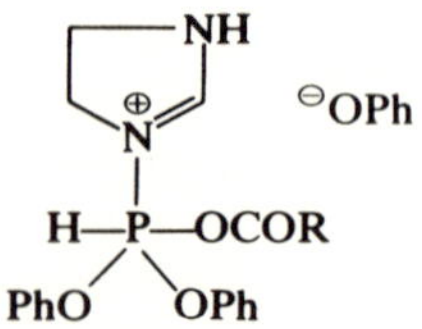

8

extent. The intramolecular phenyl esterification in **7** was affected by the type of tertiary amine used. The *N*-phosphonium salt **7** derived from pyridine was so stable that it did not easily give the phenyl ester. On the other hand, the salt **8** derived from imidazole was less stable and gave rise to the ester in a short time.[7]

The reaction mechanism involves protonation of the phosphite by a carboxylic acid to form **9**, which is transformed by pyridine into **10** and **11**. The *N*-phosphonium salt **11** reacts with the carboxylate anion to give **7** [Eq. (7.8)].

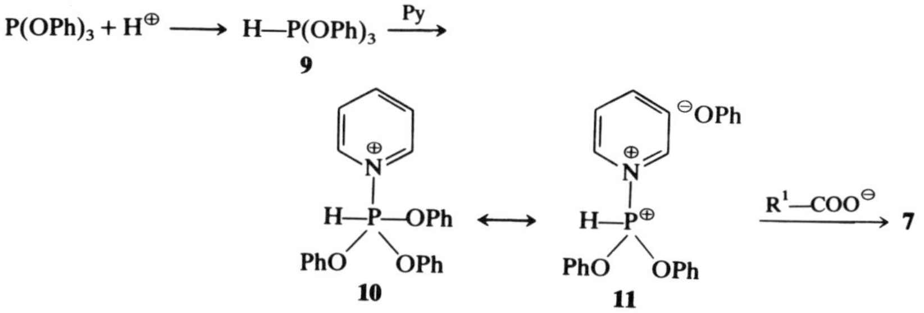

(7.8)

The weakly acidic carbon dioxide does not react with aromatic amines. However, when a catalytic amount of tertiary amine hydrochloride or amine hydrochloride was added, the reaction easily occurred to yield the corresponding ureas quantitatively [Eq. (7.9)].[8] On the other hand, the reaction with diphenyl phosphite instead of triphenyl phosphite proceeded readily even in the absence of the hydrochlorides to give quantitative yield of ureas from carbon dioxide and amines,[9] because the phosphite has a labile proton, giving rise to keto–enol tautomerization [Eq. (7.10)].

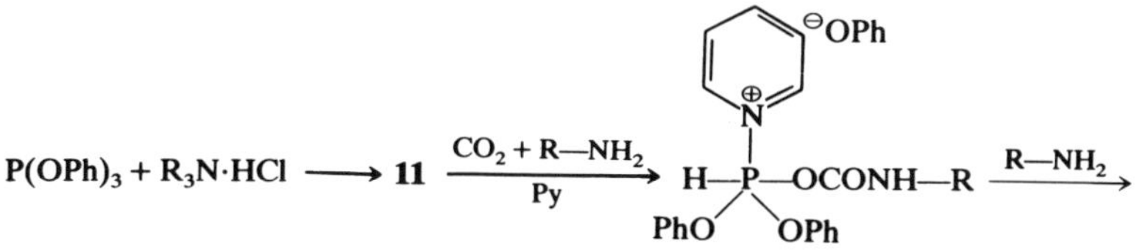

$$R\text{—}NHCONH\text{—}R + HO\text{—}P(OPh)_2 + PhOH \qquad (7.9)$$

$$\underset{\text{enol}}{HO\text{—}P(OPh)_2} \longleftrightarrow \underset{\text{keto}}{O{=}\overset{\displaystyle H}{\overset{|}{P}}\text{—}(OPh)_2} \qquad (7.10)$$

Phosphorus polyhalides such as phosphorus oxychloride have not been considered to be effective condensing agents, because they have highly reactive halogens which can cause many undesirable side reactions. When their reactivities were controlled by converting them into the N—P type intermediates, they could be effective activating agents.

Phosphorus oxychloride in pyridine formed the *N*-phosphonium salt **12**, which reacted bifunctionally with two moles of a carboxylic acid to give **14**.

Intermediate **14** reacted with two moles of an amine and a phenol to afford an amide and an ester, respectively [Eq. (7.11)].[10] The reaction with a carboxylic acid seemed to proceed sequentially: first the monoester **13** was formed, and then the diester **14**. There is a difference in reactivity between two carboxyl groups in **14**, which were aminolyzed stepwise via **15**. These stepwise reactions could be successfully used to control the sequence of the diacid in the copolyester as mentioned later.

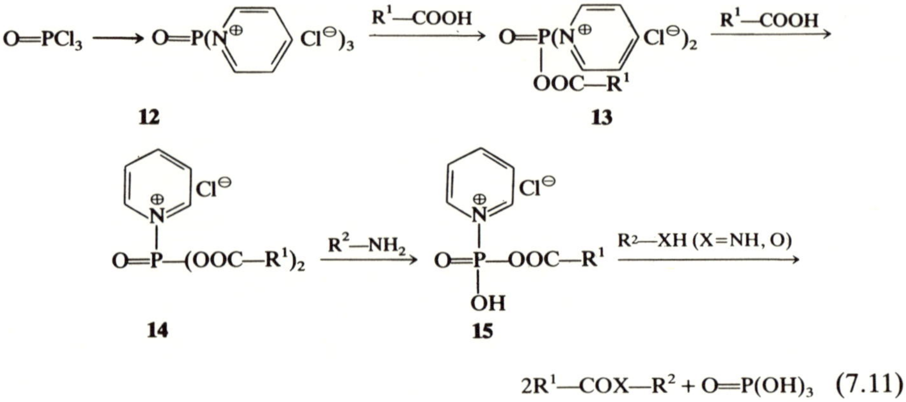

$2R^1\text{—}COX\text{—}R^2 + O{=}P(OH)_3$ (7.11)

Similarly, phenylphosphonic dichloride reacted in pyridine bifunctionally with two moles of a carboxylic acid to give a quantitative yield of the amide by reaction with two moles of an amine [Eq. (7.12)].[11] However, one of the two carboxyl groups in **17** could not react with a phenol to give the ester, because the carboxyl group in **18** is not reactive enough to undergo alcoholysis.

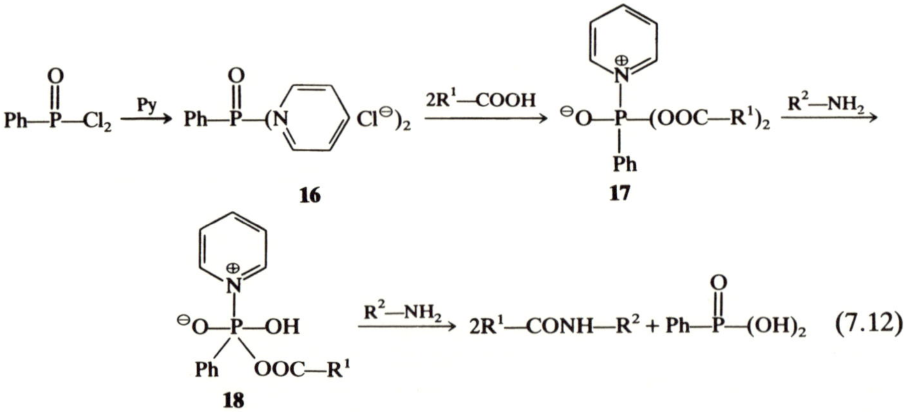

The N-phosphonium salt **19** derived from diphenyl chlorophosphate in pyridine could react monofunctionally with an equimolar amount of a carboxylic acid through the reactive intermediate **20** to yield amides and esters as shown in Eq. (7.13).[12]

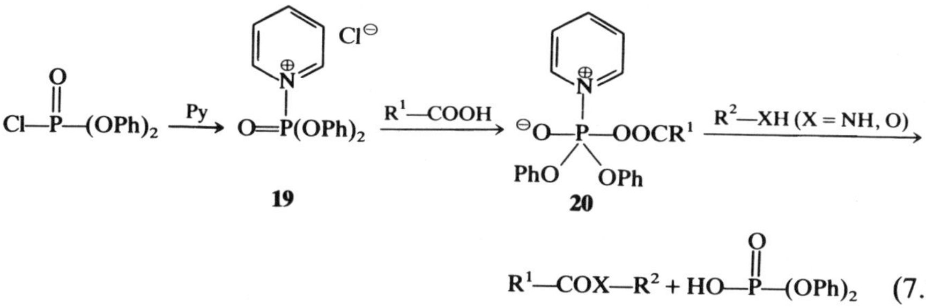

(7.13)

7.2.1.3. C—O—P Type Intermediate

Diphenyl chlorophosphate reacts with amides such as *N,N*-dimethylformamide (DMF) in pyridine to form a Vilsmeier adduct **21**. The C—O—P type intermediate could easily react with an equimolar amount of a carboxylic acid probably via **22**, which can yield an amide and an ester by aminolysis and alcoholysis, respectively [Eq. (7.14)].[13]

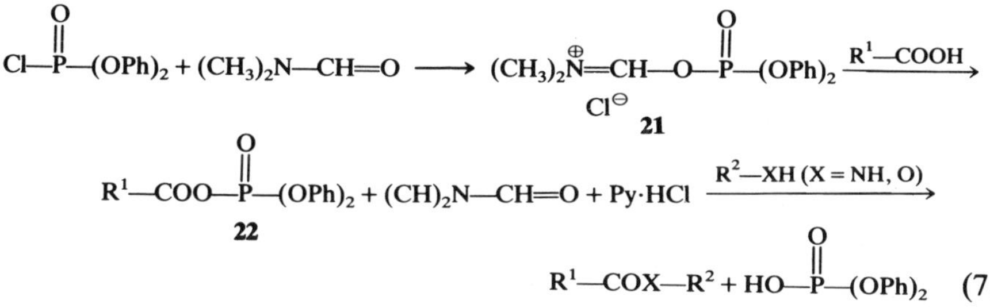

(7.14)

Triphenyl phosphite could also give an intermediate **23** similar to **21**, when the reaction of the phosphite and a carboxylic acid is carried out in aprotic amide solvents such as *N*-methylpyrrolidone (NMP) in the presence of metal salts like LiCl.[14] similar to the reaction of carbon dioxide and an amine promoted by amine hydrochloride as shown in Eq. (7.9), the reaction was considered to be initiated by complex formation with the metal ion followed by reaction with a carboxylate anion to yield **23**, which gives an amide by aminolysis [Eq. (7.15)].

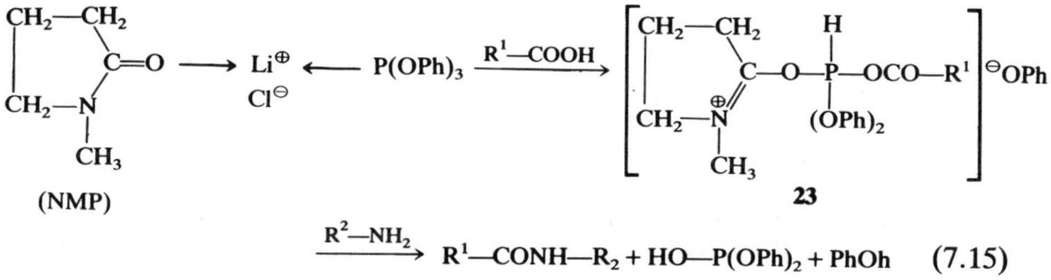

(7.15)

7.2.2. Sulfur Compounds

7.2.2.1. The N—S and C—O—S Type Intermediates

The intermediates could be derived from both thionyl chloride and arylsulfonyl chloride. It is well known that thionyl chloride reacts with a carboxylic acid to give an acid chloride, when excess thionyl chloride is used as solvent. However, no investigation has been done into the reaction stoichiometry particularly in the presence of tertiary amines like pyridine.

Thionyl chloride reacted with pyridine to form the *N*-sulfinium salt (N—S type intermediate), **24**. We found that **24** reacted with a carboxylic acid to yield **25** instead of an acyl chloride. Then, **25** gives an amide and an ester by aminolysis and alcoholysis, respectively [Eq. (7.16)].[15]

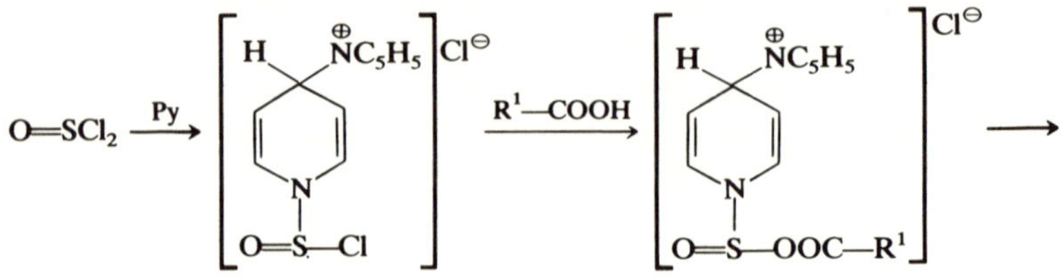

24 **25**

$$R^1\text{—CONH—}R_2 + SO_2 + HCl \qquad (7.16)$$

When reacted with amides such as *N*-methylpyrrolidone (NMP), thionyl chloride also forms a Vilsmeier adduct (C—O—S type intermediate) **26**, which could be used for activation of a carboxylic acid toward aminolysis and alcoholysis via **27** to produce an amide in nearly quantitative yield, but an ester in low yield [Eq. (7.17)]. Again, an acyl chloride is not a true intermediate in this reaction.[16]

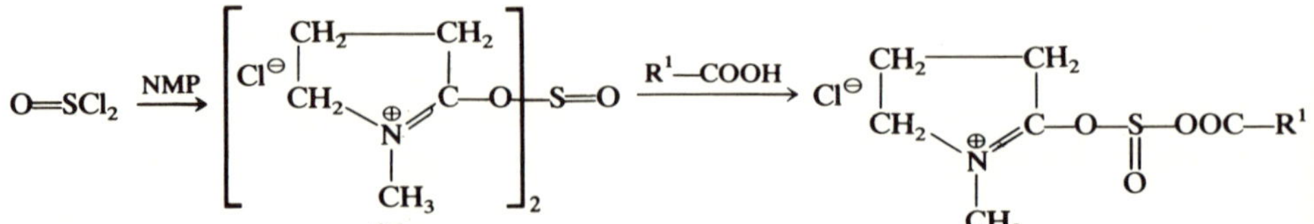

26 **27**

$$\xrightarrow{R^2\text{—XH (X = NH, O)}} R^1\text{—COX—}R^2 + SO_2 + HCl \qquad (7.17)$$

Similarly, arylsulfonyl chlorides give the N—S and C—O—P type intermediates, **28** and **30**, which were successfully used for the preparation of the ester as shown in Eqs. (7.18)[17] and (7.19).[18] However, amidation was not satisfactory due to a side reaction producing a sulfonyl amide.

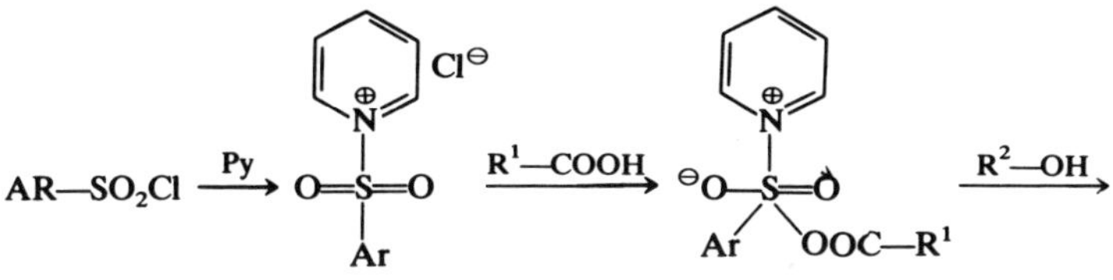

28 **29**

$$R^1\text{—COO—}R^2 + ArSO_3H + HCl \qquad (7.18)$$

$$Ar\text{—}SO_2Cl \xrightarrow{DMF} (CH_3)_2\overset{\oplus}{N}\text{=CH—O—}SO_2\text{—Ar}\ Cl^{\ominus}\ (\mathbf{30}) \xrightarrow{R^1\text{—COOH}}$$

$$[R^1\text{—COO—}SO_2\text{—Ar}]\ (\mathbf{31}) \xrightarrow{R^2\text{—OH}} R^1\text{—COO—}R^2 + ArSO_3H + HCl \qquad (7.19)$$

7.2.3. Other Compounds

Silicon tetrachloride reacted with a mixture of a carboxylic acid and an amine to form an intermediate **32**, which gave an amide in a high yield on heating [Eq. (7.20)].[19]

$$SiCl_4 + 2R^1\text{—COOH} + 2R^2\text{—}NH_2 \rightarrow (R^1\text{—COO})_n\text{—Si—}(NH\text{—}R^2)_m \quad \mathbf{32}\ (n + m = 4)$$

$$\rightarrow 2R^1\text{—CONH—}R^2 + SiO_2 + 4HCl \qquad (7.20)$$

An active ester intermediate **33** was obtained *in situ* by the reaction of a carboxylic acid and aromatic halo compounds such as picryl chloride in the presence of tertiary amines, and used for the preparation of an amide followed by aminolysis [Eq. (7.21)].[20]

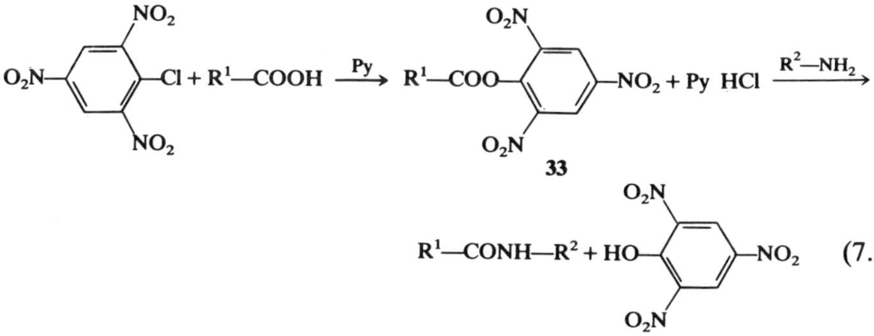

(7.21)

7.3. SCOPE AND LIMITATION OF THE POLYCONDENSATION USING ACTIVATING AGENTS

7.3.1. Polyamidation

7.3.1.1. Polyamides

Polyphosphoric acid (PPA) has been widely used as activating agent and excellent solvent for the preparation of heterocylic ladder polymers which are difficult to dissolve in ordinary solvents. Such reactions in PPA have been extensively reviewed by Iwakura.[21]

An intermediate (**2**, PPMA) derived from P_2O_5 and methane sulfonic acid promoted the reaction of dicarboxylic acid and diamines to produce polyamides, and was applied to the preparation of various heterocyclic polymers [Eq. (7.22)].[22–24] PPMA reacted with a dicarboxylic acid and a sulfonic acid to give acylating agents powerful enough to acylate an aromatic compound to polyketone **23**[25] and polysulfone **24**[26] (Eqs. (7.23) and (7.24)].

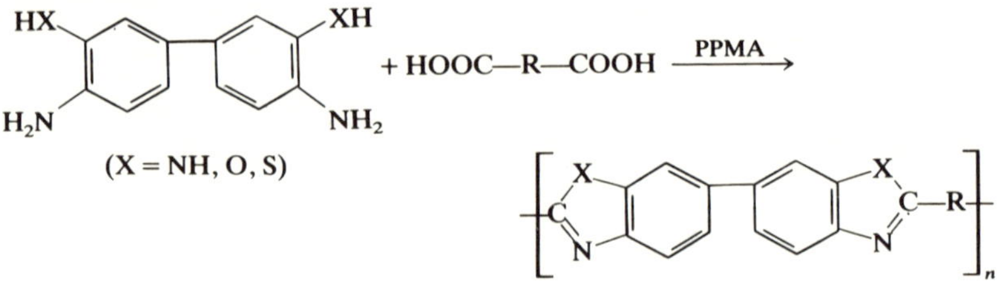

(7.22)

$\eta_{inh} = 1.0\text{–}4.7$ (H_2SO_4)

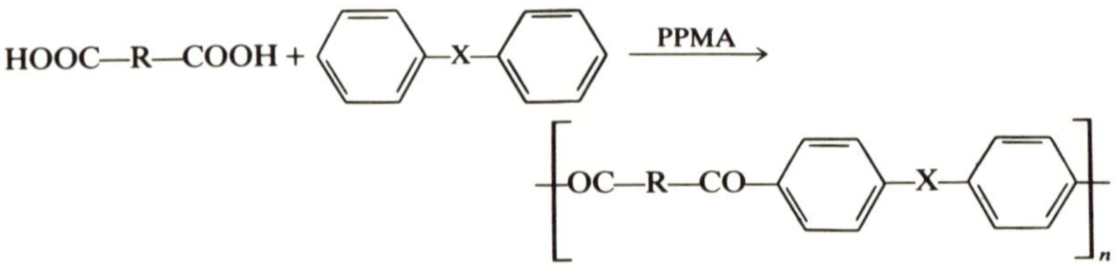

(7.23)

$\eta_{inh} = 0.5\text{–}1.1$ (H_2SO_4)

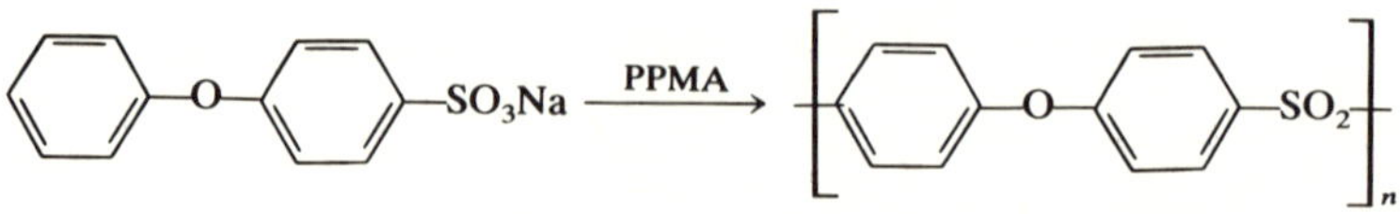

(7.24)

$\eta_{inh} = 0.85$ (DMAc)

The reaction product **4** prepared by oxidation of triphenyl phosphine with hexachloroethane was a very effective activating agent especially for the polycondensation of *p*-aminobenzoic acid in pyridine, and yielded poly(*p*-benzamide) with high molecular weight at room temperature in a short period [Eq. (7.25)].[4]

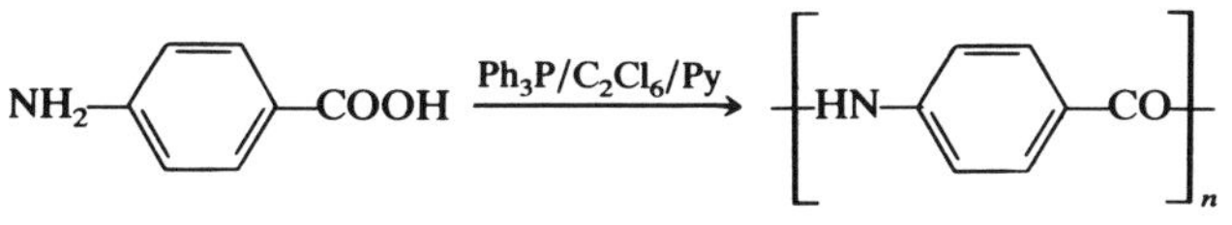

(7.25)

$\eta_{sp/c} = 5$
($M_n = 15{,}400$[27])

When a mixture of triphenyl phosphite, a dicarboxylic acid, and a diamine was heated in a mixed solvent of pyridine and *N*-methylpyrrolidone (NMP), the reaction proceeded via **7** according to Eq. (7.6), and polyamides from aromatic diamines were obtained in quantitative yields. Aliphatic diamines did not give satisfactory results. The reaction of aliphatic dicarboxylic acids and aromatic diamines proceeded homogeneously to give polymers with high molecular weight [Eq. (7.26)].[29] However, the reaction with aromatic dicarboxylic acids gave polymers of low molecular weight. The reaction was largely improved when carried out in the presence of salts such as LiCl and $CaCl_2$, which are known to promote the dissolution power of amide solvents [Eq. (7.27)].[30] The reaction has been applied to the preparation of various polymers[31] and copolymers.[32]

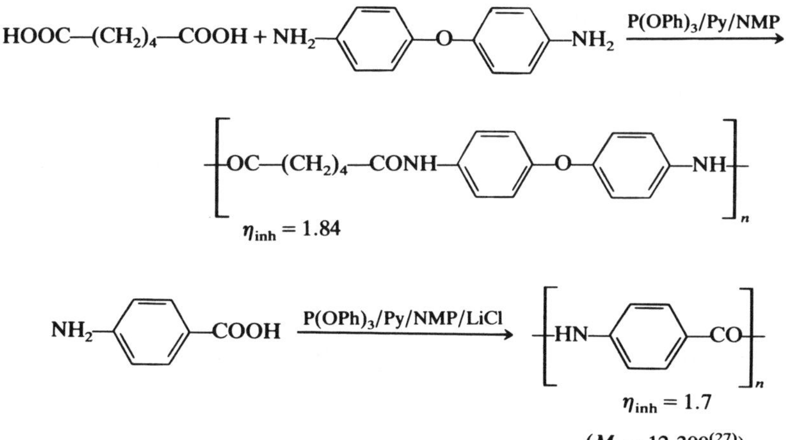

(7.26)

$\eta_{inh} = 1.84$

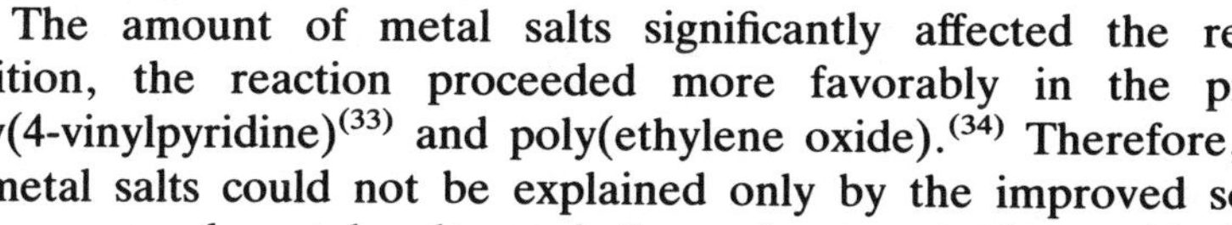

(7.27)

$\eta_{inh} = 1.7$
($M_n = 12{,}300$[27])

The amount of metal salts significantly affected the reaction. In addition, the reaction proceeded more favorably in the presence of poly(4-vinylpyridine)[33] and poly(ethylene oxide).[34] Therefore, the effect of metal salts could not be explained only by the improved solubility of polymers, and metal salts and the polymer matrices added might be involved in the reaction itself. Indeed, $CaCl_2$ capable of forming polycoordinate complexes like the well-known $CaCl_2 \cdot nNH_3$ (n = 2, 4, 8) may give similar complexes with pyridine and/or diamine, influencing the polycondensation favorably to give high-molecular-weight polyamide [Eq. (7.28)].[35] The reaction at a little higher temperature further improved the inherent viscosity to 6.2.[36]

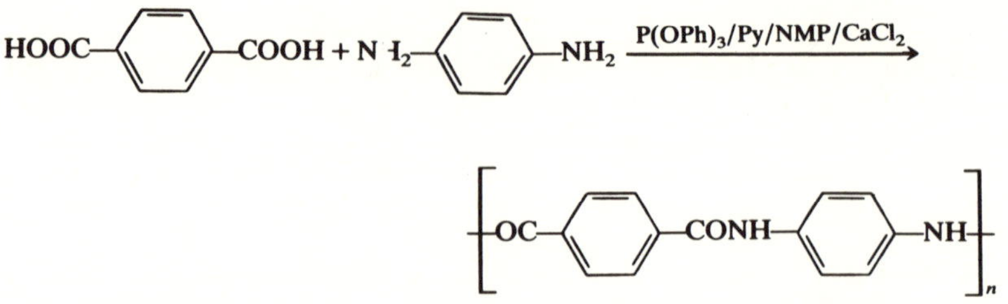

(7.28)

$\eta_{inh} = 4.5$ ($M_w = 17{,}300$[37])

The reaction was studied in more detail by modifying the rates (V_p, V_a) of forming **7** (phosphorylation) and of its aminolysis. These could be controlled by reaction conditions,[38] and by the rate of addition of a carboxylic acid in NMP to a preheated pyridine solution of a diamine and LiCl.[39] Faster addition of the acid solution was favorable for a dicarboxylic acid with lower acidity (smaller V_p), and also for a diamine with higher basicity (larger V_a). This means qualitatively that polycondensation can proceed satisfactorily to give high-molecular-weight polymers when it can satisfy Eq. (7.29).

$$V_a/V_p = (\text{const}) \tag{7.29}$$

This hypothesis appeared also to be valid in the preparation of high-molecular-weight poly(amide-hydrazide) from *p*-aminobenzhydrazide having a more basic hydrazide group. Excess of diphenyl phosphite, which is more reactive than triphenyl phosphite, was used to increase V_p [Eq. (7.30)].[40]

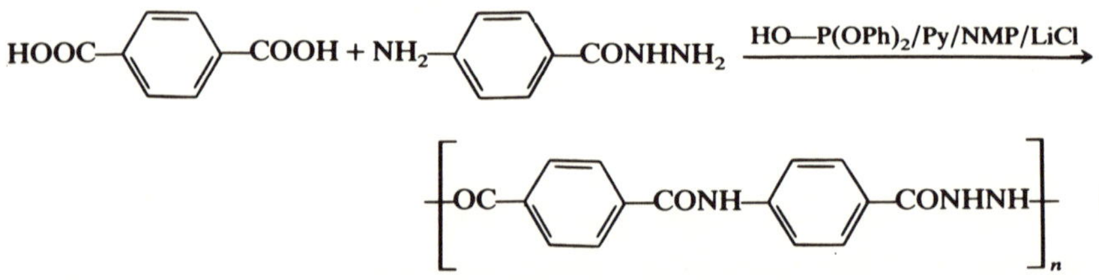

(7.30)

$\eta_{inh} = 4.7$

The reaction of triphenyl phosphite in NMP was promoted by metal salts instead of tertiary amines used according to Eq. (7.15), and utilized for the preparation of aromatic polyamides.[14] Addition of poly(*N*-vinylpyrrolidone) (PNVP) as a matrix for the metal salts in the reaction mixture increased the molecular weight of the resulting polymers.[41] Poly(α-amino acids) with high molecular weight were also obtained from unmodified α-amino acids, and their molecular weight was dependent on that of PNVP [Eq. (7.31)].[42]

$$\text{L-Leucine} \xrightarrow{\text{P(OPh)}_3\text{/LiCl/NMP/PNVP}} \text{poly(L-leucine)} \quad M_n = 35{,}000 \qquad (7.31)$$

The reaction in PNVP medium was very effective, but further modification should be necessary because isolation of the resulted polymers from PNVP was not always easy.

The Vilsmeier adduct **26** derived from thionyl chloride in NMP could activate dicarboxylic acids toward aminolysis with diamines according to Eq. (7.17), to yield polyamides of high molecular weight [Eq. (7.32)].[16]

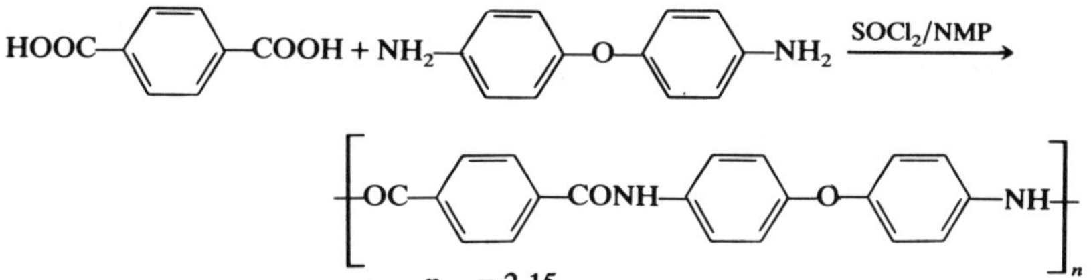

(7.32)

$\eta_{inh} = 2.15$

When carried out in NMP that contained dissolved $CaCl_2$, the reaction led to polyamide with a high inherent viscosity from terephthalic acid and *p*-phenylenediamine [Eq. (7.33)].[43] It is interesting to note that $CaCl_2$ dissolved in the NMP solution of the diamine rather than in the whole mixture promoted the reaction more significantly.

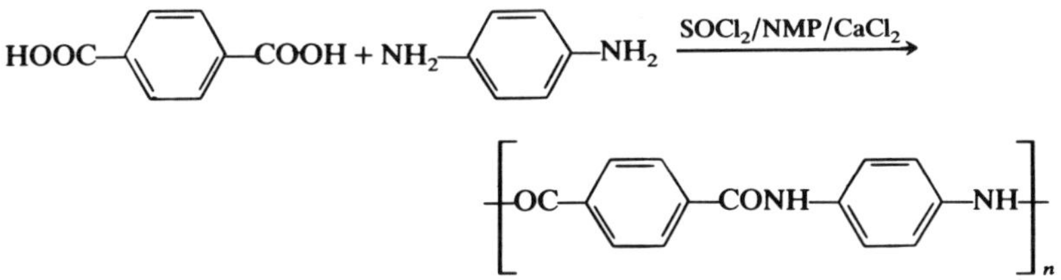

(7.33)

$\eta_{inh} = 3.8 \quad (M_w = 15{,}700$[37]$)$

When intermediate **12** prepared from phosphorus oxychloride and pyridine was reacted bifunctionally with an equivalent of dicarboxylic acid and then with diamines according to Eq. (7.11), polyamides of high inherent viscosity were obtained even from diamines with lower basicity such as 4,4′-diaminodiphenylsulfone [Eq. (7.34)].[10] A cyclic intermediate **34** has been proposed, which is aminolyzed by diamines to give linear polyamides without cycloamidation. The cyclization might be retarded, probably because of the difference in reactivity of two activated carboxyl groups on one phosphorus atom as mentioned before.

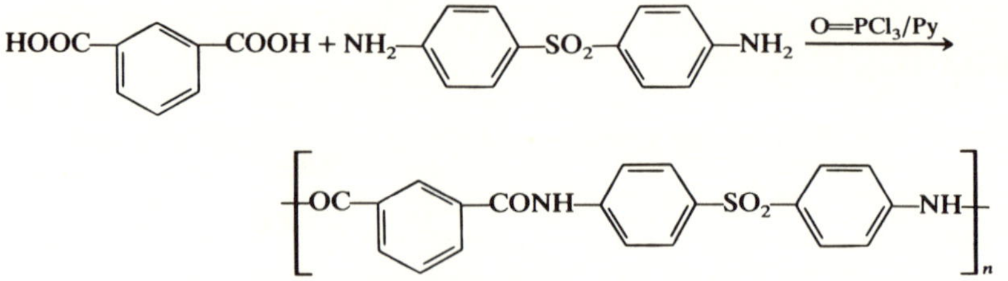

(7.34)

$\eta_{inh} = 2.24$

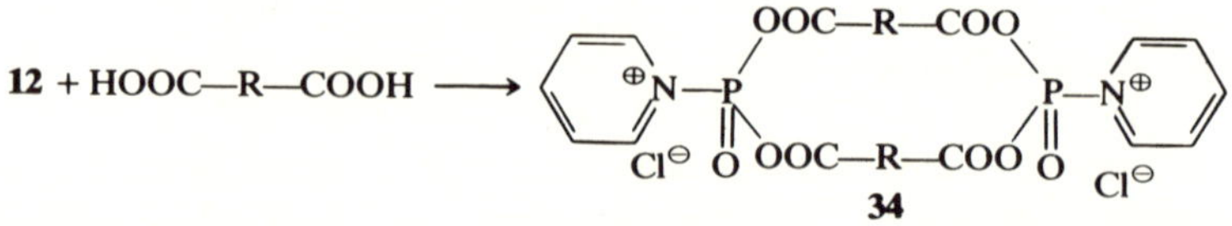

34

Phenylphosphonic dichloride in pyridine gave also linear polyamides of high inherent viscosities according to Eq. (7.11).[12] As shown in Table 7.2, these viscosity values are nearly the same or higher than those obtained by the conventional method using diacid chlorides. About 75 mol% of the *N*-phosphonium intermediate **16** with respect to a dicarboxylic acid should be used to obtain high-molecular-weight polymers, suggesting that bifunctional intermediates of possible structures of **36** and **37** and a monofunctional one **35** might be involved in the reaction.

TABLE 7.2. Polyisophthalamide Synthesis by Using **16** Derived from Phenylphosphonic Dichloride in Pyridine

$H_2N—Ar—NH_2$	η_{inh}[a]
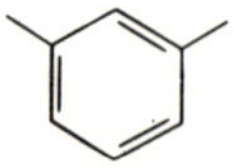	2.66 (2.47)
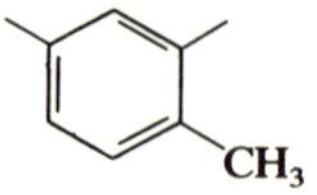	3.26 (2.50)
	3.24 (0.90)
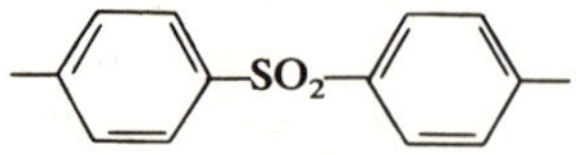	2.02 (1.53)
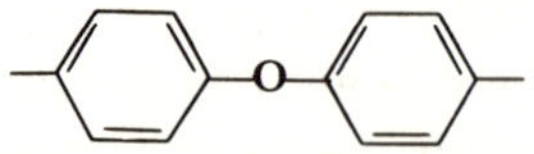	3.60 (1.20)

[a] Values in parentheses obtained by the conventional method using acid chloride.

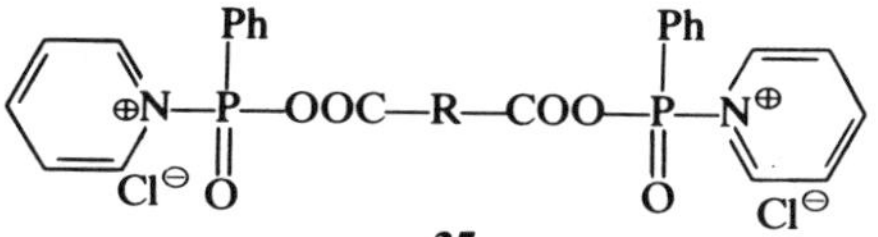

35

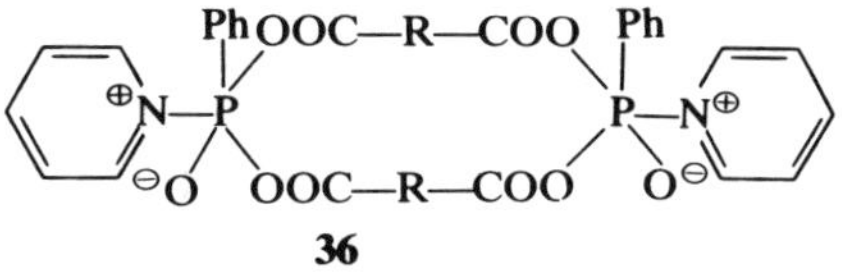

36

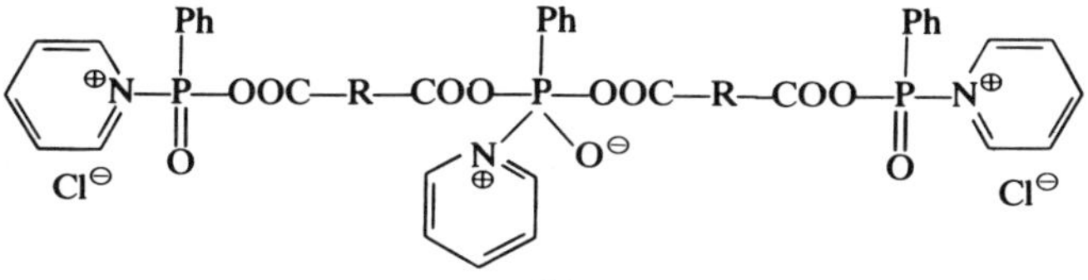

37

More than 75 mol% of **16** is needed in the case of diamines with lower basicity.[44] This is probably because one of two carboxyl groups linked to one phosphorus atom in **36** and **37** might be less reactive than the other, and more of the reactive one in **35** might be needed for the aminolysis reaction. On the other hand, the reaction of the diamines with a higher basicity gave also satisfactory results even with less than 75 mol%. Thus, the amount of **16** used might control the rate of reaction with diamines in agreement with the hypothetical equation [Eq. 7.29)].

Several condensing agents other than phosphorus and sulfur compounds have been used for polymer syntheses, and reactive intermediates **32** from silicon tetrachloride[19] and **33** from picryl chloride[20] gave polyamides according to Eqs. (7.20) and (7.21) [Eqs. (7.35) and (7.36)].

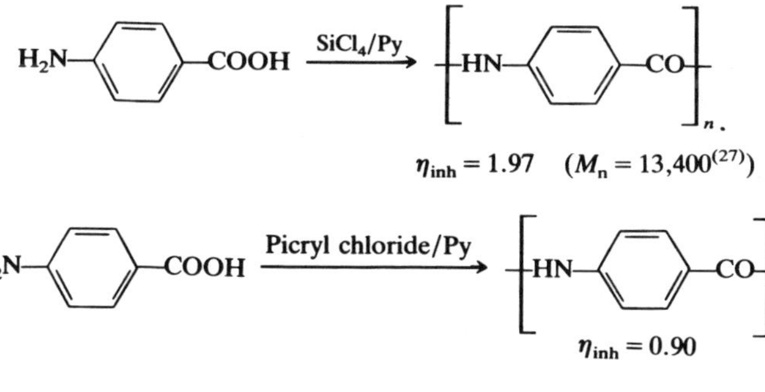

(7.35)

$\eta_{inh} = 1.97$ ($M_n = 13{,}400$[27])

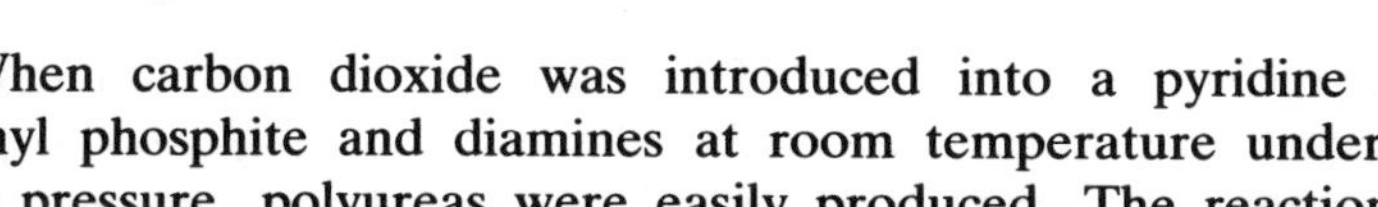

(7.36)

$\eta_{inh} = 0.90$

7.3.1.2. Polyureas

When carbon dioxide was introduced into a pyridine solution of diphenyl phosphite and diamines at room temperature under an atmospheric pressure, polyureas were easily produced. The reaction was more favorably carried out under pressure. Polymers of higher molecular weight were produced at around 20 atm [Eq. (7.37)], higher pressures not being effective.[9]

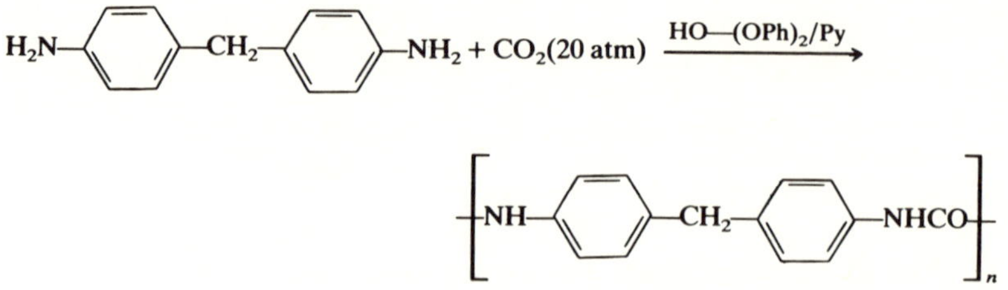

(7.37)

$\eta_{inh} = 2.2$

Triphenyl phosphite instead of diphenyl phosphite did not lead to polymerization in pyridine. However, a catalytic amount of hydrochloride from, e.g. triethylamine and pyridine promoted the reaction to give polyureas of molecular weights as high as obtained by the reaction with diphenyl phosphite [Eq. (7.38)].[8]

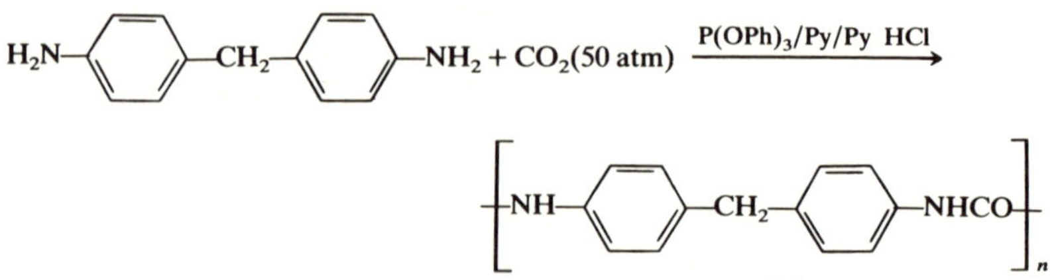

(7.38)

$\eta_{inh} = 2.6$

7.3.1.3. Typical Procedures

Preparation of Polyamides by Using Triphenyl Phosphite in the Presence of Metal Salts. A mixture of terephthalic (TPA) or isophthalic (IPA) (1.66 g, 0.01 mol), diamines (0.01 mol), and triphenyl phosphite (6.21 g, 0.02 mol) was heated at 100 °C for 3 h in a mixed solvent of NMP (20 mL) and pyridine (15 mL) containing LiCl (1.4 g, 4 wt%). Similarly, the polycondensation reaction of *p*-aminobenzoic acid (2.74 g, 0.02 mol) was carried out by using triphenyl phosphite (6.21 g, 0.02 mol). After cooling, the reaction mixture was poured into methanol (200 mL), and the precipitated polymer was separated by filtration, ground to a powder, washed well with methanol, and dried.

Preparation of Polyterephthalamides in NMP Solution Containing Dissolved $CaCl_2$ and Pyridine. A mixture of TPA (5 mmol, 0.831 g) and diamines (5 mmol) or of *p*-aminobenzoic acid (10 mmol, 1.371 g) and triphenyl phosphite (10 mmol, 3.10 g) in NMP (50 mL) that contained dissolved $CaCl_2$ (3 g) and pyridine (4.3 mL, $Py/CaCl_2$ = 2 mol/mol) was maintained at 100 °C for 4 h. The resulting gel was ground in methanol to a powder, separated by filtration, washed in boiling methanol, and dried. Similarly, the reaction of TPA (10 mmol, 1.66 g) and diamines (10 mmol) or of *p*-aminobenzoic acid (20 mmol, 2.74 g) and triphenyl phosphite (20 mmol, 6.20 g) was conducted at 100 °C for 4 h in NMP (50 mL) that contained $CaCl_2$ (3 g) and pyridine (4.3–13.0 mL, molar ratios to $CaCl_2$ = 2–6) or picoline (2.6–10.6 mL, molar ratios to $CaCl_2$ = 1–4).

Preparation of Polyisophthalamides by Using Phenylphosphonic Dichloride in Pyridine. To a pyridine (5 mL) solution of phenylphosphonic dichloride (1.50 g, 7.7 mmol) maintained at room temperature for 30 min, a pyridine (5 mL) solution of IPA (5 mmol) was added, and the reaction mixture was kept at room temperature for 20 min and then heated at 120 °C for 10 min. During aging at 120 °C precipitates initially formed disappeared, and a clear solution was obtained. To the hot solution, a pyridine (15 mL) solution of diamines (5 mmol) was added, and the mixture was further heated at 120 °C for 3 h. *m*-Phenylenediamine hydrochloride (MPD·2HCl), insoluble in pyridine, was added in solid followed by pyridine (15 mL). The resulting mixture was diluted with pyridine and then poured into methanol to precipitate polymers.

Preparation of Polyureas from Carbon Dioxide and Diamines by Using Triphenyl Phosphite. A mixture of triphenyl phosphite (10 mmol), a diamine (10 mmol), and pyridine hydrochloride (20 mmol) in a mixed solvent of pyridine (20 mL) and NMP (20 mL) was placed in a 100-mL autoclave, and then gaseous carbon dioxide was introduced. The reaction was carried out under an initial pressure of carbon dioxide (10–50 atm) at a given temperature for several hours with stirring. The pressure of carbon dioxide was allowed to drop during polycondensation. The reaction mixture was poured into methanol and the precipitated polymer was collected by filtration, washed in boiling methanol followed by filtration, and dried.

7.3.2. Polyesterifications

7.3.2.1. Polyesters

Polycondensation of aromatic diacids and dipehnols can be carried out by using reactive intermediate **4** divided from phosphine oxide and oxalyl chloride or phosgene in the presence of tertiary amines.[5]

The cyclic intermediate **34** in the polyamide synthesis prepared by the reaction of phosphorus oxychloride, pyridine, and a dicarboxylic acid was not effective in polyesterification probably because of the lower reactivity of one of two activated carboxyl groups on one phosphorus atom of **33**. However, when about an equivalent amount of LiCl with a diphenol in pyridine was added to **34**, polyesterification proceeded smoothly to give high-molecular-weight polyesters [Eq. (7.39)].[45] The salt should be used together with the diphenol, since complexation of a diphenol with LiCl in pyridine will enhance the reactivity of phenolic hydroxyl groups. A characteristic feature of the reaction with **34** will be described later in the preparation of sequential copolymers.

$$O{=}PCl_3 \xrightarrow{Py} \mathbf{12} \xrightarrow{R^1{-}(COOH)_2/Py} \mathbf{34} \xrightarrow{R^2{-}(OH)_2/LiCl/Py} {+}OC{-}R^1{-}COO{-}R^2{-}O{+}_n \qquad (7.39)$$

The reaction of diphenyl chlorophosphate in pyridine [Eq. (7.13)] has been applied to the preparation of polyesters from dicarboxylic acids and diphenols,[12] and polyesteramides from dicarboxylic acids and aminophenols,[46] dicarboxylic acids, diphenols, and diamines,[47] and aminobenzoic acid and hydroxybenzoic acids[47] [Eqs. (7.40)–(7.43)].

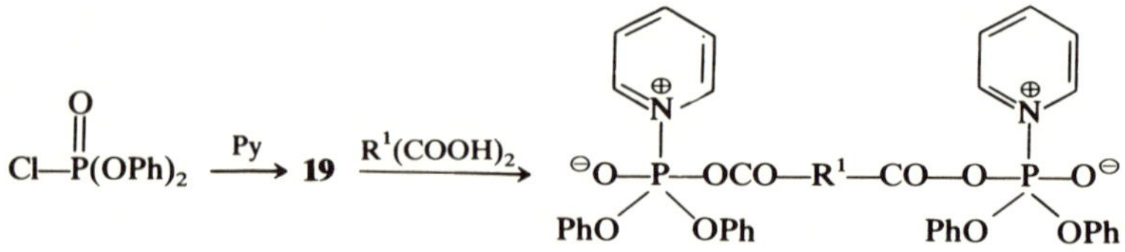

38

$$\mathbf{38} \xrightarrow{HO{-}R^2{-}OH} {+}OC{-}R^1{-}COO{-}R^2{-}O{+}_n \qquad (7.40)$$

$$\mathbf{38} \xrightarrow{HO{-}R^3{-}NH_2} {+}OC{-}R^1{-}CONH{-}R^3{-}O{+}_n \qquad (7.41)$$

$$\mathbf{38} \xrightarrow{HO{-}R^2{-}OH/NH_2{-}R^4{-}NH_2} {+}OC{-}R^1{-}CONH{-}R^4{-}NH{+}_m{+}OC{-}R^1{-}COO{-}R^2{-}O{+}_n \qquad (7.42)$$

$$\mathbf{19} \xrightarrow{NH_2{-}R^5{-}COOH/HO{-}R^6{-}COOH} {+}NH{-}R^5{-}CO{+}_m[O{-}R^6{-}CO{+}_n \qquad (7.43)$$

The reaction is further improved when **19** is transformed into **39** by complexing with lithium halides like LiBr, and polyesters of high molecular weight were obtained from hydroxybenzoic acids [Eq. (7.44)] as well as dicarboxylic acids and diphenols.[48]

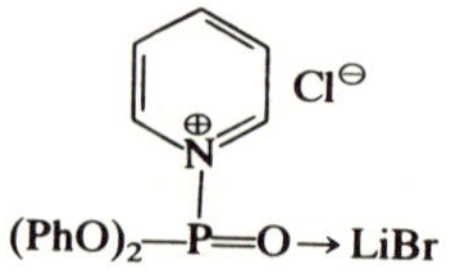

39

$$\mathbf{39} \xrightarrow{HO{-}R^6{-}COOH} {+}O{-}R^6{-}CO{+}_n \qquad (7.44)$$

The Vilsmeier adduct **21** derived from diphenyl chlorophosphate and DMF was also useful for polyesters from dicarboxylic acids and diphenols as well as hydroxy benzoic acids,[13] but this C—O—P type intermediate **21** was less effective than the N—P type intermediate **19**.

The reaction of arylsulfonyl chlorides in pyridine [Eq. (7.18)] could not be used for polyesterification because a side reaction forming sulfonates took place. When the reaction was carried out in the presence of LiCl, the side reaction was retarded to a large extent, and high-molecular-weight polymers were obtained along with a small amount of oligomers.[17] However, the reaction was significantly improved to give high-molecular-weight polymers without oligomers by dropwise addition of diphenols dissolved in pyridine and carboxamides such as *N,N*-diethylformamide [Eq. (7.45)].[49] In the case of hydroxybenzoic acids, **29** in pyridine [Eq. (7.46)] was added to pyridine/carboxamide solutions of these acids.

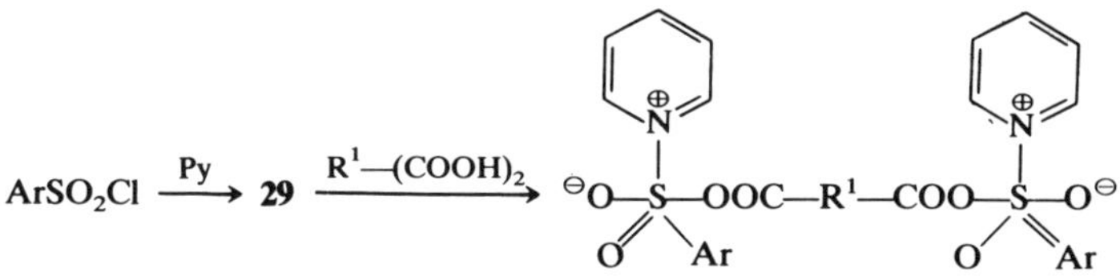

40

$$\xrightarrow{R^2\text{—}(OH)_2/\text{carboxamides}} \left[OC\text{—}R^1\text{—}COO\text{—}R^2\text{—}O \right]_n \qquad (7.45)$$

$$HO\text{—}R^3\text{—}COOH/Py/\text{carboxamides} \xrightarrow{29/Py} \left[O\text{—}R^3\text{—}CO \right]_n \qquad (7.46)$$

The molecular weight of polymers obtained by the reaction with arylsulfonyl chlorides could be more effectively improved by using the Vilsmeier adduct **30** derived from sulfonyl chlorides and DMF [Eq. (7.19)]. High-molecular-weight polyesters were obtained by heating a mixture of **30** and hydroxybenzoic acids, and by stepwise reaction of **30** with dicarboxylic acids and diphenols, respectively.[50]

Among all activating agents developed so far, the *N*-sulfinium salt **24** prepared from thionyl chloride in pyridine [Eq. (7.16)] was the most effective for polyesterification, and polyesters of the highest molecular weight were obtained. The reactions were conducted by heating a mixture of **24** and hydroxybenzoic acid,[51] [Eq. (7.48)] and by initial reaction of **24** with dicarboxylic acids followed by alcoholysis with diphenols[15] [Eq. (7.47)]. The reaction was used for the preparation of the copolymers with relatively highly ordered sequences of monomers as mentioned later in Section 7.3.3.

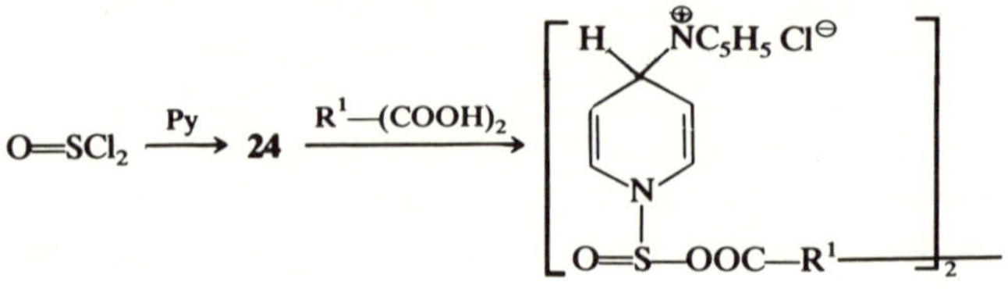

41

$$\xrightarrow{R^2\text{—}(OH)_2} \left[OC\text{—}R^1\text{—}COO\text{—}R^2\text{—}O \right]_n \tag{7.47}$$

$$\mathbf{24} \xrightarrow{HO\text{—}R^3\text{—}COOH} \left[O\text{—}R^3\text{—}CO \right]_n \tag{7.48}$$

Results of the polycondensations with several activating agents (active intermediates) are shown in Tables 7.3 and 7.4, which show that polyesters of sufficiently high molecular weight were obtained.

7.3.2.2. *Typical Procedures*

Polyesterification using Thionyl Chloride in Pyridine. To cold thionyl chloride (11 mmol) in an ice-bath, pyridine (10 mL) was added slowly for 10 min to keep the reaction temperature low, and the reaction mixture was stirred for 30 min. To the cold solution, isophthalic acid (IPA, 5 mmol) and terephthalic acid (TPA, 5 mmol) in pyridine (10 mL) were added dropwise for 20 min, and the mixture was kept at room temperature for 20 min. Bisphenols (5 mmol) in pyridine (10 mL) was added all at once, and the whole mixture was heated at 80 °C for 4 h. The resulting viscous solution was diluted with pyridine, and poured into methanol to precipitate polymer,

TABLE 7.3. Polycondensation of Isophthalic and Terephthalic Acids (50/50) with 2,2-Bis(4-hydroxyphenyl) Propane by Using Several Active Intermediates

	Polyester	
Active intermediate	η_{inh}	M_n
24	2.28	>100,000
39	1.48	86,000
30	1.39	78,000

TABLE 7.4. Copolycondensation of 4-Hydroxybenzoic Acid and 3,5-Dimethoxy-4-hydroxybenzoic Acid (50/50) by Using Several Active Intermediates

	Copolymer
Active intermediate	η_{inh} [a]
24	3.78
39	2.95
28	2.99

[a] Measured in *p*-chlorophenol at 50 °C.

which was washed in boiling methanol and dried. Similarly, to a cold pyridine (10 mL) solution of thionyl chloride (11 mmol), prepared as described above, hydroxybenzoic acid (10 mmol) in pyridine (20 mL) was added dropwise for 20 min, and the resulting mixture was maintained at room temperature for 20 min, and then heated at 80 °C for 3 h. Polymers were isolated by working up as described above.

Polyesterification using Tosyl Chloride. A solution of tosyl chloride (TsCl, 13 mmol) in pyridine (10 mL) and DMF (3 mmol) was maintained at room temperature for 30 min and added to IPA (5 mmol), TPA (5 mmol), or their mixture of equimolar amounts (2.5 mmol) in pyridine (10 mL). The reaction mixture was maintained at room temperature for 10 min and then at 120 °C for 10 min. To this mixture bisphenols (5 mmol) in pyridine (10 mL) were added over 10 min and the whole solution was heated at 120 °C for 3 h. Polymers were separated by working up as above. Similarly, polycondensation of hydrobenzoic acids was carried out as follows. TsCl (13 mmol) was dissolved in pyridine (10 mL), and the solution was maintained at room temperature for 30 min. The reaction mixture was treated with DMF (10–50 mmol) for 30 min at room temperature, and then added all at once at 120 °C to a warmed solution of hydroxybenzoic acids (10 mmol) in a mixed solvent (20 mL) of pyridine and an amide solvent such as DMF preheated at 120 °C for 5 min. The whole solution was maintained at 120 °C for 3 h.

Polycondensation using Diphenyl Chlorophosphate. A lithium halide (10 mmol) dissolved in pyridine (10 mL) was treated with diphenyl chlorophosphate (DPCP, 13 mmol) at room temperature for 30 min, followed by the addition of IPA (5 mmol), TPA (5 mmol), or their mixture of equimolar amounts (2.5 mmol) in pyridine (10 mL). An exothermic reaction took place that yielded precipitates, and the reaction mixture was maintained at room temperature for 10–30 min, and then at 120 °C for 5–20 min. To this mixture was added bisphenols (5 mmol) in pyridine (10 mL) for 10 min and the whole mixture was heated at 120 °C for 3 h. Similarly, polycondensation of hydroxybenzoic acids was carried out as follows. A pyridine (20 mL) solution of DPCP (13 mmol) and LiBr (0–15 mmol), prepared as described above, was added dropwise over 0–40 min to a hot solution of equal parts (5 mmol) of hydroxybenzoic acids in pyridine (10 mL) preheated at 120 °C for 5 min and the whole solution was heated at 120 °C for 3 h.

7.3.3. A Possible Control of Monomer Sequences in Copolymers

A characteristic feature of the activated polycondensations is the simplicity of the process in which polymers with high molecular weight can be obtained by mixing monomers and activating agents in a solvent, and then heating at temperatures up to 120 °C for several hours. Another feature is that monomer sequences in copolymers can possibly be controlled by sequential reaction of two kinds of monomers such as diamines and diphenols with the activated dicarboxylic acids prepared *in situ*.

When methylhydroquinone (MeHQ) and 20 mol% of phenylhydroquinone (PhHQ) were added to a mixture of isophthalic (IPA) and 50 mol% terephthalic acid (TPA) activated by diphenyl chlorophosphate in pyridine, the reaction proceeded according to Eq. (7.13), and a thermotropic liquid crystalline copolymer was obtained quantitatively. The transition temperature above which the copolymer exhibited a complete anistropic melt varied with time and mode of addition of diphenols (Table 7.5).[52] This transition temperature above which liquid crystalline phases were observed may be interpretated as the melting point of crystalline sequences of MeHQ, PhHQ, and TPA in the copolymer. The longer these segments, the higher the transition temperature (melting point) due to higher crystalline order. Addition of the diphenols to IPA/TPA/diphenylchlorophosphate in pyridine all at once resulted in a homogeneous reaction mixture from which a copolymer showing an anisotropic melt above 270 °C was obtained. Similar effects were found by sequential addition of monomer

TABLE 7.5. Copolycondensation of IPA/TPA (50/50) with MeHQ/PhHQ (80/20) by Using **19** Prepared from Diphenyl Chlorophosphate in Pyridine

Initial reaction monomer (min)	Subsequent reaction monomer (min)	State	Optical anisotropy (>°C)
MeHQ (5)	PhHQ (5)	Ppt	270
PhHQ (5)	MeHQ (5)	Soln	210
MeHQ/PhHQ (mixture)	(0)	Soln	210
	(10)	Ppt	270

TABLE 7.6. Copolycondensation of IPA/TPA (50/50) with MeHQ/ODA (80/20) by Using **19** Prepared from Diphenyl Chlorophosphate in Pyridine

Ageing after the reaction with MeHQ (min)	Optical anisotropy (>°C)
0	225
10	260
20	290

instead of simultaneous addition. Analogously, copolyesteramides have been prepared from MeHQ, IPA, TPA, and 20 mol% of 4,4′-oxydianiline (ODA) using diphenylchlorophosphate in pyridine. Polycondensation was carried out sequentially, first with MeHQ and then with ODA. Transition temperatures to anisotropic melts varied with the time of the initial reaction with MeHQ ("ageing," see Table 7.6).[53] Higher transition temperatures are probably due to longer liquid crystalline polyester segments in the copolymer.

Variation in time and mode of addition of IPA and TPA also affected sequences in copolyesters with MeHQ. As mentioned before in the reaction with phosphorous oxychloride in pyridine [Eq. (7.11)], active intermediate **12** reacted stepwise with two moles of a carboxylic acid to form initially **13** and then **14**, and with a dicarboxylic acid to give a cyclic intermediate **34**. The reactivity of the two activated carboxyl groups on one phosphorus atom is different from each other. Therefore, when **12** was treated with equal parts of isophthalic (I) and terephthalic (T) acids, **42** and **43** would be initially formed, and then three types of intermediates, **44**, **45**, and **46**, were expected to be produced by further reaction with the acids.

Addition of isophthalic acid first and then terephthalic acid might favor the formation of **44**, which might give the copolymer of alternating sequence

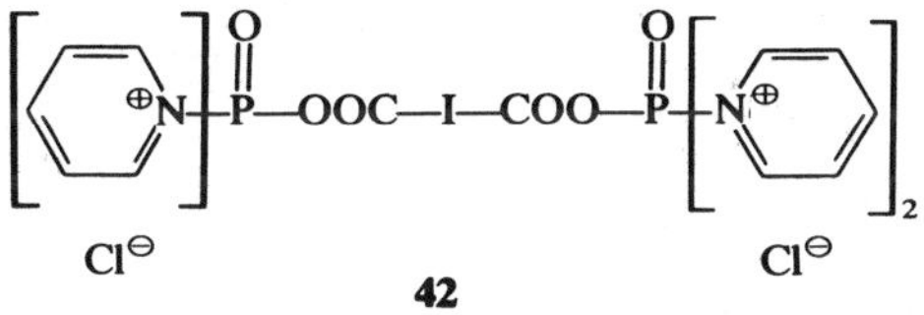

42

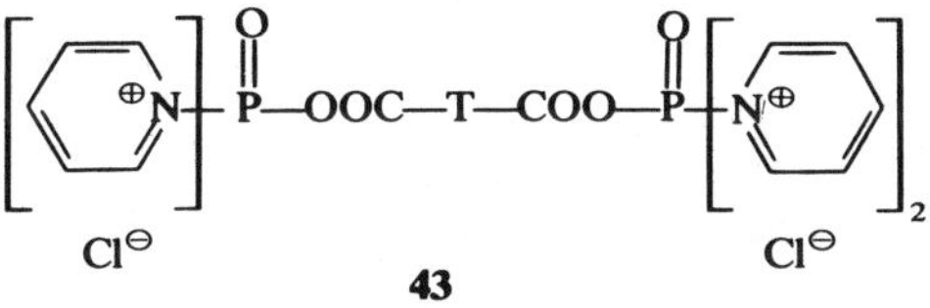

43

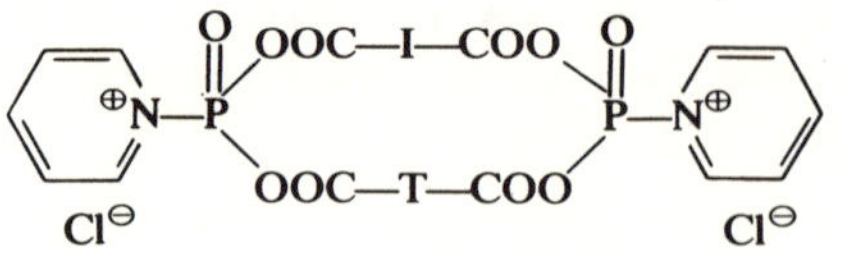

44

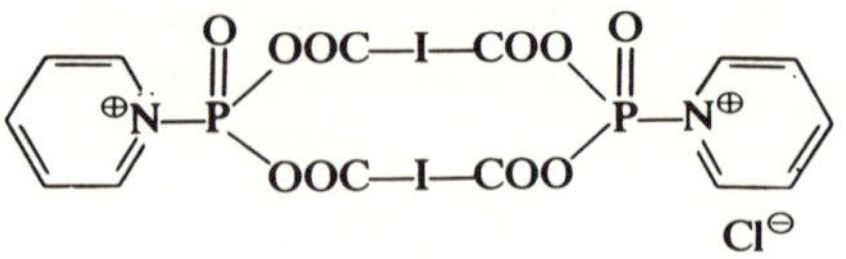

45

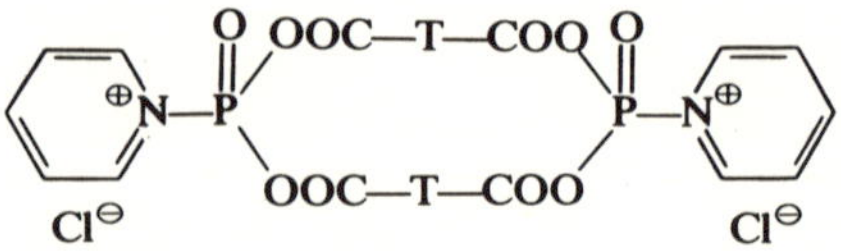

46

of isophthalic and terephthalic acids by alcoholysis with diphenols because of the different reactivity of the two activated carboxyl groups. When a mixture of the diacids was added, **45** and **46** would be produced along with **44**, and a random copolymer might be obtained. The changes in monomer sequences in the copolymers were again analyzed by recording liquid crystalline transition temperatures (see Table 7.7).[(45)]

When a mixture of the diacids was added in 10 min the transition temperature was 275 °C. Slower addition (30 min), more strongly involving intermediates **45** and **46** in the copolymerization, decreased this to 258 °C. Sequential addition, IPA–TPA, increased this temperature again to 284 °C. No attempt has been made to discuss the relationship between anisotropic characteristics and mesogenic segment length quantitatively.[(53)]

The inherent viscosity of the copolymer prepared by sequential addition is much higher than that prepared by simultaneous addition (Table 7.7). This may be due not so much to molecular weight differences, but rather also to different monomer sequences in the copolymers, giving rise to different conformations in the solvent used. Slower addition of isophthalic acid increased the viscosity, but not the liquid crystalline transition temperature.

TABLE 7.7. Copolycondensation of IPA/TPA (50/50) with MeHQ by Using **12** Derived from Phosphorus Oxychloride in Pyridine

Initial reaction: addition time of monomer (min)	Initial reaction: ageing (min)	Optical anisotropy (>0°C)	η_{inh}[a]
IPA (0)[b]	10	283	3.31
	30	284	2.84
(10)[b]	10	284	4.86
IPA/TPA(10)	0	275	1.13
(30)		258	1.09

[a] Measured in *p*-chlorophenol at 50 °C.
[b] Sequential addition of IPA during time indicated, then TPA.

Control of monomer sequences in copolymers was attempted also in the reaction using **24** derived from thionyl chloride in pyridine.[54] Copolymerization of 25 mol% of terephthalic acid (TPA), 50 mol% of *p*-hydroxybenzoic acid (PHB), and PhHQ proceeded more smoothly and favorably, when carried out by reacting initially with TPA and PHB and then with PhHQ in order to produce intermediate **47** and to prevent formation of the intermediate **48** from PHB alone. The high-molecular-

47

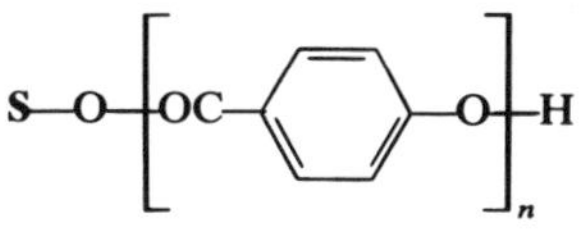

48

S =

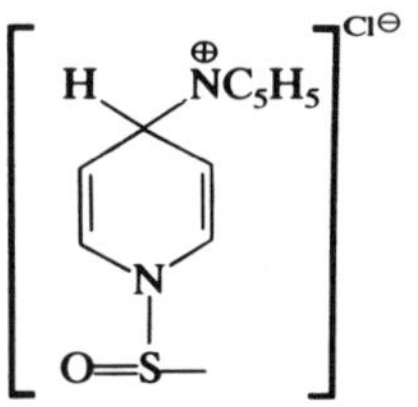

weight copolymer with a relatively highly ordered sequence of monomers was obtained. This was also the case for the copolycondensation of *m*-hydroxybenzoic acid (MHB) and PHB.

REFERENCES

1. M. Kurihara, H. Saito, K. Nukada, and N. Yoda, *J. Polym. Sci., A-1* **7**, 2897 (1969).
2. G. Schramm, H. Grotsch, and W. Pollmann, *Angew. Chem., Int. Ed. Engl.* **1**, 1 (1962).
3. M. Ueda, *Kobunshi* **35**, 128 (1986).
4. G. C. Wu, H. Tanaka, K. Sanui, and N. Ogata, *Polym. J.* **14**, 571 (1982).
5. S. Kitayama, K. Sanui, and N. Ogata, *J. Polym. Sci., Polym. Chem. Ed.* **22**, 2705 (1984).
6. N. Yamakawa and F. Higashi, *Tetrahedron* **30**, 1319 (1974).
7. F. Higashi and A. Mochizuki, unpublished results.
8. F. Higashi, T. Murakami, and Y. Taguchi, *J. Polym. Sci., Polym. Chem. Ed.* **20**, 103 (1982).
9. N. Yamazaki, F. Higashi, and T. Iguchi, *J. Polym. Sci., Polym. Chem. Ed.* **13**, 785 (1975).
10. F. Higashi and A. Kobayashi, *J. Polym. Sci., Polym. Chem. Ed.* **27**, 507 (1988).
11. F. Higashi, Y. N. Lee, and A. Kobayashi, *J. Polym. Sci., Polym. Chem. Ed.* **26**, 2077 (1988).
12. F. Higashi, A. Hoshio, and J. Kiyoshige, *J. Polym. Sci., Polym. Chem. Ed.* **21**, 3241 (1983).
13. F. Higashi, Y. Yamada, and A. Hoshio, *J. Polym. Sci., Polym. Chem. Ed.* **22**, 2181 (1984).
14. F. Higashi, M. Goto, and H. Kakinoki, *J. Polym. Sci., Polym. Chem. Ed.* **18**, 1711 (1980).
15. F. Higashi, T. Mashimo, and I. Takahashi, *J. Polym. Sci., Polym. Chem. Ed.* **24**, 97 (1986).
16. F. Higashi and T. Nishi, *J. Polym. Sci., Polym. Chem. Ed.* **24**, 701 (1986).
17. F. Higashi, N. Akiyama, and T. Koyama, *J. Polym. Sci., Polym. Chem. Ed.* **21**, 3233 (1983).

18. F. Higashi, N. Akiyama, I. Takahashi, and T. Koyama, *J. Polym. Sci., Polym. Chem. Ed.* **22**, 2181 (1984).
19. P. Strohriegl, W. Heitz, and G. Weber, *Makromol Chem., Rapid Commun.* **6**, 111 (1985).
20. H. Tanaka, G. Wu, Y. Iwanaga, K. Sanui, and N. Ogata, *Polym. J.* **14**, 635 (1982).
21. Y. Iwakura, *Kobunshi* **17**, 130 (1968).
22. M. Ueda, M. Sato, and A. Mochizuki, *Macromolecules* **18**, 2723 (1985).
23. M. Ueda, H. Sugita, and M. Sato, *J. Polym. Sci., Polym. Chem. Ed.* **24**, 1019 (1986).
24. M. Ueda, S. Yokote, and M. Sato, *Polym. J.* **18**, 117 (1986).
25. M. Ueda and T. Kanno, *Makromol. Chem., Rapid Commun.* **5**, 833 (1984).
26. M. Ueda, *Makromol. Chem., Rapid Commun.* **6**, 271 (1985).
27. According to the relation[28]; $[\eta] = 1.9 \times 10^{-7} \cdot M^{1.7}$.
28. J. R. Schaefgen, and V. S. Foldi, *Polym. Prepr., Am. Chem. Soc., Div. Polym. Chem.*, **17**(1), 69 (1976).
29. N. Yamazaki, F. Higashi, and J. Kawabata, *J. Polym. Sci., Polym. Chem. Ed.* **12**, 2149 (1974).
30. N. Yamazaki, F. Higashi, and M. Matsumoto, *J. Polym. Sci., Polym. Chem. Ed.* **13**, 1373 (1975).
31. J. Preston, W. R. Krigbaum, and R. Kotek, *J. Polym. Sci., Polym. Chem. Ed.* **20**, 3241 (1982).
32. J. Asrar, J. Preston, A. Ciferri, and W. R. Krigbaum, *J. Polym. Sci., Polym. Chem. Ed.* **20**, 373 (1982).
33. F. Higashi, M. Goto, Y. Nakano, and H. Kakinoki, *J. Polym. Sci., Polym. Chem. Ed.* **18**, 851 (1980).
34. F. Higashi, Y. Nakano, M. Goto, and H. Kakinoki, *J. Polym. Sci., Polym. Chem. Ed.* **18**, 1099 (1980).
35. F. Higashi, N. Akiyama, and S. Ogata, *J. Polym. Sci., Polym. Chem. Ed.* **21**, 913 (1983).
36. W. R. Krigbaum, R. Kotek, and Y. Mihara, *J. Polym. Sci., Polym. Chem. Ed.* **23**, 1907 (1985).
37. According to the relation[28]: $[\eta] = 2.8 \times 10^{-7} \cdot M^{1.7}$.
38. F. Higashi, Y. Taguchi, N. Kokubo, and H. Ohta, *J. Polym. Sci., Polym. Chem. Ed.* **19**, 2745 (1981).
39. F. Higashi and Y. Taguchi, *J. Polym. Sci., Polym. Chem. Ed.* **19**, 3345 (1981).
40. F. Higashi and M. Ishikawa, *J. Polym. Sci., Polym. Chem. Ed.* **18**, 2905 (1980).
41. F. Higashi and Y. Taguchi, *J. Polym. Sci., Polym. Chem. Ed.* **18**, 2875 (1980).
42. F. Higashi, K. Sano, and H. Kakinoki, *J. Polym. Sci., Polym. Chem. Ed.* **18**, 1841 (1980).
43. F. Higashi and T. Nishi, *J. Polym. Sci., Polym. Chem. Ed.* **26**, 3235 (1988).
44. F. Higashi and Y. N. Lee, *J. Polym. Sci., Polym. Chem. Ed.* **26**, 3123 (1988).
45. F. Higashi, H. Shimazaki, and A. Kobayashi, *Polym. Prepr. Jpn.* **38**, 2 (1989).
46. F. Higashi, A. Hoshio, and H. Ohtani, *J. Polym. Sci., Polym. Chem. Ed.* **21**, 3241 (1983).
47. F. Higashi, M. Ozawa, A. Hoshio, and A. Mochizuki, *J. Polym. Sci., Polym. Chem. Ed.* **23**, 1699 (1985).
48. F. Higashi, A. Hoshio, Y. Yamada, and M. Ozawa, *J. Polym. Sci., Polym. Chem. Ed.* **23**, 69 (1985).
49. F. Higashi and T. Mashimo, *J. Polym. Sci., Polym. Chem. Ed.* **23**, 2999 (1985).
50. F. Higashi, I. Takahashi, N. Akiyama, and T. C. Chang, *J. Polym. Sci., Polym. Chem. Ed.* **22**, 3607 (1984).
51. F. Higashi and T. Mashimo, *J. Polym. Sci., Polym. Chem. Ed.* **24**, 1697 (1986).
52. F. Higashi, *Nihon Settchaku Kyokai* **24**, 504 (1988).
53. F. Higashi, H. Arikawa, and A. Mochizuki, *J. Polym. Sci., Polym. Chem. Ed.* **26**, 3071 (1988).
54. F. Higashi, S. Sugimori, and T. Mashimo, *J. Polym. Sci., Polym. Chem. Ed.* **26**, 1003 (1988).

8

Polymer Synthesis via Metathesis Chemistry

E. A. Ofstead and K. B. Wagener

8.1. INTRODUCTION

8.1.1. Background

Metathesis chemistry, now more than twenty years old, pertains to a unique reaction of olefins that can be used to synthesize both small molecules and a broad spectrum of polymers. Early researchers, impressed with the beauty and elegance of this chemistry, frequently used terms such as "fascinating," "intriguing," "exciting," and even "doubly exciting" (commercially and academically), to express their sentiments about the subject.

A great deal has been learned about all aspects of the process, in part because it is accessible using both the techniques of polymerization chemistry, which are very sensitive to minor factors as they affect chain processes, and the techniques of small molecule chemistry, which can be ideal for studying model reactions. The opportunity to effectively employ both approaches, an opportunity which is relatively rare in the synthesis of hydrocarbon polymers, has provided a powerful combination of tools for unraveling this chemistry. As Ken Ivin has written, "We now know quite a lot about this beautiful woman, metathesis, but she still has a few secrets locked away."[1] Productive discoveries continue unabated in this field of research.

E. A. Ofstead • Research Division, The Goodyear Tire Rubber Company, Akron, Ohio 44316 *K. B. Wagener* • Department of Chemistry and Center for Macromolecular Science and Engineering, University of Florida, Gainesville, Florida 32611.

New Methods for Polymer Synthesis, edited by W. J. Mijs. Plenum Press, New York, 1992.

There now exists a vast literature on the olefin metathesis reaction covering catalysis, mechanisms, and applications to synthesis in both organic and polymer chemistry. This work is well documented through about 1984 in two indispensible books on the subject by Ivin[1] and by Dragutan, Balaban, and Dimonie.[2] The basic features of this chemistry, which are only briefly presented in this chapter, are extensively reviewed in these books and additional sources cited therein. More recent developments (through 1989) have been summarized in other reviews.[3–6] In this chapter we will focus on the metathesis of cyclic and acyclic olefins as used to create polymers, but will not cover the related metathesis of alkynes.

8.1.2. Metathesis Chemistry and Polymerization

The word "metathesis" is derived from the Greek "meta" (change) and "tithemi" (place), and grammatically it refers to a transposition of sounds or letters in a word. The term "olefin metathesis" was originally adapted for this organic reaction by Calderon and co-workers in 1967.[7] As shown in Eq. (8.1) for the example of the reversible metathesis of 2-pentene to form

$$2\ CH_3{-}CH{=}CH{-}C_2H_5 \overset{\text{catalyst}}{\rightleftharpoons} CH_3{-}CH{=}CH{-}CH_3 + C_2H_5{-}CH{=}CH{-}C_2H_5 \tag{8.1}$$

2-butene and 3-hexene, this definition is appropriate for the chemistry at hand since it refers to the interchange of moieties, linked in this case by an olefinic bond. When sufficiently active catalysts are used with acyclic olefins, the reaction can go to an equilibrium position which is determined primarily by entropy considerations, since there is no net change in the numbers or types of bonds.

In 1964, Banks and Bailey first reported the disproportionation (also called dismutation) of olefins over high-temperature metal oxide catalysts to give higher and lower alkenes.[8] Thus, propylene afforded primarily ethylene and 2-butene. It was not until 1967, however, that the reaction was shown to proceed uniquely via scission of the strongest bond in the molecule, namely, the olefinic carbon–carbon bond.[7,9] These results were subsequently verified by others,[10,11] and direct confirmation that double bond scission was indeed the site of ring cleavage during the polymerization of cycloalkenes was provided by Dall'Asta.[12]

When cyclic olefins are employed, cleavage of the ring leads to the formation of a difunctional moiety which effectively forms the building block for making a polymer chain (Scheme 8.1). In the absence of some method for generating chain ends, molecular weights presumably would approach infinity, but a variety of sources of chain ends exist to limit molecular weights, which will be discussed later. The acronym ROMP has recently become a familiar term for ring-opening metathesis polymerizations.

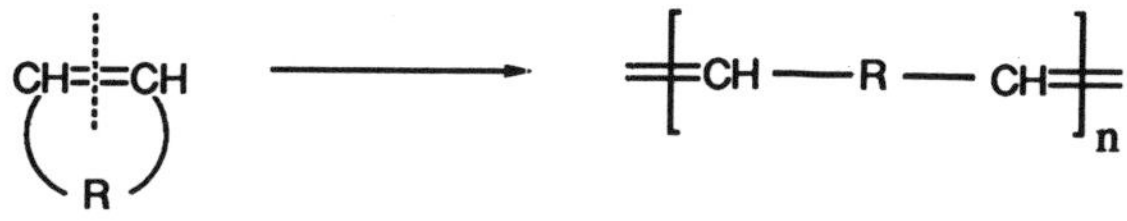

SCHEME 8.1.

Cyclic olefins differ in one major respect from their acyclic counterparts in that the reaction is usually (but not necessarily) accompanied by the release of ring strain. Thus, an additional driving force exists which results in shifting the reaction equilibrium in favor of the polymer. However, ring strain is not a prerequisite, since entropy alone can drive the polymerization of strain-free monomers to high polymer if the ring size is sufficiently large. This effect has been demonstrated by the polymerization of the 24-carbon cyclic trimer of cyclooctene, cyclotetraeicosa-1,9,17-triene.[13]

The first cycloolefin ring-opening polymerizations were reported in a patent filed by Eleuterio in 1957, which included descriptions of the preparation of polypentenamer and polynorbornene.[14] Further details regarding polynorbornene were published by Truett and co-workers in 1960,[15] but both these reports were based on the use of inefficient catalysts, and conditions unfavorable for polymer synthesis. Fortunately, research efforts were spurred on in 1964 by the key discovery by Natta and Dall'Asta of soluble, highly active catalysts which were effective for the solution polymerization of low-strain monomers such as cyclopentene under mild conditions.[16]

Unfortunately, these early major discoveries dealing with disproportionation and polymerization were isolated in two disparate fields of chemistry, namely, petrochemical processes and polymerization chemistry. As a result, the common basis for these two chemistries was not recognized until the discovery by Calderon and co-workers that a catalyst system based on tungsten hexachloride, ethanol, and ethylaluminumdichloride could bring about not only the rapid polymerization of cycloolefins such as cyclooctene and 1,5-cyclooctadiene,[17] but also the facile disproportionation of 2-pentene at room temperature.[7,18] Progress since then in both areas of metathesis chemistry has been intimately linked, to the mutual advantage of each field.

The vast body of published work has focused on the use of cyclic olefins as a convenient route to high polymers. However, an alternative approach exists, analogous to polycondensation, which utilizes an elimination reaction as shown in Eq. (8.2). This process allows the molecular weight of the

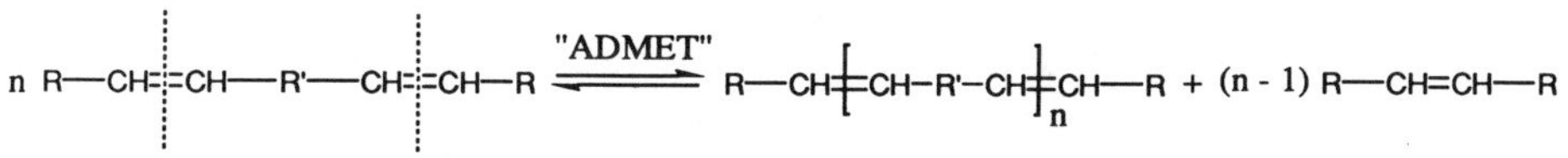

(8.2)

product to be controlled if the olefinic byproduct is selectively removed during the reaction. However, efficient removal of the olefin becomes a very challenging requirement in order to drive the reaction toward high-molecular-weight products. In addition, the starting diene must be entirely free of monoolefin. Finally, some care is required in the choice of the middle alkylene link in order to avoid the formation of stable small-ringed products. For example, 1,7-octadiene produces cyclohexene rather than linear oligomers.[19] Unfortunately, early workers only achieved low-molecular-weight oligomers by this process.[19,20] with molecular weights not exceeding about 500.

The successful attainment of high-molecular-weight products via this process[21] did not occur until the development of a new family of well-defined catalyst systems. These had evolved from basic studies of the nature of metathesis catalysis, and had been successfully introduced for the preparation of a remarkable series of living metathesis polymers through the work of Schrock and Grubbs.[4] These catalysts possessed the critical characteristics of being long-lived, highly active, and essentially free of undesirable side reactions.

Improved procedures utilizing these catalysts were developed recently by Wagener and co-workers[21,22] which, for the first time, completely eliminated all competing reactions and achieved the required efficient removal of byproduct olefin, leading to the formation of high molecular weights comparable to those from ROMP. In laboratory studies, this has been accomplished through the use of high-vacuum techniques. The designation ADMET was coined for this reaction, signifying acyclic diene metathesis. The ADMET process now provides a viable alternative to the use of cyclic olefins for the synthesis of metathesis polymers with desirably high molecular weights. From a practical point of view, the ADMET process is unique in that it is essentially thermoneutral, with no requirement for cooling during the reaction. Most polymerizations require the removal of heat, a factor which contributes to the cost and complexity of commercial processes.

8.2. GENERAL POLYMERIZATION FEATURES

8.2.1. The Central Metathesis Step

Regardless of whether polymerization is done by ring-opening (ROMP) or step polycondensation (ADMET) techniques, a central metathesis step exists. Early proposals for the mechanism of ring-opening metathesis polymerization suggested that a four-centered process was involved,[23] wherein a cycloaddition reaction of two alkenes was promoted by coordination to the transition metal catalyst as in Scheme 8.2.[18] Theoretical arguments were subsequently developed to support the notion that a

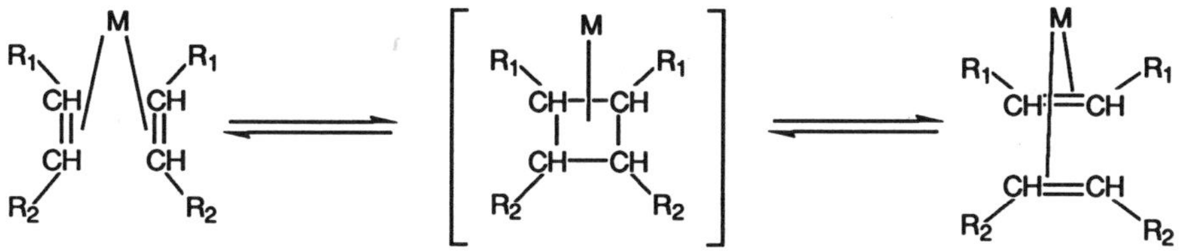

SCHEME 8.2. Bis–Olefin mechanism.

normally forbidden (2 + 2) cycloaddition could be transformed into an allowed process in the presence of a transition metal having the appropriate oxidation state. However, this scheme proved to be short-lived.

The bis-olefin scheme gave way during the 1970s to another intriguing mechanism involving a carbene-metal species. In 1970, Herisson and Chauvin observed that some metathesis products could not be adequately accounted for by the bis-olefin scheme, and offered an alternative pathway involving the transformation of a carbene-bearing metal species into an intermediate four-membered metallacyclobutane ring,[24] as shown for the acyclic olefin example in Scheme 8.3. Each step in the process is reversible. The symmetric metallacycle can revert to the starting complex, or can transform to the new complex with equal ease. For most catalyst systems, this process is not highly stereospecific, and thus provides a pathway for the interconversion of *cis* and *trans* vinylene units. This concept has since been thoroughly validated in a variety of elegant ways,[25] and is universally accepted for most catalyst systems.

The remaining requirement for this process to operate catalytically is the olefin exchange step. The product olefin, which is presumably coordinated to the metal, must undergo facile dissociation and exchange with a new one from the reaction medium in order to propagate the reaction (Scheme 8.4).

8.2.2. Distinctions in ROMP and ADMET Propagation

While metallacycle formation is a central feature in both ROMP and ADMET mechanisms, these two polymerizations differ in many respects, particularly in the nature of the propagation reaction.

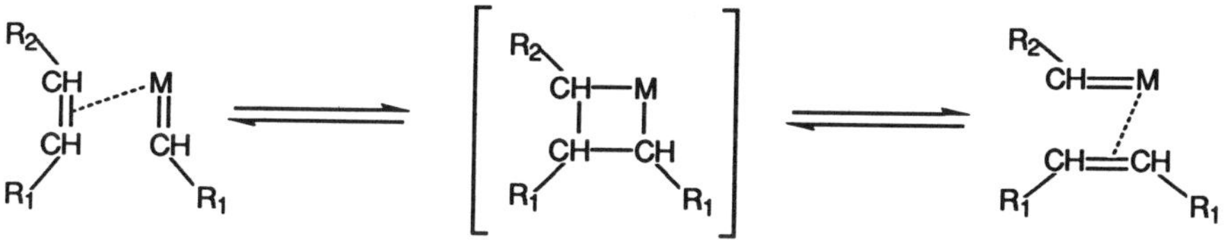

SCHEME 8.3. Carbene mechanism.

SCHEME 8.4. Olefin exchange.

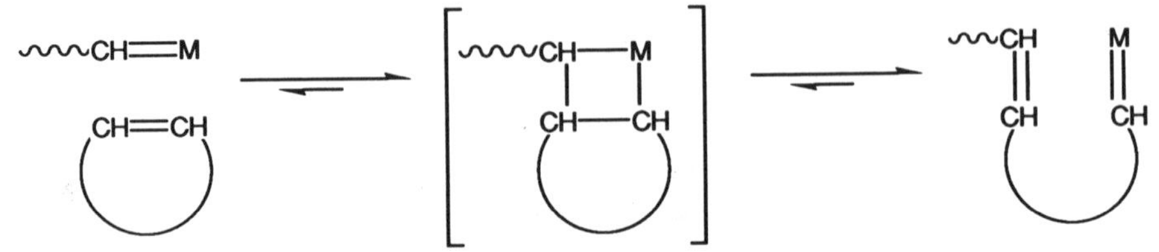

SCHEME 8.5. ROMP propagation.

In ROMP polymerization, the reaction described in Scheme 8.3 becomes the chain growth process as in Scheme 8.5, whereby a monomer undergoes ring opening when it inserts at the reactive chain end. For monomers having low ring strain such as cyclopentene, this process can become easily reversible, resulting in monomer–polymer equilibrium effects.

With asymmetric olefins, two reaction pathways exist, and this becomes a particularly important factor in ADMET polymerization. These two pathways are illustrated below for small molecule olefins. This so-called productive reaction (Scheme 8.6) provides for the formation of a new olefin, while the degenerate process (Scheme 8.7) leads to products which have the same structure as the starting material. Normally the degenerative process is transparent, although with internal olefins it may be manifested by *cis–trans* isomerization. Nevertheless, the existence of this nonproductive redistribution has been clearly demonstrated by the use of isotopically labeled olefins.[26,27]

The relative rates of productive and degenerate processes can be fairly similar in reactions of internal olefins such as 2-pentene. However, with α-olefins, the degenerate process is highly favored, and the productive reaction becomes much slower. Consequently, the preparation of new

SCHEME 8.6. Productive metathesis.

SCHEME 8.7. Degenerate metathesis.

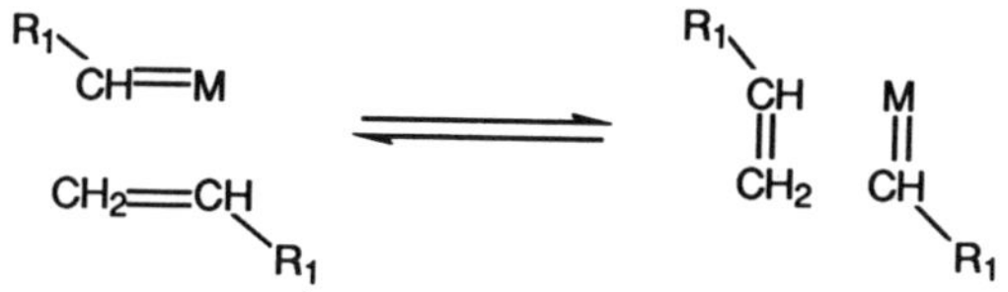

SCHEME 8.8. Degenerate metathesis of α-olefins.

olefins from α-olefins is a more challenging task than if internal olefins are used. The active catalyst appears to bear the substituted carbene most of the time (Scheme 8.8).

Degenerate metathesis plays an important role during the ADMET process if α,ω-dienes are employed, principally by dramatically reducing reaction rates. This fact increases the need to use long-lived, clean-reacting catalysts in order to promote the reluctant productive metathesis reaction, as in Scheme 8.9, which in fact is the only possible ADMET propagation reaction. This points emphasizes the importance of the catalyst developments referred to in the discussion of ADMET above.

8.3. SPECIFIC FEATURES OF ROMP CHEMISTRY

8.3.1. Polymerizability of Cycloolefins

It is generally known that the ring-opening polymerization of simple monocyclic monomers is highly favored thermodynamically for three- and four-membered rings, and for eight-membered and all higher rings.[28] Rings with five and seven members are borderline cases, which will show marked reversibility and significant monomer concentrations at equilibrium. Cyclo-

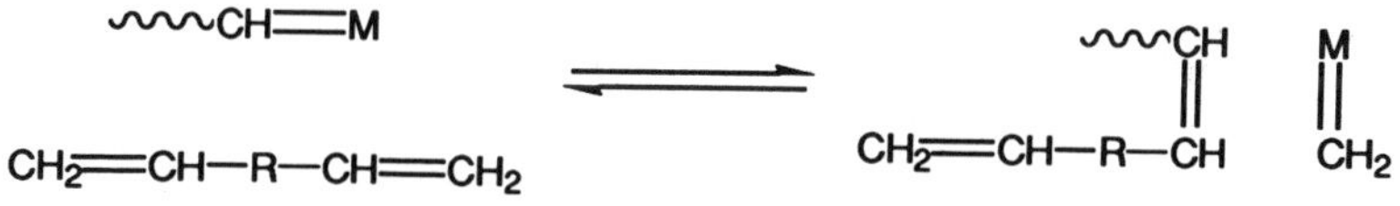

SCHEME 8.9. ADMET propagation.

hexene and nearly all its derivatives are thermodynamically non-polymerizable, although bicyclo-2,2,2-oct-2-ene, which contains only six-membered rings, has been polymerized.[13] Substituents on rings generally reduce polymerizability somewhat and, in the case of cyclopentene, may render the monomer unpolymerizable.[29] When monomer–polymer equilibrium occurs for small-ring monomers such as cyclopentene, the equilibrium is usually shifted in favor of polymer by lowering the polymerization temperature.

Polymerizability derives from the net combination of entropic and enthalpic factors. Ring strain (enthalpic) is the most common driving force for rings of less than about 15 atoms, and is especially dominant with highly strained bicyclic systems such as the norbornenes. However, entropy alone can serve as the driving force for the polymerization of large, strain-free rings. The attainment of low but measurable equilibrium concentrations of macrocyclic oligomers in the polymerization of simple cycloolefins reflects the polymerizability of these same rings. Experiments which actually used macrocyclic oligomers as the starting point for polymerizations have been reported.[13,30,31] However, for rings of less than about nine carbons, the entropy contribution is unfavorable for polymerizability, thus ring strain becomes a prerequisite.

8.3.2. Initiation of Metathesis Reactions

Because of the remarkable diversity of catalyst systems which are capable of initiating metathesis reactions, there apparently exist many processes which can effectively generate the active carbene species. In some instances, if an organometallic compound bearing a well-defined carbene ligand is used, the source of the initiation step is relatively clear. For example, Casey and Burkhardt reported the use of a diphenylcarbene–tungsten complex to initiate metathesis, and showed that the initiating carbene was found as expected in products from olefin scission reactions.[32] Many similar reports have followed.

Recently, the work of Schrock and co-workers[33] has led to the evolution of a series of Lewis acid-free catalysts which have proven very useful in ROMP for making living polymers, and in ADMET reactions as well. Versions based on both tungsten (**1**) and molybdenum (**2**) have been prepared. Molybdenum derivatives appear to be preferred if the monomer to be used possesses polar substituents, as in the example of the ester-containing monomer 5,6-dicarbomethoxynorbornene.[34]

In a variation on the use of carbene–metal complexes, precursors known to generate carbene ligands thermally have also been used. For example, a titanacyclobutane complex derived from the reaction of Tebbe's reagent with an olefin has been employed in reactions with terminal olefins,

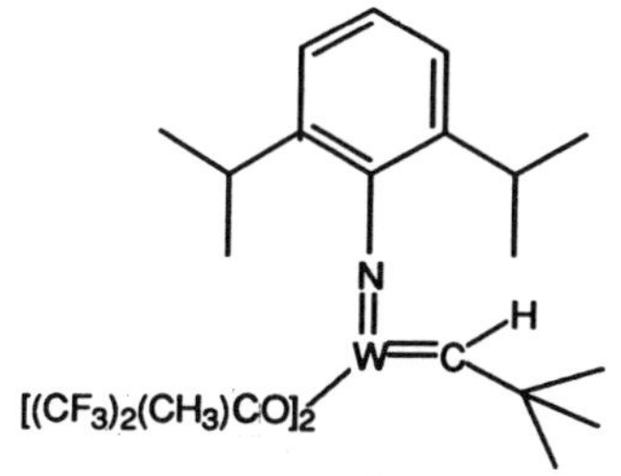

1

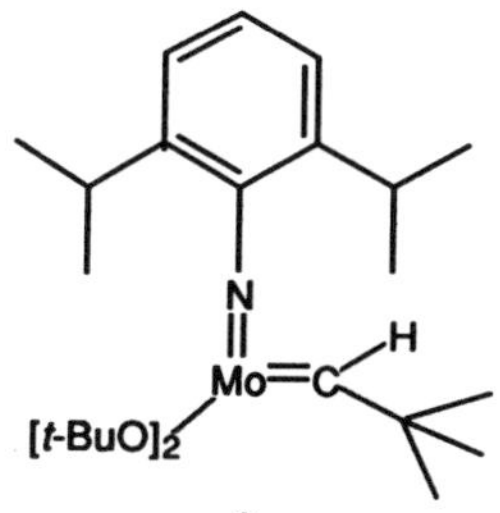

2

although only degenerate metathesis has been observed with that system due to extreme steric hindrance about the transition metal.[35]

More effective metallacycle systems have been developed based on highly strained titanacyclobutanes possessing a fused cyclopropane unit,[36] as in Scheme 8.10. These have shown broader activity for promoting productive metathesis reactions. Their reactivity is driven in part by the added relief of ring strain which accompanies the first metathesis step. Catalysts of this type were used in the early development of living metathesis polymerization systems.

Quite often, however, the mechanism of formation of the catalyst is never resolved, but merely speculated on. This is particularly true with the older catalyst systems based on Ziegler–Natta-type combinations of transition metal salts with organometallic reducing agents. According to several researchers, carbene formation likely results from alkylation of the transition metal, followed by the removal of HX as in Eq. (8.3). This is probably assisted by reaction of excess cocatalyst with the elements of HX to generate alkane byproducts. An alternative pathway may involve dialkylation of the transition metal, followed by an elimination involving an α-hydrogen [Eq. (8.4)].

$$WX_6 + (RCH_2)_3Al \longrightarrow [RCH_2WX_5] \xrightarrow{-HX} [RCH{=}WX_4] \qquad (8.3)$$

$$WX_6 + (RCH_2)_4Sn \longrightarrow [(RCH_2)_2WX_4] \xrightarrow{-RCH_3} [RCH{=}WX_4] \qquad (8.4)$$

Questions related to initiation processes for several catalyst systems have been discussed recently.[37] These papers reveal the complexity of this

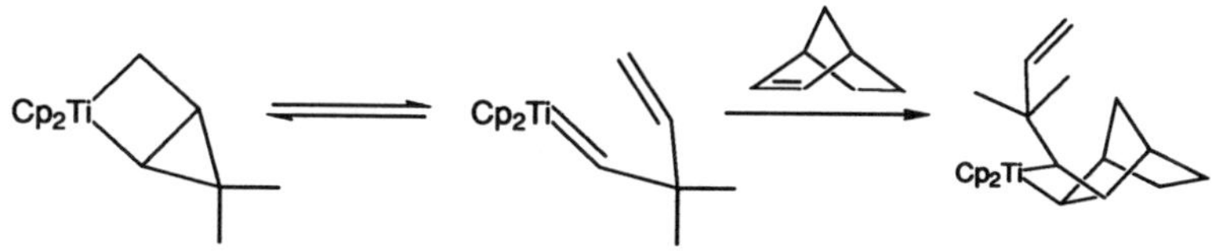

SCHEME 8.10. Initiation of norbornene polymerization.

subject. For example, tungsten hexachloride apparently can initiate the polymerization of cyclopentene in the absence of other agents. Reduction of WCl_6 to WCl_5 can be observed, along with formation of halogenated derivatives of the monomer. It has also been speculated that cyclopentene can react directly to form a cyclopentyl tungsten species, which then reacts further to generate the active carbene. Many of these issues remain unresolved.

8.3.3. ROMP Propagation and Secondary Metathesis Reactions. Consequences of the Carbene Mechanism

Several general features of metathesis polymerizations should be anticipated based on the carbene mechanism. Each of the following primary metathesis steps must be considered when dealing with ROMP polymerizations:

A. Initiation (a source of chain ends, see above).

B. Propagation (including monomer–polymer equilibrium, see Scheme 8.11).

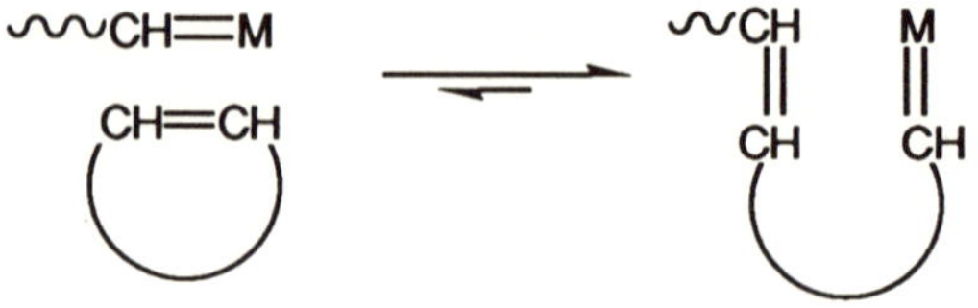

SCHEME 8.11.

C. Chain transfer (e.g., during propagation, see Scheme 8.12).

SCHEME 8.12.

D. Chain scission (see Scheme 8.13).

SCHEME 8.13.

E. Ring-chain equilibrium (see Scheme 8.14).

SCHEME 8.14.

F. Redistribution of molecular weights (see Scheme 8.15).

SCHEME 8.15.

G. Redistribution of monomer units in copolymers (see Scheme 8.16).

SCHEME 8.16.

H. *cis–trans* interconversions.

(possible in all steps B–G)

Certainly each of these processes would be expected to exhibit somewhat different kinetic behavior. Rates of the various steps are actually quite dependent on both the catalyst system used, and the structure of the monomer and the corresponding polymer repeat unit. With most monomers possessing substantial ring strain, the propagation step can go to completion before the other processes have a chance to reach equilibrium. As a consequence, a great diversity of polymerization data has been reported, wherein many different and seemingly conflicting combinations of the above processes have been observed. These differences can often be attributed to reactions having been terminated at various stages along the complex path to equilibrium. However, since metathesis equilibrium is readily achievable with strain-free acyclic olefins when using sufficiently active catalysts, it would be expected that equilibrium behavior would be attainable in ROMP systems as well. Indeed, this has been demonstrated for some of the simple unsubstituted, monocyclic monomers.

In 1968, Scott and co-workers first described cycloolefin polymerizations as equilibrium reactions.[30] They demonstrated the reversible formation of macrocyclic oligomers from high polymer (E, above), as well as the converse, and they proposed that these systems appeared to obey the laws of ring-chain equilibrium with regard to the distribution of both linear and cyclic species, in accordance with Jacobson–Stockmayer theory.[38] It was shown that the fraction of the product which consists of cyclic oligomers depends on the overall concentration of polymer in the system. Beyond a certain critical concentration of cyclic oligomers (amounting to about 20 g per liter for the case of cyclooctene), all additional monomeric units go into the formation of the high-molecular-weight polymer.

Others have reported similar results. For example, Chauvin and co-workers quantified the position of equilibrium for 1,5-cyclooctadiene polymerizations with regard to the cyclic oligomer content, independent of whether the starting material was monomer, oligomer, or high polymer.[31] Hocker determined the equilibrium dependence of oligomer concentration on ring size for several cycloolefin systems.[39]

From a practical point of view, cyclic oligomers are unlikely to contribute usefully to the physical properties of a polymer, and it would be best to choose reaction conditions which would minimize the amount of oligomers formed. Methods based on catalyst control of this process are very difficult to define. However, one obvious approach, based on the equilibrium concepts discussed above, is simply to use the maximum permissible concentration of monomer during polymerization in order to minimize the fraction converted to oligomers.

Monomer–polymer equilibrium (Scheme 8.11) is generally an important phenomenon only with rings having low strain energy. As a result, this subject has been explored with only a few monomers. With cyclopentene, e.g., the reversible formation of monomer from purified high polymer was demonstrated in 1971.[13] It has been possible to establish the basic thermodynamic characteristics of the polypentenamer system from equilibrium conversion data.[13,40,41]

Recently, Drapeau and Leonard elegantly refined our understanding of the equilibrium behavior of polypentenamer.[42] They identified the contribution of randomness in the distribution of *cis* and *trans* units along the polymer chain to the entropy of the polymer, and its effect on equilibrium polymerization. Experimental data nicely verified the model they proposed.

Equilibration of microstructure has been reported with certain cycloolefins, although nonequilibrium results are more commonly observed. With cyclopentene, for example, several catalysts which showed diverse stereospecificity during early stages of polymerization all led to approximately the same microstructure by the end of the reaction, namely 85% *trans* content.[43] Commercially available crystalline polyoctenamer (Vestenamer 8012, Chemishe Werke Huls) has a *trans* content of about 80%,[44] quite close to the expected equilibrium value, suggesting that the polymerization may be one which is simply carried to near-equilibrium in order to achieve the desired microstructure.

In a unique demonstration that some equilibria can be achieved independently of others, it has been found that equilibrium conversion of monomer to polymer can occur with cyclopentene when using catalysts which are very highly *cis*-stereospecific.[41] The amount of monomer present at equilibrium is much higher when the polymer has the *cis* structure in contrast to the *trans*, reflecting the relative stabilities of the *cis* and *trans* polymers. These experiments demonstrated the unusual attainment of monomer–polymer equilibrium (Scheme 8.11) under conditions which denied *cis–trans* equilibration (H). One can easily imagine other possible combinations of strange behavior as well.

In extreme contrast to the above equilibrium results, the nearly total lack of equilibrium behavior is seen in living ROMP polymerizations,[4] when norbornene-type monomers are used (Scheme 8.17). Here, the only process which can be considered to go to equilibrium is the total conversion of the monomer to polymer. All processes listed above, with the possible exception of the chain transfer process (Scheme 8.12), appear to be virtually inaccessible in these polymerizations. (However, chain transfer during propagation appears not to have been examined, since the addition of acyclic olefins to these polymerizations would be counterproductive for making living polymers.) It is likely that the hindered structure of the norbornene polymer chain fortuitously plays a major role in inhibiting secondary metathesis reactions.

Other polymerization systems can demonstrate behavior which reflects some extent of contribution of the various processes listed above. Chain transfer and post-polymerization chain scission have been well documented. In fact, the chain scission process (alternating combination of Scheme 8.12 and Scheme 8.13) forms the basis of polymer analysis methods developed by Hummel and co-workers,[45] who used large excesses of olefin to break down unsaturated polymers into small, characterizable fragments.

Acyclic olefins having different structures vary in their reactivity during the course of ROMP polymerization, which affects the molecular weight of the polymer prior to attainment of equilibrium. This reflects a dependence of the chain transfer rate (Scheme 8.12) on the structure of the olefin. In the case of cyclopentene polymerization,[29] the order of reactivity of olefins is: C═CR > RC═CR ≫ C═CR_2, RC═CR_2, R_2C═CR_2. In some situations, α-olefins can be more reactive than the cyclic monomer, resulting in molecular weights which are lower initially than they are at the end of the reaction. The opposite effect can be observed with less reactive olefins.

Adventitious amounts of acyclic olefins are commonly found in

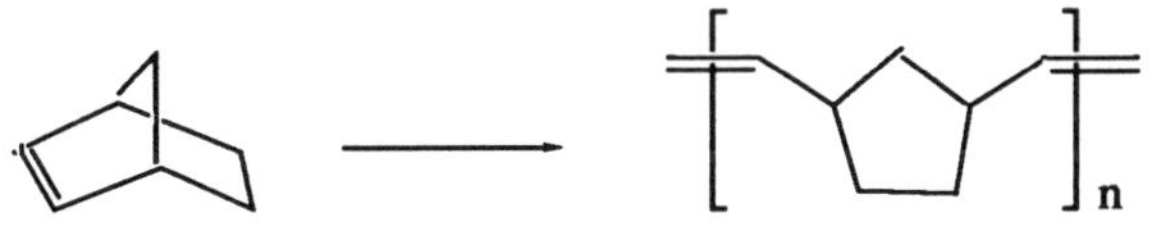

SCHEME 8.17.

commercially available monomers. Linear olefins such as 2-pentene are normally present in cyclopentene, and 4-vinylcyclohexene appears to be an unavoidable impurity in 1,5-cyclooctadiene. It is important to monitor the levels of minor impurities such as these in order to control the molecular weight of polymers.

The formation of most-probable molecular weight distributions for the linear, high-molecular-weight fraction of ROMP polymers (Scheme 8.15) is common in the polymerization of the simpler cycloolefin monomers[46] although, unfortunately, molecular weight distribution data are rarely reported in the literature for systems othan than living polymers. The narrow distributions seen in preparations of living polymers are indeed unique for metathesis polymerizations.

8.3.4. ROMP Termination

Hard data related to termination reactions remain largely unavailable, similar to the situation with the initiation process. The subject remains filled with much speculation, sometimes supported by indirect evidence. Any catalyst/monomer system would have to be evaluated individually if an attempt to establish likely termination processes were to be made. An abbreviated listing of some of the proposed termination reactions is given below:

- cyclopropanation, involving reaction of the carbene–metal species with the olefinic substrate,
- formation of a relatively unreactive carbene–metal species via reaction with an impurity in the system such as a conjugated diene,
- elimination of the organic portion of the intermediate metallocycle, with direct formation of an olefin,
- reductive coupling of two carbene–metal species, leading to an olefin and dimerization of the metal catalyst,
- β-hydrogen transfer from the carbene to the metal, accompanied by olefin elimination,
- reaction with miscellaneous adventitious impurities in the system. Unknown polar, possibly protic entities, which may contain oxygen or nitrogen, for example, often terminate the active catalyst or, at the very least, modify its behavior.

The use of controlled termination reactions can be an effective route to well-defined polymer chain ends. For example, the addition of an aldehyde leads to a Wittig-type coupling of the aldehyde to the active carbene.[47] Techniques developed by Grubbs and co-workers have opened the door to a most interesting host of block- and graft-copolymers.

The virtual absence of termination processes can be demonstrated when ultrahigh-molecular-weight polymers are obtained. Certain monomer/catalyst combinations have been shown to produce such materials. For example, polynorbornene (Norsorex) having a molecular weight greater than 2,000,000 is produced and sold commercially.[48] Similar results have been obtained with polypentenamer, where inherent viscosities greater than 10 have been measured when *cis*-directing catalysts have been used.[43] Lack of termination is also a critical requirement which has been achieved in the preparation of living polymers via ROMP, and in ADMET polymerizations as well.

8.4. SPECIFIC FEATURES OF ADMET CHEMISTRY

8.4.1. Polymerizability of Dienes

Less detail can be provided for ADMET polymerization, given that its discovery is relatively new. In principle, any diene capable of ejecting a small molecule which can be removed to drive the polymerization should be amenable to ADMET polymerization. Exceptions, of course, are monomers more susceptible to cyclization (intramolecular chemistry) rather than polymerization (intermolecular chemistry), which is a typical constraint for any step polymerization reaction monomer.

The polar (or nonpolar) nature of the diene must also be considered, since as previously stated a metathesis catalyst can be poisoned by the presence of polar functional groups. Thus far, only hydrocarbon and aromatic α,ω-dienes have been polymerized successfully by ADMET chemistry, though there are indications that dienes possessing silicon and oxygen may also polymerize.

While α,ω-dienes have been investigated most thoroughly to date, other dienes should polymerize also, and in fact should do so more rapidly. This research is being undertaken very actively at the moment.

8.4.2. The ADMET Polymerization Cycle. Consequences of the Carbene Mechanism

With the exception of the propagation step in ADMET chemistry, all the processes shown for ROMP reactions (Schemes 8.11–8.16) should be operative in ADMET polymerization as well. However, because of the very different starting point in ADMET reactions, the mechanistic steps of major importance early on are equivalent to the reverse steps in chain transfer and chain scission (Schemes 8.12 and 8.13 above), since olefin is eliminated during ADMET, rather than being incorporated into the polymer.

The ADMET process is a classical example of a polycondensation polymerization, promoted in this case by metathesis chemistry. It might be expected that the equilibrating processes D–H (where applicable) would be occurring simultaneously with olefin elimination, since the "polymerization" step does not have any unique driving force, such as cycloolefins provide, which would accelerate chain growth relative to other processes. In fact, molecular distributions observed with ADMET products are close to 2 as would be expected, and equilibrium *trans* microstrucuctures can also be achieved.[22] Reversibility of the process is confirmed by the drop in molecular weight which occurs if ethylene is reintroduced to the reaction.

The polymerization cycle as envisioned for ADMET chemistry is shown in Fig. 8.1. In this scheme, a diene monomer forms an initial metallacyclobutane with the catalyst carbene, which then cleaves to attach the catalyst moiety at the monomer chain end. Propagation then occurs (clockwise in the scheme) where another monomer molecule or polymer forms another metallacyclobutane leading to the elimination of a methylene carbene, resulting in the first connection in a growing chain. The methylene carbene encounters another monomer or polymer unit, followed by the evolution of ethylene in the case for an α,ω-diene. This polymerization cycle then continues in a stepwise manner to produce a high-molecular-weight polymer if ethylene is efficiently removed from the system. Thus, in ADMET polymerization the propagation step involves two different metallacyclobutane intermediates in the complete cycle, as compared with only one for ROMP.

8.5. RECENT TRENDS IN METATHESIS POLYMERIZATION

Several major areas of activity stand out. These topics have led metathesis research into interesting territories considered inaccessible only a decade ago. These activities have benefited significantly from recent major advances in catalyst development.

8.5.1. Aylic Diene Metathesis Polymerization (ADMET)

As previously stated, ADMET polymerization chemistry proceeds by a condensation polymerization mechanism in which a diene is condensed via the formation of two metallacyclobutanes to produce repeat units identical to those as described for ROMP chemistry. Since many cycles of degenerate steps must occur for each productive cycle, and since mechanistically two metallacyclobutane formations are required in each propagation reaction, the demands on the catalysts are more stringent. As a consequence, until recently no successful ADMET polymerization had been observed. Research in the 1970s on the ADMET polymerization of 1,4-pentadiene was discontinued since high polymer could not be formed.[19,49,50]

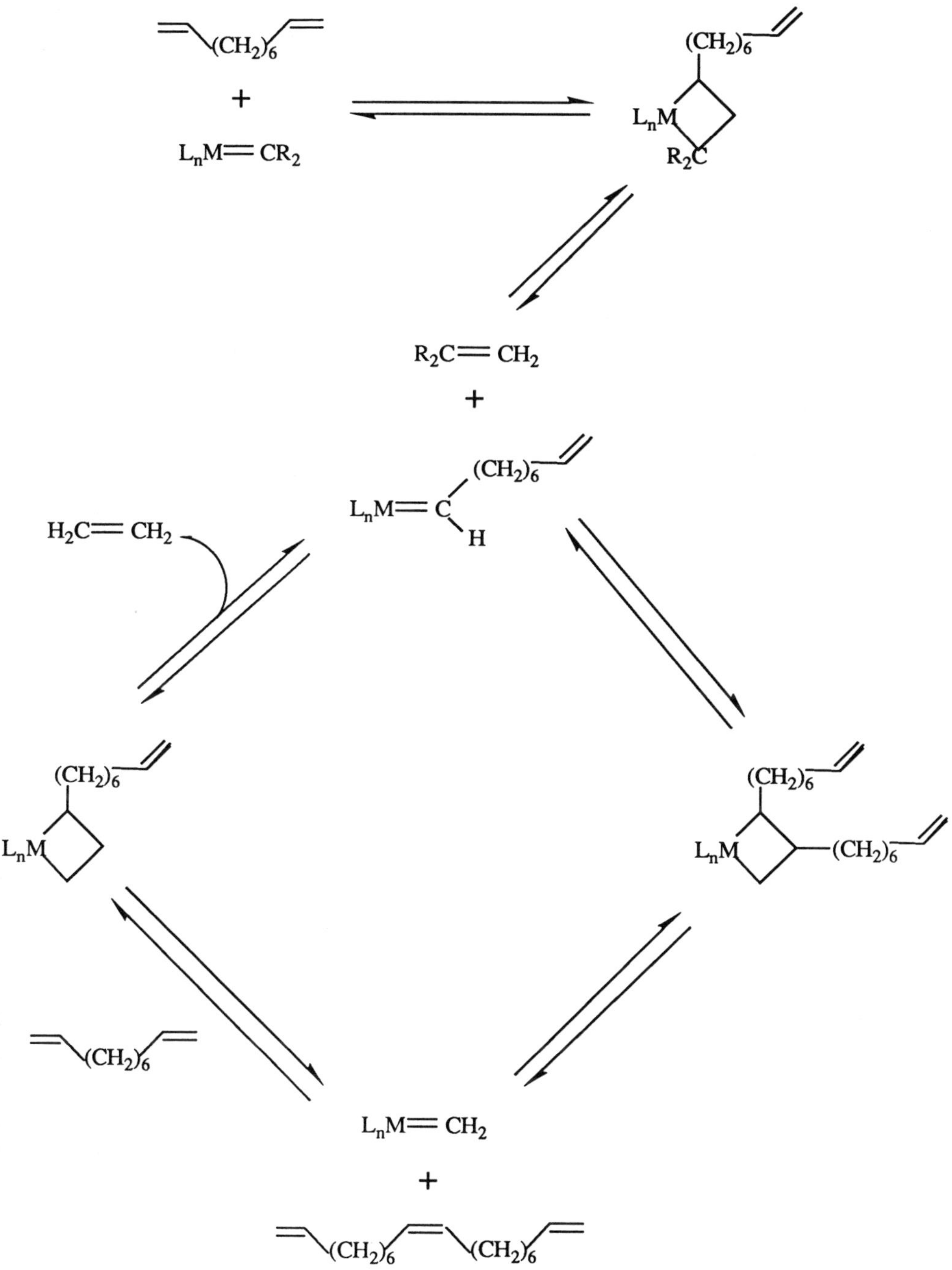

FIG. 8.1. The ADMET polymerization cycle.

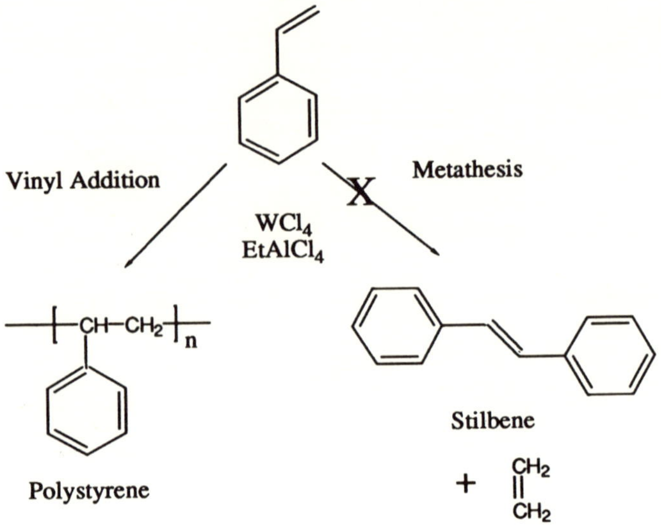

SCHEME 8.18.

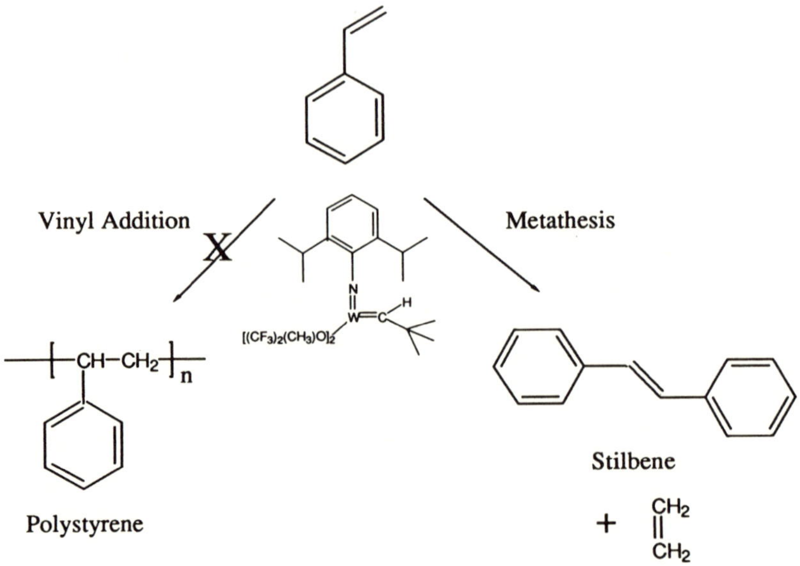

SCHEME 8.19.

Recently it has been demonstrated that the key to success in ADMET polymerization consists of obviation of the formation of cations during the metathesis reaction. This is illustrated in Schemes 8.18 and 8.19 with styrene.[51,52] It became apparent that the attempted metathesis of styrene to produce *trans*-stilbene was not possible in the presence of Lewis acid-containing catalyst systems. Instead, vinyl addition chemistry via the formation of cations led to polymerization generating polystyrene, and as a consequence, Lewis acid-free catalyst systems were considered instead.

The nature of the styrene reaction can be dictated completely by the choice of the type of catalyst employed where a Lewis acid-free catalyst, in particular Schrock's catalyst, results in exclusive metathesis generating *trans*-stilbene in quantitative yields (Scheme 8.19). This has proven to be a general reaction for a variety of substituted styrenes.

With this new information in hand, ADMET polymerization has been successfully demonstrated by condensing 1,9-decadiene to polyoctenamer, as illustrated in the reaction in Scheme 8.20.[21] The polymerization was done in vacuum under bulk conditions initially, but toluene was added later in order to maintain liquid state at a temperature of less than 50 °C. Polymerization initiated instantaneously upon contact of Schrock's catalyst with the monomer decadiene (1:5000), and ethylene evolved during the entire course of the reaction. The polymer which resulted produced a carbon NMR as illustrated in Fig. 8.2 wherein all resonances can be assigned on the basis of polyoctenamer produced by ring-opening metathesis polymerization chemistry. The polymer is pure and is high in molecular weight (intrinsic viscosity of 0.95, correlating to a number average molecular weight having a minimal value of 50,000).

Further evidence for the step nature of the ADMET polymerization is shown in Fig. 8.3 which presents a gel permeation chromatogram of polyoctenamer made by this technique. ADMET polymerization generates a molecular weight distribution value of 2, which is statistically predicted for

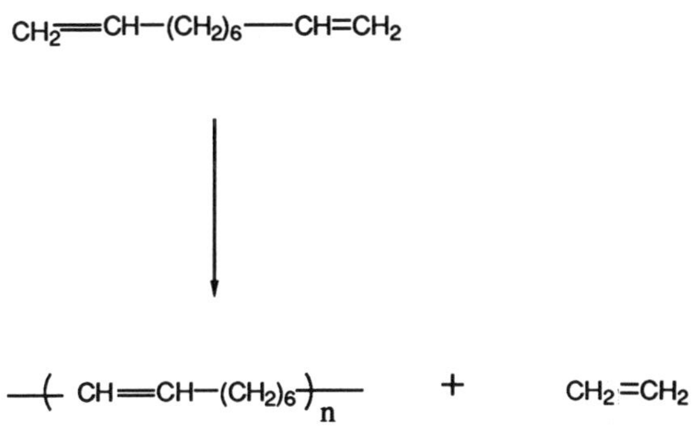

SCHEME 8.20.

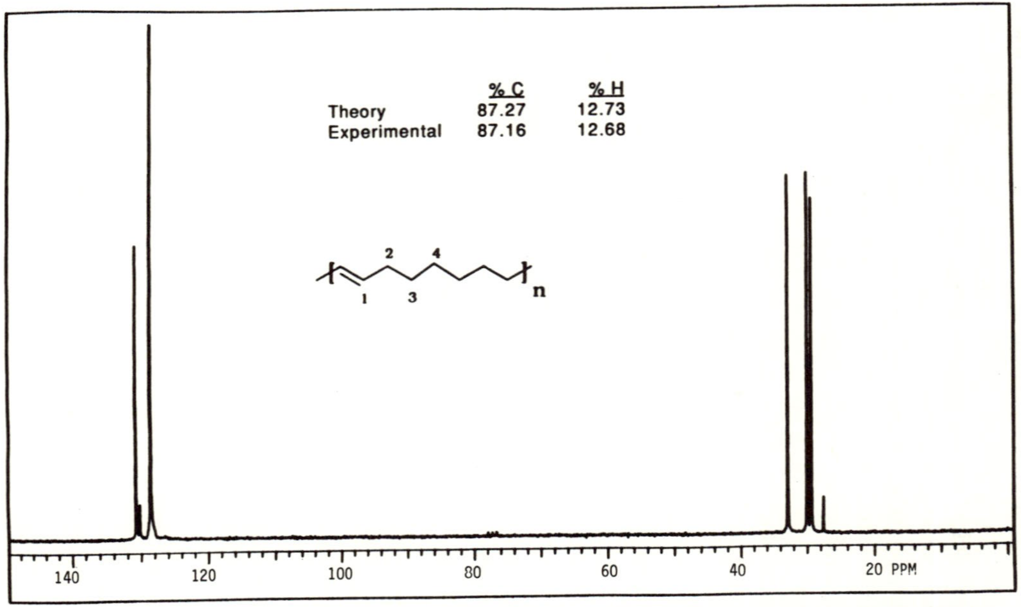

FIG. 8.2. ^{13}C NMR of 92% trans polyoctenamer formed by acyclic diene metathesis polymerization. IV = 0.95 dL/gram; conversion = 99 + %; C_1(*cis*) = 129.8; C_1(*trans*) = 130.4, C_2(*cis*) = 27.3; C_2(*trans*) = 32.7, C_3 = 29.8, C_4 = 29.1; solvent = 128.

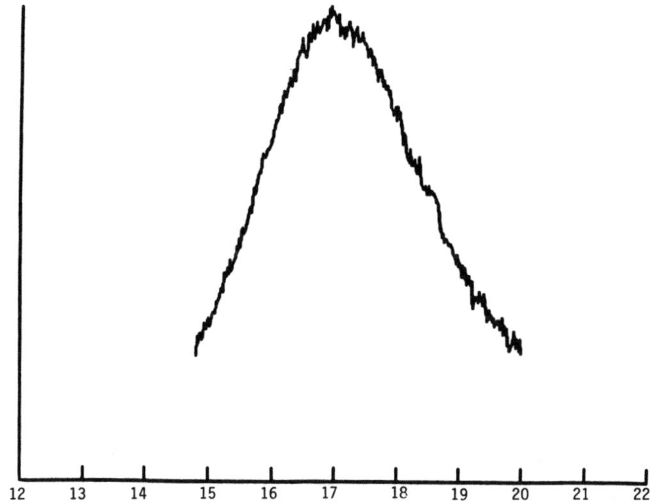

FIG. 8.3. Size exclusion chromatography curve of linear polyoctenamer produced by ADMET polymerization of 1,9-decadiene. $M_w = 16352$; $M_n = 8258$; $M_w/M_n = 1.98$.

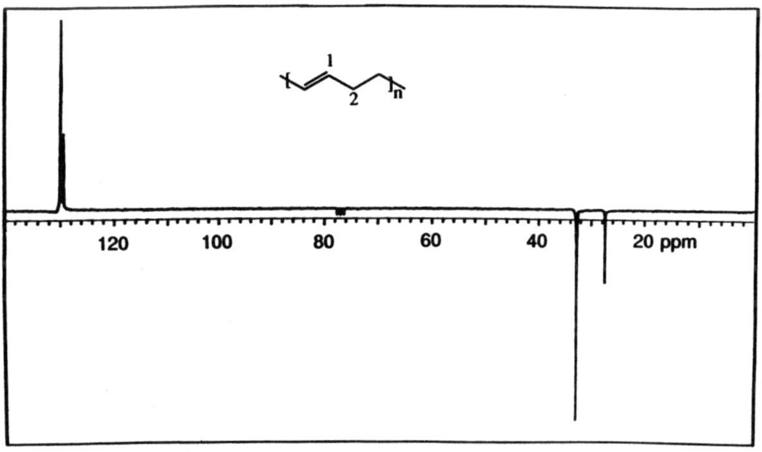

FIG. 8.4. ^{13}C NMR (attached proton test) of linear 1,4-polybutadiene produced by ADMET polymerization of 1,5-hexadiene. C_1(*cis*) = 129.4, C_1(*trans*) = 130.0, C_2(cis) = 27.5, C_2(trans) = 32.7.

any step polymerization reaction. Note that the distribution is uniform in its shape and illustrative of the random nature of the polymerization itself.

Other homopolymers that have been prepared by the ADMET technique include polyphenylene vinylene (from β,β'-dipropenylbenzene), 1,4-polybutadiene (from 1,5-hexadiene),[22] and a variety of silicon- and ether-containing polymers.[53]

Figure 8.4 also shows the carbon NMR spectra of 1,4-polybutadiene produced by ADMET polymerization which illustrates how clean the polymerization is itself.

Copolymers of hexadiene and decadiene have been made following the technique illustrated in Scheme 8.20. Since these monomers possess olefins of equal reactivity, random copolymers would be predicted to be formed, and this turns out to be the case.[54] The carbon NMR spectrum (Fig. 8.5) of the copolymer which is formed shows the sp^2 carbon region to be more complicated as would be anticipated by different types of connections between copolymer repeat units. The expanded carbon region (Fig. 8.6) has been completely assigned (Table 8.1); Table 8.2 shows that the composition of the polymer is dictated by monomer feed, as would be expected in any step polycondensation reaction. Only ethylene is released, with the remaining fragments of the monomers being placed into the polymer chain itself.

8.5.2. Ring-Opening Metathesis Polymerization (ROMP)

Since this metathesis route to producing polymers is more mature than ADMET polymerizations, a number of different areas of ROMP research have advanced quite rapidly. Each is discussed in some detail below.

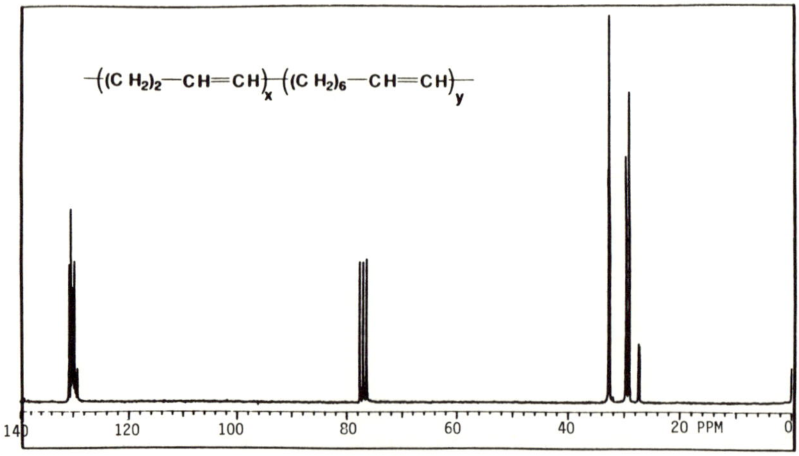

FIG. 8.5. ^{13}C NMR of copolymer.

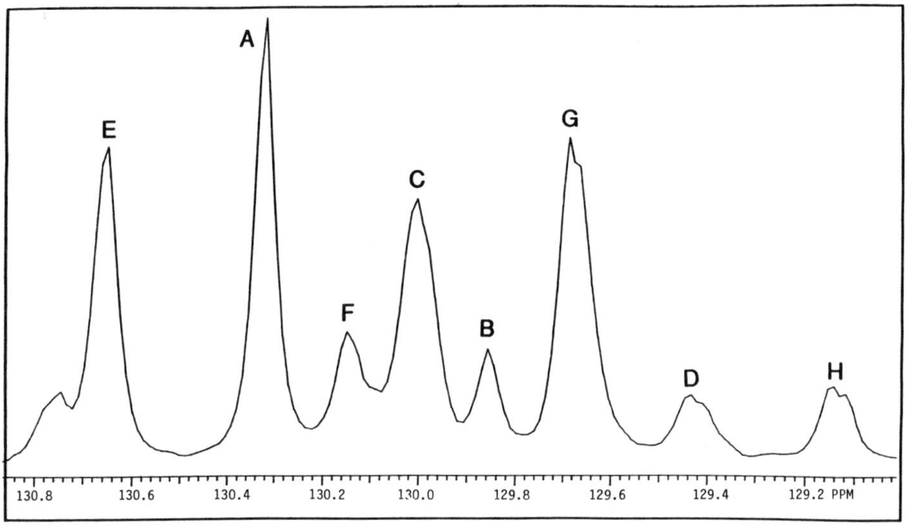

FIG. 8.6. ^{13}C NMR expanse of sp^2 region of polymer.

8.5.2.1. Functionalized Cycloolefin Monomers for Making Reactive Polymers

Early metathesis catalyst systems tended to be very sensitive to polar entities such as ester or alcohol groups. Thus, attempts to polymerize polar-substituted monomers were relatively unsuccessful initially, although a great many monomers were studied. However, the new generation of stable, relatively nonacidic catalysts based on W, Mo, Ti, and Ta has significantly broadened the scope of materials which may be considered for use as feasible monomers.

TABLE 8.1. Copolymer (50:50) Quantitative Carbon NMR Peak Assignments

Peak	ppm	Monomer unit linkages		*Relative # of carbons* observed	*Relative # of carbons* statistical
A	130.32	*Trans*	1,9–1,9		
B	129.86	*Cis*	1,9–1,9	1.00	1.00
C	130.00	*Trans*	1,5–1,5		
D	129.44	*Cis*	1,5–1,5	0.87	1.00
E	130.64	*Trans*	1,9–1,5		
F	130.14	*Cis*	1,9–1,5	2.10	2.00
G	129.68	*Trans*	1,5–1,9		
H	129.14	*Cis*	1,5–1,9	2.00	2.00

TABLE 8.2. Monomer Ratios (1,9-Decadiene : 1,5-Hexadiene)

Monomer feed	^{1}H NMR[a]	^{13}C NMR	
		Methylene[b]	Alkene[c]
50:50	48:52	51:49	49:51
75:25	72:28	75:25	75:25
95:5	95:5	92:8	92:8

[a] Determined by integration of methylene protons (1.8–2.0 ppm).
[b] Determined by integration of methylene carbons (26–34 ppm).
[c] Determined by integration of alkene carbons (129–131 ppm).

Over the years, many substituted norbornenes possessing a wide range of polar substituents have been subjected to metathesis polymerization, often with some measure of success. These monomers were of interest because they could be readily prepared through the Diels–Alder condensation of cyclopentadiene with diverse functional olefins. As tabulated by Ivin,[(55)] the remarkable range of substituents tested includes halogen, nitrile, alcohol, ether, ester, carboxylic acid and anhydride, ketone, amide, and various silanes! Tungsten-based catalysts were often employed, but more tolerant catalysts based on Re, Ru, Os, or Ir were more often required. In general, these experiments did not produce very practical results; high levels of expensive catalysts were required, yields were often poor, and control of the process was generally not satisfactory.

One effective approach to the preparation of a hydroxyl-bearing polymer is to block the functional site with a substituent which can be tolerated by the catalyst. Chung and Ramakrishnan prepared a precursor polymer using an organoborane derivative of norbornene, which could then be oxidized with hydrogen peroxide to the desired hydroxy polymer,[(56)] as in Scheme 8.21. A variety of tungsten catalysts were effective for this polymerization.

The new generation of carbene-bearing catalysts shows some promise in the preparation of certain classes of polar-substituted norbornenes. Aldehydes (and presumably ketones) may not be tolerated with these systems, since they terminate polymerizations via a Wittig-type exchange. However, a variety of cyclic amines (**3**) and ethers (**4**)[(57)] were polymerized

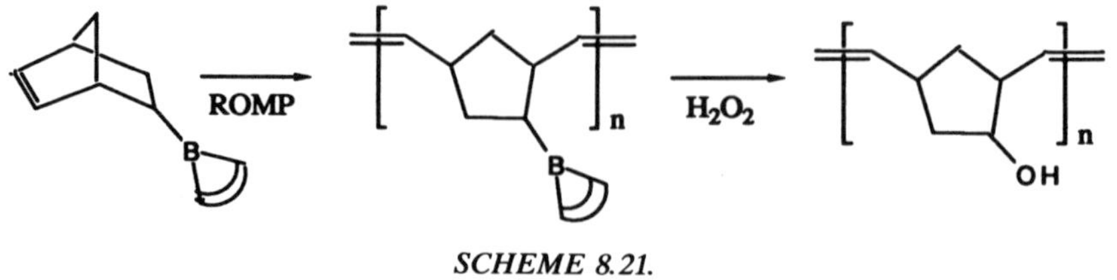

SCHEME 8.21.

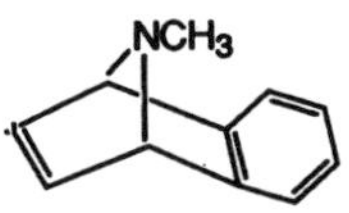

3

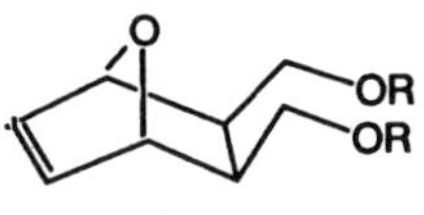

4

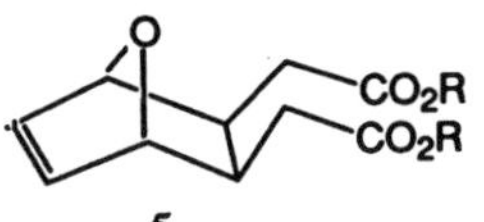

5

with a tungsten catalyst. The more stable molybdenum catalyst was preferred for the polymerization of the ester (**5**).[58]

Very early studies demonstrated that Ru and Os salts could affect the ring opening of norbornene under highly protic conditions in the presence of water or alcohol.[59,60] Recently it was found that Ru salts showed very high activity in aqueous polymerizations of monomer **4**, and that on reuse of the residual aqueous layer, catalyst activity increased remarkably.[61] The polymers which resulted were water soluble, and possessed unique ionophoric properties similar to crown ethers.

8.5.2.2. New Routes to Conducting Polymers

Conducting polymers, usually variations on the polyacetylene structure, have been prepared by several different approaches involving metathesis chemistry. The most direct approach has been via the ROMP polymerization of cyclooctatetraene,[62] which produces a fully conjugated structure directly as in Scheme 8.22. The previously-described tungsten carbon complex **1** has been very effective for this polymerization. The solubility and processing characteristics of this polymer can be improved by incorporating other structural units into the polymer chain, for example, by copolymerization with cyclooctadiene.[63]

An unusual route to a cross-conjugated, potentially conducting polymer involves the polymerization of 3,4-diisopropylidenecyclobutene, as shown in Scheme 8.23.[64] Polymerization was initiated using a titanacene–carbene precursor similar to that described in Scheme 8.10 above.

Other major routes to conducting polymers have involved the preparation of precursor polymers which have the capability to be converted readily to the conjugated structure by means of thermal or other activation processes. The first approach to be developed was the Durham route, involving the use of "Feast" monomers, as in Scheme 8.24.[65,66] A conventional WCl_6/Me_4Sn catalyst was effective for this class of monomers,

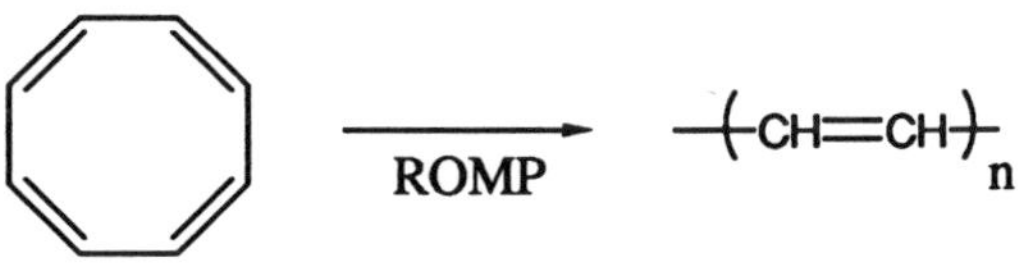

SCHEME 8.22.

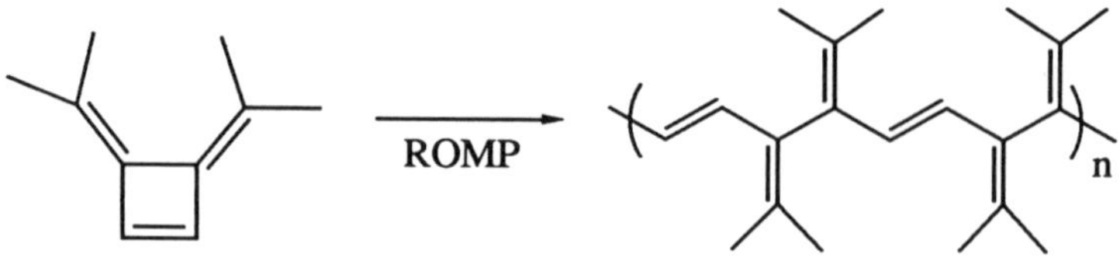

SCHEME 8.23.

although the new carbene catalysts **1** and **2** above have also been used to prepare polyacetylenes of low polydispersity.[67]

In this case, a substituted tricyclodecatriene polymerizes exclusively through the cyclobutene ring, leading to a soluble, well-characterized precursor to polyacetylene. After the intermediate polymer is converted to the desired form (usually a film), heating results in elimination of a stable, volatile aromatic compound to form the desired conjugated structure. Modifications of this general structure have allowed the preparation of more stable precursor polymers and provided for a more convenient decomposition temperature, increasing the utility of the scheme.[66]

Another example of the precursor approach to polyacetylene is illustrated in Scheme 8.25 where benzvalene is polymerized to an unstable polymeric structure. The desired conjugated polymer is then obtained by isomerization induced by $HgCl_2$.[68] Here again, catalyst **1** was employed.

It should be noted that polyacetylenes have also been prepared by the direct polymerization of acetylene with many catalysts, including metathesis-type catalysts.[69] However, these preparations do not appear to involve conventional metathesis chemistry.

8.5.2.3. Living Polymers

Recent progress in developing techniques to synthesize living polymers from certain types of cycloolefins has created a number of fascinating opportunities for making new functional or block- and graft-copolymers by metathesis chemistry. These results are based on the use of the new-generation carbene catalysts referred to previously, in combination with norbornene derivatives primarily.

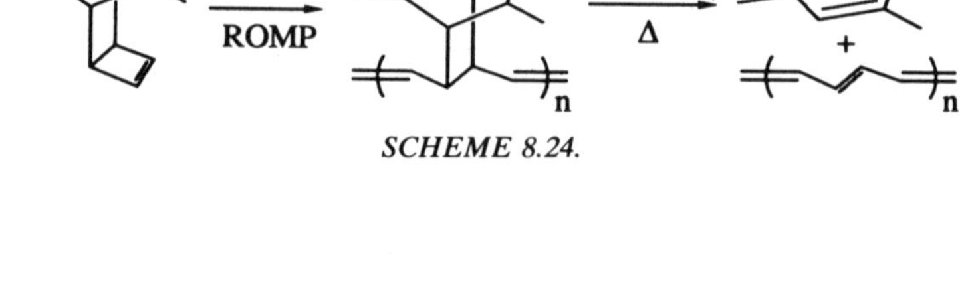

SCHEME 8.24.

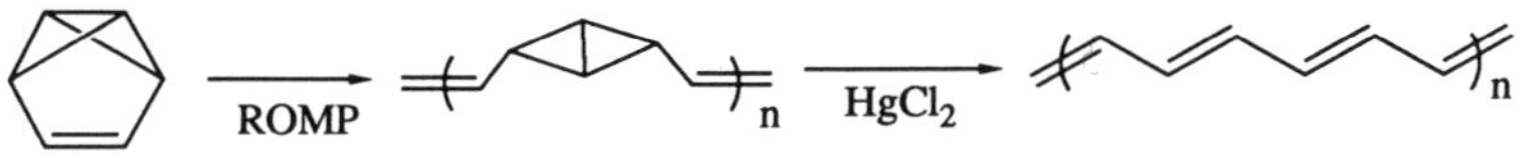

SCHEME 8.25.

Catalysts such as that shown in Scheme 8.10, and the more recently-developed ones of type **1** and **2**, which evolved from early carbene-related catalyst investigations of Schrock,[70] provide some of the requirements needed for the attainment of monodisperse living polymers, namely, rapid initiation relative to propagation, freedom from multiple types of active species during polymerization, and lack of termination or chain transfer reactions.

In metathesis polymerizations, however, lack of interchain transfer reactions is dependent also on the nature of the monomer–polymer system being investigated. For example, monodisperse polymers have not been attainable with cyclopentene-based systems, presumably due to the interchain reactions (Schemes 8.15 and 8.16). Norbornene-type monomers are much preferred because they offer significant advantages here. They possess the unique combination of enhanced polymerizability due to ring strain, coupled with a greatly reduced tendency to undergo undesired secondary metathesis of the polymeric vinylene units, inhibited by steric hindrance due to branching at the α-carbons. Consequently, chain transfer steps can usually be ignored in these systems.

The first living polymerization of a cycloolefin was reported by Grubbs and Gilliom in 1984 for norbornene.[71] Classical verification of this achievement was demonstrated by the linear increase in molecular weight with conversion, by the direct correspondence of molecular weight with the amount of catalyst used, and by the narrow polydispersities obtained of less than 1.1. These living polymers possessed excellent stability of the active chain end if stored protected at room temperature.

Polymerizations could be reinitiated after several months by the addition of fresh monomer. These polymerizations are quite temperature sensitive. Below a certain threshold temperature of about 50–70 °C, the catalyst becomes dormant, but can be reinitiated immediately on rewarming.

Living polymerizations have also been carried out using tantalum-based complexes. The tantallacyclobutane initiator was prepared by the low-temperature coupling of norbornene with a carbene–tantalum complex.[72] However, polymerization of norbornene at 50 °C resulted in a polymer with a relatively broad molecular weight distribution, and there was also some evidence of secondary interchain metathesis as indicated in a slight change in the *cis–trans* content of the polymer with time. These results emphasize the fact that a stable, living catalyst can still lead to broad polydispersities with metathesis chemistry if secondary reactions can occur.

Living diblock metathesis copolymers involving polar-substituted monomers have been prepared by using the molybdenum catalyst **2** above. Thus, sequential addition of monomers was employed in order to prepare a block of polynorbornene, followed by a block of copolymer based on the polar monomer 5,6-dicarbomethoxynorbornene.[34]

An intriguing variety of applications of living metathesis polymer technology for the preparation of block- and graft-copolymers has been described by Risse and Grubbs.[73] Many of these involve termination of the active carbene by coupling with various aldehydes, followed in some cases by a second-stage process. For example, termination of a norbornene polymerization by coupling with a polyether ketone afforded a graft-copolymer in Scheme 8.26. Similar coupling chemistry was used to generate block copolymers by reaction of the living polymer with a block of another polymer which possessed terminal ketone units.

One fascinating version of the diblock preparation process involved the transformation of the terminal carbene to a carbonyl end group through reaction with an excess of a dialdehyde.[73] As shown in Scheme 8.27, the resulting aldehyde-terminated polymer was then used as the initiator for the polymerization of *t*-butyldimethylsilyl vinyl ether by means of the Aldol-Group Transfer Polymerization process,[74] activated by the Lewis acid $ZnCl_2$. The resulting polymer could be hydrolyzed to generate a hydrophilic polyvinyl alcohol block. Other indirect routes to metathesis block polymers have also been reported. For example, Amass used a living polystyryl lithium block as the initiator for the metathesis polymerization of cyclopentene, and found that some diblock polymer was obtained.[75] However, the catalyst transformation process was not very efficient, and much of the product was homopolymer with typical broad polydispersity.

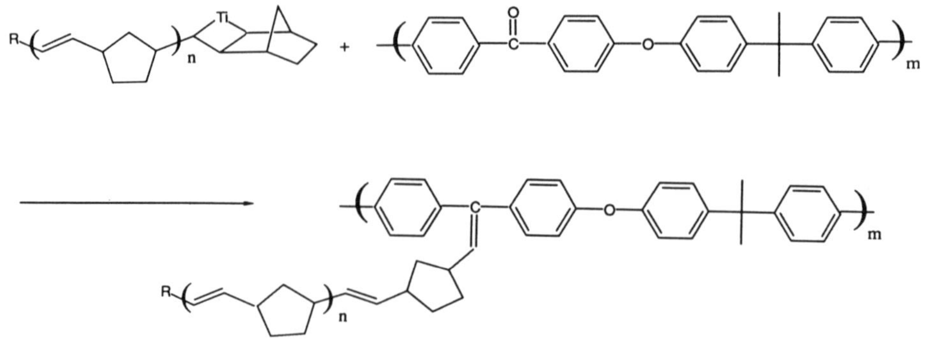

SCHEME 8.26.

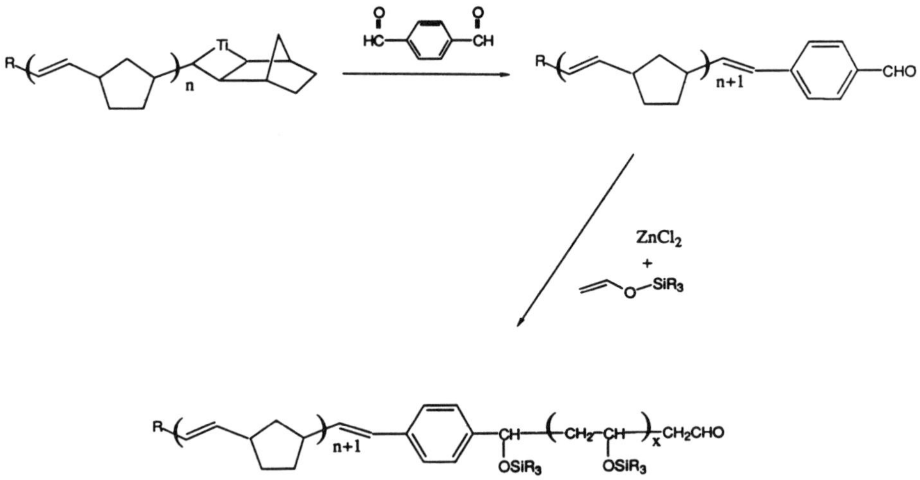

SCHEME 8.27.

8.5.2.4. Commercial Polymers

Three types of ROMP polymers have been commercialized, although several other materials have been intensively investigated as well. The products are Norsorex (from norbornene), Vestenamer (cyclooctene), and Metton (dicyclopentadiene).

Norsorex was the first product to be introduced. Production was begun by CdF Chimie in France in 1975. Production capacity is substantial, and currently believed to be about 5 MM lb per yr.[76] This polymer has certain unique properties, including a glass transition temperature of about 35 °C, and an extremely high molecular weight of about 2 MM.[48] It is sold in the form of a high surface-area powder which can be easily incorporated into rubber formulations. Because its T_g is somewhat above room temperature, this polymer is not strictly an elastomer. However, it is customarily used in formulations which contain high levels of plasticizing oils. This effectively lowers the T_g of the blend, imparting elastomeric properties to the product. These formulations are usually highly hysteretic, and thus find use in vibration-damping applications, for example.

Norbornene polymers have played a major role in more academic areas of metathesis chemistry as well. The critical examination of polymers obtained from norbornene and its substituted derivatives has been an invaluable tool in elucidating intricate features of the stereochemistry of ring opening for a variety of catalyst systems, in extensive studies conducted by Ivin, Rooney, and co-workers.[77]

Trans-polyoctenamer was introduced as Vestenamer 8012 by Chemische Werke Huls in 1980.[44] This polyalkenamer has a high *trans* content of about 80% and is a hard, semicrystalline material at room temperature, but melts near 55 °C. It has a moderately low molecular weight for easy processability when melted, and it is sulfur vulcanizable. This combination of properties has made Vestenamer attractive as a blending agent for rubber compounds to improve processsability and dimensional stability of uncured parts.

Polymers from dicyclopentadiene (DCPD) have been explored extensively in several laboratories, primarily because of the low cost of the monomer. In 1982, B. F. Goodrich announced the development of a thermoplastic version of poly-DCPD which they named Telene,[78] although its commercialization has not yet been realized.

In 1984, Hercules Inc. announced the development of a RIM (reaction injection molding) technology based on DCPD.[79,80] The process involves the rapid mixing and injection into the mold of two monomer streams which separately contain the two catalyst components. Polymerization initiates instantly, and goes to completion rapidly. The product is called Metton. This material has a T_g well above 100 °C, and also has high strength, rigidity, and impact resistance, making it attractive for a variety of engineering applications. The structure of the polymer is represented in Scheme 8.28 where only one of the rings is opened, although there has been speculation in the literature that the other ring may also undergo metathesis and ring opening.

The polymer of cyclopentene, known as polypentenamer, was the first metathesis polymer to be seriously considered for commercial production. Many corporations had active projects to develop this elastomer. It possesses an attractive combination of properties as an elastomer,[16] and the monomer was expected to be low-cost since it is derived from the petrochemical byproduct cyclopentadiene. However, changes in feedstock availability and elastomer demand resulted in discontinuation of its development.

Two general forms of polypentenamer are known. These are the *trans* form, which has a crystalline melting point just below room temperature and thus is a strain-crystallizing elastomer, and the *cis* form, which is normally amorphous (melting point, −40 °C) and has an unusually low T_g of −114 °C.[81] The *trans* polymer is readily prepared using a wide variety of catalyst systems, such as alcohol-modified tungsten salts in combination with organoaluminum cocatalysts.[29] Catalyst activity is improved with the use of

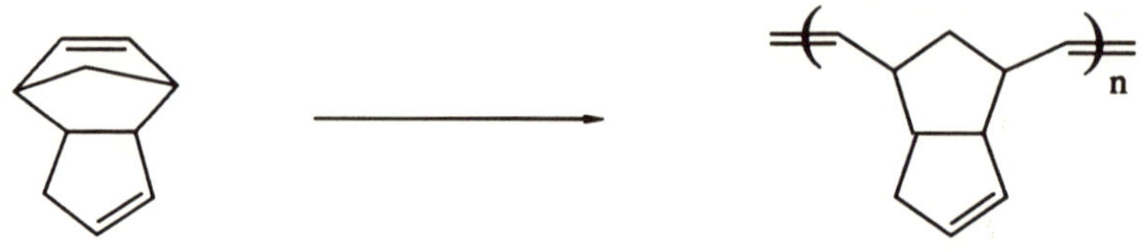

SCHEME 8.28.

β-halo alcohols. The preparation of the *cis* polymer is more difficult, with relatively few catalysts being effective. In some cases, low polymerization temperatures improve the *cis*-stereospecificity of the catalyst, as with $WF_6/EtAlCl_2$,[24,29] $MoCl_5/Et_3Al$,[12] or WCl_6/Et_3Al/dibenzoyl peroxide.[82]

8.6. PROCEDURES FOR METATHESIS POLYMERIZATION

8.6.1. Ring-Opening Metathesis Polymerization (ROMP)

In general, relatively simple reaction conditions are acceptable as long as the reaction can be protected adequately from moisture and air, the two most detrimental factors for most polymerization systems. Reaction temperatures for polymerizations usually range from ambient to about 75 °C. Furthermore, pressures are typically atmospheric or slightly positive, to reduce chances of air being drawn into the reaction mixture. Depending on the monomer/catalyst combination, reactions may take several hours and not reach completion, or may be complete in a few minutes.

Many procedures for achieving these conditions have been used. The simplest procedures utilize ordinary glass bottles or flasks, into which the monomers, solvent, and catalysts are introduced. Materials can be transferred by hypodermic syringe, for example, or even by being dispensed from reservoirs using simple valves and tubing, as long as a positive flow or pressure of an inert gas is maintained to prevent admission of air. More elaborate procedures have also been used, for example, involving Schlenke-type apparatus, or even high-vacuum rack methods for ultracontrol of impurities.

Purification of materials is a prime consideration, since traces of the wrong contaminants can prove disastrous. For this reason, it is often advisable to monitor monomer purity with sensitive gas chromatographic techniques. In polymerizations of cyclopentene, for example, levels of allenes as low as 10 ppm can be extremely detrimental to catalyst activity. Impurities such as cyclopentadiene are also to be avoided, and acyclic olefins must be monitored if molecular weight control is important. Obviously, amounts of only a few 100 ppm can have a significant effect on molecular weight, depending on the structure of the acyclic olefin. In the case of norbornene, commercially available monomer is very high in quality, and detrimental impurities other than potential oxidation products are less of a consideration.

Removal of polar impurities may require special treatment of the monomer, since olefinic monomers are particularly prone to oxidation over time, leading to various oxygenated species. Treatment of a solution of the monomer, by passing it through a purification column containing an absorbent such as silica gel, alumina, or molecular sieve, is usually a very effective process. When using high-vacuum equipment, distillation of monomers from sodium mirrors is also effective when working with nonpolar monomers.

In carrying out reactions, catalysts are often handled as solutions, and may be conveniently introduced into the reaction mixture by hypodermic syringe. When relatively high concentrations of monomer are used, it is advisable to use some sort of temperature control, such as a constant temperature bath, to prevent fast reactions from overheating. If it is intended to conduct experiments on neat reagents, it is best that small quantities be tested to verify that the process is safe to scale up to larger amounts.

Reactions can be terminated by a variety of techniques, including addition of polar protic substance such as alcohols (for most catalyst systems). Oxygen may be one of the most effective short-stop agents, but is not often used intentionally. In general, it is advisable to add some sort of stabilizer to the terminated reaction mixture to serve as an antioxidant for the unsaturated polymer after it has been dried. 2,6-Di-*t*-butyl-*p*-cresol is commonly used for this purpose, but there are many other effective stabilizers as well.

8.6.2. Acyclic Diene Metathesis Polymerization (ADMET)

Metathesis polymerizations are conducted using fairly standard synthetic procedures as found in polymer chemistry. Generally speaking, ROMP chemistry is done using solution polymerization techniques, while ADMET polymerization is conducted under bulk reaction conditions. Additionally, the dicyclopentadiene polymerization generating thermoset polymers is conducted under conditions required for reaction injection molding (RIM reaction).

Syntheses carried out using ADMET chemistry have all been accomplished on a small scale using vacuum line techniques to achieve high-molecular-weight polymers. In a typical experiment, monomer is rigorously purified to remove oxygen and water, then sealed in glass ampules which are then attached to the vacuum line. The catalyst is also crystallized several times before being dissolved in a nonpolar solvent which is free of water and oxygen, and then placed in a reaction ampule. This ampule is also sealed to the vacuum line. The whole apparatus is dried thoroughly, followed first by introduction of the catalyst, then introduction of ampules of monomer. Polymerization begins instantaneously and ethylene is collected in a liquid nitrogen trap mounted above the reaction vessel on the vacuum line. Polymerizations are done over a period of time where ethylene evolution is monitored until no additional ethylene is collected. The polymerizations are terminated simply by exposure to air, followed by characterization of the polymer with the usual techniques.

Though metathesis polymerization has seen commercial utility in only three or four cases, it is obvious that the renaissance of the reaction will lead to increased activity on a larger scale to generate new classes of polymers by this intriguing reaction.

REFERENCES

1. K. J. Ivin, *Olefin Metathesis* (Academic Press, London, 1983).
2. V. Dragutan, A. T. Balaban, and M. Dimonie, *Olefin Metathesis and Ring-Opening Polymerization* (Wiley, Chichester, 1985).
3. *Comprehensive Polymer Science,* ed. by G. Allen and J. Bevington, (Pergamon, Oxford, 1989), Vol. 4, Chap. 6 and 7.
4. R. H. Grubbs and B. M. Novak, in *Encyclopedia of Polymer Science and Engineering* (Wiley, New York, 1990), Suppl. I, p. 20.
5. K. J. Ivin, in *Encyclopedia of Polymer Science and Engineering* (Wiley, New York, 1988) Vol. 9, p. 634; E. A. Ofstead, in *Encyclopedia of Polymer Science and Engineering* (Wiley, New York, 1988), Vol. 11, p. 287.
6. R. H. Grubbs and W. Tumas, *Science* **243**, 907 (1989).
7. N. Calderon, H. Y. Chen, and K. W. Scott, *Tetrahedron Lett.* 3327 (1967).
8. R. L. Banks and G. C. Bailey, *Ind. Prod. Res. Dev.* **3**, 170 (1964).
9. J. C. Mol, J. A. Moulijn, and C. Boelhouwer, *J. Chem. Soc., Chem. Commun.* 633 (1968).
10. J. Levisalles, H. Rudler, and D. Villemin, *J. Organomet. Chem.* **193**, 235 (1980).
11. K. Tanaka, K. Tanaka, and K. Miyahara, *J. Chem. Soc., Chem. Commun.* 666 (1980).
12. G. Dall'Asta and G. Motroni, *Eur. Polym. J.* **7**, 707 (1971).
13. E. A. Ofstead and N. Calderon, *Makromol. Chem.* **154**, 21 (1972).
14. H. S. Eleuterio, U.S. Pat. 3,074,918, Jan. 22, 1963 (filed June 30, 1957).
15. W. L. Truett, D. R. Johnson, I. M. Robinson, and B. A. Montague, *J. Am. Chem. Soc.* **82**, 2337 (1960).
16. G. Natta, G. Dall'Asta, and G. Mazzanti, *Angew. Chem., Int. Ed. Engl.* **3**, 723 (1964).
17. N. Calderon, E. A. Ofstead, and W. A. Judy, *J. Polym. Sci.* **A-I**(5), 2209 (1967).
18. N. Calderon, E. A. Ofstead, J. P. Ward, W. A. Judy, and K. W. Scott, *J. Am. Chem. Soc.* **90**, 4132 (1968).
19. E. A. Zuech, W. B. Hughes, D. H. Kubicek, and E. T. Kittleman, *J. Am. Chem. Soc.* **92**, 528 (1970).
20. G. Dall'Asta, G. Stigliani, A. Greco, and L. Motta, *Chim. Ind.* (*Milan*) **55**, 142 (1973).
21. K. B. Wagener, J. M. Boncella, R. P. Duttweiler, and J. G. Nel, *Polym. Prepr., Am. Chem. Soc.* **30**(1), 283 (1989).
22. K. B. Wagener, J. G. Nel, and J. M. Boncella, *Polym. Prepr., Am. Chem. Soc.* **30**(2), 130 (1989).
23. C. P. C. Bradshaw, E. J. Howman, and L. Turner, *J. Catal.* **7**, 269 (1967).
24. J. L. Herisson and Y. Chauvin, *Makromol. Chem.* **141**, 161 (1970).
25. For example, see review in Ref. 1, pp. 43–78.
26. W. J. Kelly and N. Calderon, *J. Macromol. Sci., Chem.* **A9**, 911 (1975).
27. J. McGinnis, T. J. Katz, and S. Hurwitz, *J. Am. Chem. Soc.* **98**, 605 (1976).
28. K. J. Ivin, in *Reactivity, Mechanism and Structure in Polymer Chemistry,* ed. by A. D. Jenkins and A. Ledwith. (Wiley–Interscience, London, 1974), Chap. 16.
29. P. Gunther, F. Haas, G. Marwede, K. Nutzel, W. Oberkirch, G. Pampus, N. Schon, and J. Witte, *Angew. Makromol. Chem.* **14**, 87 (1970).
30. K. W. Scott, N. Calderon, E. A. Ofstead, W. A. Judy, and J. P. Ward, paper presented at Div. Ind. Eng. Chem., 155th National Meeting of the American Chemical Society, April 1968, Abstract L54; K. W. Scott, N. Calderon, E. A. Ofstead, W. A. Judy, and J. P. Ward, *Adv. Chem. Ser.* **91**, 399 (1969).

31. Y. Chauvin, D. Commereuc, and G. Zaborowski, *Recl. Trav. Chim. Pays-Bas* **96**, M131 (1977); Y. Chauvin, D. Commereuc, and G. Zaborowski, *Makromol. Chem.* **179**, 1285 (1978).
32. C. P. Casey and T. J. Burkhardt, *J. Am. Chem. Soc.* **96**, 7808 (1974).
33. R. R. Schrock, *J. Organomet. Chem.* **300**, 249 (1986); J. S. Murdzek and R. R. Schrock, *Organometallics* **6**, 1373 (1987).
34. J. S. Murdzek and R. R. Schrock, *Macromolecules* **20**, 2640 (1987).
35. T. R. Howard, J. B. Lee, and R. H. Grubbs, *J. Am. Chem. Soc.* **102**, 6876 (1980).
36. L. R. Gilliom and R. H. Grubbs, *J. Am. Chem. Soc.* **108**, 733 (1986).
37. E. Ceausescu, A. Cornilescu, E. Nicolescu, M. Popescu, S. Coca, and M. Cuzmici, *J. Mol. Catal.* **46**, 415 (1988).
38. H. Jacobson and W. H. Stockmayer, *J. Chem. Phys.* **18**, 1600 (1950).
39. H. Hocker, W. Reimann, L. Reif and K. Reidel, *Recl. Trav. Chim. Pays-Bas* **96**, M47 (1977); H. Hocker, W. Reimann, L. Reif, and K. Riebel, *J. Mol. Catal.* **8**, 191 (1980).
40. D. Kranz and M. Beck, *Angew. Makromol. Chem.* **27**, 29 (1972).
41. K. L. Makovetskii and L. I. Red'kina, *Dokl. Akad. Nauk SSSR* **231**, 143 (1976).
42. A. Drapeau and J. Leonard, *Macromolecules* **18**, 144 (1985).
43. E. A. Ofstead, J. P. Lawrence, M. L. Senyek, and N. Calderon, *J. Mol. Catal.* **8**, 227 (1980).
44. A. Draxler, *Kautsch. Gummi. Kunstst.* **34**, 185 (1981).
45. K. Hummel, *Pure Appl. Chem.* **54**, 351 (1982).
46. E. A. Ofstead, unpublished results.
47. W. Risse and R. H. Grubbs, *Macromolecules* **22**, 4462 (1989); W. Risse and R. H. Grubbs, *Polym. Prepr., Am. Chem. Soc.* **30**(1), 193 (1989).
48. R. F. Ohm, *Chemtech,* 183 (1980).
49. G. Dall'Asta, *Pure Appl. Chem.* **1**, 133 (1973).
50. J. Doyle, *J. Catal.* **30**, 118 (1973).
51. M. L. Hamberg and K. B. Wagener, *Macromolecules* **20**, 2949 (1987).
52. K. B. Wagener, J. M. Boncella, J. G. Nel, R. P. Duttweiler, and M. A. Hillmyer, *Makromol. Chem.,* **191**, 365 (1990).
53. K. B. Wagener, J. M. Boncella, K. Brzezinska, and D. W. Smith, Jr., manuscript in preparation.
54. K. B. Wagener, J. G. Nel, J. Konzelman, and J. M. Boncella, *Macromolecules,* to appear; *Polym. Prepr., Am. Chem. Soc.* **31**(1), 464 (1990).
55. K. J. Ivin, *Olefin Metathesis* (Academic Press, London, 1983), Chap. 13.
56. T. C. Chung and S. R. Ramakrishnan, *Polym. Prepr., Am. Chem. Soc.* **31**(1), 55 (1990).
57. L. F. Cannizo, L. R. Gilliom, and R. H. Grubbs, unpublished results cited in Ref. 6.
58. J. S. Murdzek and R. R. Schrock, *Macromolecules* **20**, 2460 (1987).
59. F. W. Michelotti and W. P. Keaveney, *J. Polym. Sci.* **A3**, 395 (1965).
60. R. E. Rhinehart, *J. Polym. Sci.* **C27**, 7 (1969).
61. B. M. Novak and R. H. Grubbs, *J. Am. Chem. Soc.* **110**, 7542 (1988).
62. F. L. Klavetter and R. H. Grubbs, *Polym. Prepr., Am. Chem. Soc.* **28**(2), 425(1987); F. L. Klavetter and R. H. Grubbs, *J. Am. Chem. Soc.* **110**, 7807 (1988).
63. F. L. Klavetter and R. H. Grubbs, *Polym. Mater. Sci. Eng.* **58**, 855 (1988); F. L. Klavetter and R. H. Grubbs, *Synth. Met.* **28**, D99 (1989).
64. T. M. Swager and R. H. Grubbs, *J. Am. Chem. Soc.* **109**, 894 (1987).
65. J. H. Edwards and W. J. Feast, *Polym. J.* **21**, 595 (1980).
66. W. J. Feast and J. N. Winter, *J. Chem. Soc., Chem. Commun.,* 202 (1985).
67. S. A. Krouse and R. R. Schrock, *Macromolecules* **21**, 1885 (1988).
68. T. M. Swager and R. H. Grubbs, *J. Am. Chem. Soc.* **111**, 4413 (1989).
69. T. J. Katz, S. J. Lee, and M. A. Shippey, *J. Mol. Catal.* **8,** 219 (1980); A. J. Amass, M. S. Beevers, T. R. Farren, and J. A. Stowell, *Makromol. Chem., Rapid Commun.* **8**, 119 (1987).

70. R. R. Schrock, *J. Am. Chem. Soc.* **96**, 6796 (1974); R. R. Schrock, *J. Am. Chem. Soc.* **98**, 3478 (1976).
71. R. H. Grubbs and L. R. Gilliom, *Proceedings of the 4th Int. Symp. on Homogeneous Catalysis,* 24–28 Sept. 1984, Leningrad, USSR (Gordon and Breach, New York, 1985); see also Ref. 36.
72. K. C. Wallace and R. R. Shrock, *Macromolecules* **20**, 450 (1987).
73. W. Risse and R. H. Grubbs, *Polym. Prepr., Am. Chem. Soc.* **30**(1), 193 (1989).
74. D. Y. Sogah and O. W. Webster, *Macromolecules* **19**, 1775 (1986).
75. A. J. Amass, S. Bas, D. Gregory, and M. C. Mathew, *Macromol. Chem.* **186**, 325 (1985).
76. *Plast. Rubber Wkly.* **1030**, p. 20, Mar. 24, 1985.
77. See Chap. 11 in Ref. 1.
78. L. P. Tenney, paper presented at Soc. Plast. Eng. Ann. Tech. Conf., ANTEC '82, San Francisco, 1982.
79. *Plast. World,* p. 22, May 1985.
80. R. P. Geer, *Soc. Plast. Eng. Regional Tech. Conf. Prepr.,* Detroit, MI, 1983, p. 104.
81. G. Dall'Asta and P. Scaglione, *Rubber Chem. Technol.* **42**, 1235 (1969).
82. R. T. Minchak and H. Tucker, *Polymer Prepr., Am. Chem. Soc.* **13**, 885 (1972).

9

Synthesis of Polymers of Isocyanides

Roeland J. M. Nolte and Wiendelt Drenth

9.1. INTRODUCTION

9.1.1. Synthesis of Isocyanides

In 1868 two articles appeared in *Annalen der Chemie und Pharmacie* on "a new series of homologues of hydrogen cyanide." In the first article A. W. Hofmann[1] reported on an extension of his earlier work[2] in which he described the formation of isomers of nitriles in a reaction between a primary amine, chloroform, and alcoholic potassium hydroxide. In the second article Gautier[3] reported on the synthesis of a series of compounds, hinted at in an article of his in 1867,[4] and apparently identical to Hofmann's. Gautier, however, obtained these compounds in a different way, namely, in a reaction between an alkyl iodide and silver cyanide. Thus isocyanide chemistry was born,[5,6] although in the preceding decade several chemists, trying to prepare alkyl cyanides from alkyl iodides and silver cyanide, had isolated considerable amounts of substances with a horrifying odor.

Isocyanides have the general structure R—N=C, or, written in another resonance form, $R{-}N^{+}{\equiv}C^{-}$. They are also known as isonitriles and carbylamines, but the name isocyanide is preferred.[7] They are derivatives of the extremely unstable hydroisocyanic acid, H—N=C.

Roeland J. M. Nolte • Department of Organic Chemistry, University at Nijmegen, Toernooiveld, 6525 ED Nijmegen, The Netherlands. *Wiendelt Drenth* • Emeritus Professor of Organic Chemistry, University at Utrecht, Ovidiuslaan 4, 3584 AW Utrecht, The Netherlands.

New Methods for Polymer Synthesis, edited by W. J. Mijs. Plenum Press, New York, 1992.

The early history of isocyanide chemistry is reviewed by J. A. Green II and P. T. Hofmann in the classical text *Isonitrile Chemistry*, edited by I. Ugi.[5]

Gautier prepared isocyanides by reaction of silver cyanide with alkyl iodides,

$$\mathrm{AgNC + RI \rightarrow AgI + RNC}$$

In Hofmann's carbylamine reaction a primary amine is treated with chloroform and a strong base. Dichlorocarbene is an intermediate in this reaction,

$$\mathrm{CHCl_3 + KOH \rightarrow KCl + H_2O + CCl_2}$$

$$\mathrm{CCl_2 + R{-}NH_2 \rightarrow R{-}N^+H_2{-}C^-Cl_2 \rightarrow R{-}N{=}C + 2\,HCl}$$

The original methods of Gautier and Hofmann were not generally applicable for synthesis in quantity. Because of this drawback, for almost one century isocyanides remained an uncommon class of compounds.

A breakthrough in isocyanide synthesis occurred when Hagedorn[8] and a little later Corey[9] and Ugi[10] described the preparation by dehydration of *N*-monosubstituted formamides. In this way larger amounts can be obtained in good yield;

$$\mathrm{R{-}NH_2 \rightarrow R{-}NH{-}C(O)H \rightarrow R{-}N{=}C + H_2O}$$

Since then, the chemistry of isocyanides and their use as ligands in metal complexes has rapidly expanded.[11] A frequently applied dehydrating agent is phosphorus oxychloride in combination with pyridine, quinoline, or triethylamine. Also diphosgene, toluenesulfonyl chloride, and a variety of other dehydrating agents have been applied.

More recently, Gautier's method was improved by Songstad *et al.*,[12,13] who replaced silver cyanide by an onium dicyanoargentate such as triphenylmethylphosphonium or tetramethylammonium dicyanoargentate. With this modification, the Gautier method in some cases gives a quantitative yield of alkyl isocyanide. Hofmann's synthetic procedure is still being applied, in particular after it was improved by using a phase transfer catalyst.[14,15]

9.1.2. Polymerization of Isocyanides

Since their discovery in the nineteenth century, it has been known that isocyanides easily polymerize. Systematic study of this polymerization was first made by Yamamoto and other Japanese workers[16–18] and in particular by Millich.[19–21]

Yamamoto tested the activity of radical generating compounds, anions, and Lewis acids as initiators in the polymerization reaction. Only Lewis acids such as boron trifluoride appeared to be catalytically active. Yamamoto's experiments have partly been duplicated by Stackman.[22] Saegusa *et al.*[23] observed the almost quantitative polymerization of cyclohexyl and isopropyl isocyanide when treated with a catalytic amount of a Lewis acid such as $BF_3 \cdot OEt_2$. Under the same conditions, *tert*-alkyl isocyanides gave the isomeric cyanides as well as isocyanide oligomers.

Yamamoto also described the polymerization of cyclohexyl isocyanide in the presence of the transition metal complexes nickel tetracarbonyl and dicobalt octacarbonyl, while the same polymer was found by Saegusa *et al.*[24] under the influence of copper(II) chloride. Otsuka *et al.*[18] polymerized *tert*-butyl and other isocyanides in the presence of a complex formed by alkylation of $Ni(C{=}N{-}t\text{-}C_4H_9)_4$ with CH_3I.

Millich and Sinclair[25] developed a catalyst for the polymerization of, in particular, 1-phenylethyl isocyanide, which consisted of a strong acid, a radical initiator, and a glass surface. Although the polymerization proceeds slowly, a high molecular weight is obtained. These authors[26] also investigated the structure of the polymer, which is the subject of Section 9.3.

In a study of metal ion catalyzed additions to isocyanides, Nolte[27–29] observed the formation of polymers. He then tested approximately twenty different metal ions on their suitability as polymerization catalysts and found nickel(II) to be by far the best. In the succeeding years polymers of approximately one hundred different isocyanides have been prepared in particular by the group of Drenth and Nolte employing the nickel(II) method.

9.2. SCOPE OF THE POLYMERIZATION

9.2.1. General

Polymerization of isocyanides is most conveniently carried out with the help of nickel(II) salts such as $NiCl_2 \cdot 6H_2O$, $Ni(ClO_4)_2 \cdot 6H_2O$, or nickel acetylacetonate. The solvent is methanol, ethanol, or a mixture of these solvents with chloroform. In many cases it is preferable to omit the solvent, for instance, when bulky isocyanides (*tert*-butyl isocyanide or 2,6-dichlorophenyl isocyanide) are polymerized or when high molecular weights are required. The reaction temperature is 25 °C.

Other metal complexes have been tested as polymerization catalysts. Some results obtained with metal acetylacetonates are compiled in Table 9.1, which indicates that cobalt(II), palladium(II), copper(II), and platinum(II) also show some activity. Excellent catalysts are also $Ni(CO)_4$

TABLE 9.1. Polymerization of Ethyl Isocyanide by Various Metal Catalysts in Chloroform[a]

Catalyst	Polymer yield (%)
Ti(IV)$(Acac)_2Cl_2$	12
Zr(IV)$(Acac)_4$	0
V(III)$(Acac)_3$	0
Cr(III)$(Acac)_3$	8
Fe(III)$(Acac)_3$	0
Co(III)$(Acac)_3$	9
Rh(III)$(Acac)_3$	2
Al(III)$(Acac)_3$	0
Be(II)$(Acac)_2$	0
Mu(II)$(Acac)_2$	1
Co(II)$(Acac)_2$	61
Ni(II)$(Acac)_2$	100
Pd(II)$(Acac)_2$	53
Pt(II)$(Acac)_2$	20
Cu(II)$(Acac)_2$	29
Zn(II)$(Acac)_2$	0
Co(II)$(Acac)_2$	0
$Co_2^0(CO)_8$	100
$Ni^0(CO)_4$	100
BF_3	55

[a] Taken from Refs. 27 and 28; Acac denotes acetylacetonate anion; reaction time 5 h; T = 61 °C.

and $Co_2(CO)_8$. These complexes, however, are very toxic and are more difficult to handle than nickel(II) salts.

Almost any isocyanide can be polymerized with nickel(II) salts, the exceptions being monomers that contain strongly metal binding properties, such as imidazolyl groups. Addition of a small amount of acid usually neutralizes these binding properties and polymerization of the isocyanide can take place.[30,31]

The polymerization rate of a number of isocyanides has been measured.[29] The rates depend on the bulkiness of the group R in R—N═C and on the presence of electron-donating or electron-withdrawing substituents in R. Some first-order rate constants for monomers polymerized with nickel chloride and nickel acetylacetonate are given in Table 9.2.

In general, the molecular weights of the polymers depend on the type of monomer used and on the applied reaction conditions.[32–38] Under standard conditions (temperature 20 °C, neat or methanolic solution, 1 mol% of $NiCl_2$) $\bar{M}_v$ values of approximately 65,000 are obtained.[33] Lowering the temperature and decreasing the amount of catalyst to 0.1 or 0.01 mol% gives $\bar{M}_v$ values up to 250,000.[37] Isocyanides with bulky substituents generally yield oligomers with $\bar{M}_v$ values of 1500–3000.[38]

TABLE 9.2. First-Order Rate Constants of the Polymerization of Various Isocyanides by Nickel Chloride and Nickel Acetylacetonate[a]

	$k_1 \times 10^5$ (s^{-1})	
R in RNC	$NiCl_2$	$Ni(Acac)_2$
CH_3	8	0.8
C_2H_5	36	7.0
n-C_4H_9	57	8.2
i-C_4H_9	60	9.7
i-C_3H_7	12.4	7.0
s-C_4H_9	10	6.0
cycloC_6H_{11}	4.4	
$CH_3CH(C_6H_5)$	1.7	2.6
t-C_4H_9	0.0011	0.00
C_6H_5	21.6	
p-$CH_3OC_6H_4$	5.3	
p-$CH_3C_6H_4$	8.5	
p-ClC_6H_4	60	

[a] Taken from Ref. 29. Reaction conditions: $T = 30\,°C$; initial [R—N═C] 0.9 mol dm^{-3}; [$NiCl_2$] 4.5×10^{-4} mol dm^{-3}; [$Ni(Acac)_2$] 6.4×10^{-3} mol dm^{-3}; solvent: ethanol ($NiCl_2$) or *sec*-butanol [$Ni(Acac)_2$].

9.2.2. *Diastereoselective Polymerization of Chiral Isocyanides*

In Section 9.3 it will be shown that the polymers of most isocyanides have a rigid helical configuration and that polymerization of isocyanides proceeds stereoselectively with respect to screw sense. When the monomer is achiral a racemic mixture of right-handed (*P*) and left-handed (*M*) helices is formed. However, when the monomer is a homochiral isocyanide (one enantiomer of an isocyanide), its polymer can be expected to be a mixture of diastereomers, and right-handed (*P*) and left-handed (*M*) helices will not be obtained in equal amounts. In order to test this hypothesis different optically active isocyanides were synthesized and polymerized with nickel chloride as catalyst;[32,34–38] some data are given in Table 9.3. In general, polymers are obtained with optical rotations quite different from those of the corresponding monomers. The screw senses of the polymers were deduced from careful analysis of the CD spectra.[32,35] They were also derived from the optical rotation data. The molar optical rotations [M] in Table 9.3 are the sum of two contributions: one from the helical main chain and a second from the optically active side chains. The optical rotation due to the side chains can be estimated from model compounds, e.g., *N*-neopentylidene derivatives of optically active amines.[32,35]

$$[R^*\text{—N═C} <]_n \qquad\qquad R^*\text{—N═CH—}t\text{-}C_4H_9$$

polymer — model compound

TABLE 9.3. Diastereoselective Polymerization of Chiral Isocyanides R*NC[a]

Monomer	Polymer	
*R**	$[M]_{578}^{20}$ [b]	Screw sense
(*R*)-$CH_3(CH_2)_5CH(CH_3)$	+47.3	*M*
(*R*)-$(CH_3)_2CHCH_2CH(CH_3)$	+68.9	*M*
(*R*)-$(CH_3)_2CHCH(CH_3)$	−16.2	*M*
(*S*)-$CH_3CO_2CH_2CH(CH_3)$	−190	*P*
(*S*)-$CH_3CO_2CH_2CH[CH(CH_3)_2]$	−172	*P*
(*S*)-$C_2H_5OC(O)CH(CH_3)$	−355	*P*
(*S*)-$PhCH(CH_3)$	−458	*P*
(*S*)-$PhCH_2CH(CH_3)$	+22	*P*
(*R*)-$CH_3CO_2CH_2CH(Ph)$	−156	*P*
(*S*)-$H_2C{=}CHCH(C_2H_5)$	+27.6	*M*
(*S*)-$HC{\equiv}CCH(C_2H_5)$	−102	*P*
(*S*,*R*)-$ImCH_2CH(COOH)NHCOCH(CH_3)$	0	*P*
(*S*,*S*)-$ImCH_2CH(COOH)NHCOCH(CH_3)$	+1025	*M*
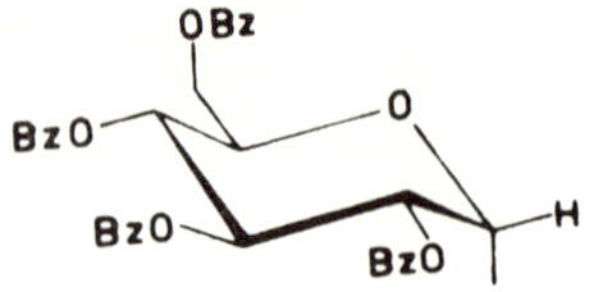	>+514	*M*
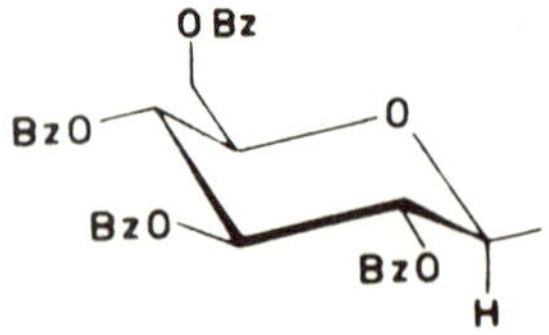	0	*P*

[a] Data have been taken from Refs. 32, 34–38.
[b] [M] is molar optical rotation, expressed per polymer repeating unit. Solvent $CHCl_3$ or CH_3OH, *c* 0.005–1.0.

For instance, for R* = (*R*)-(*i*-C_3H_7)CH_3CH the difference between the molar optical rotation value of the polymer (−16.2°, Table 9.3) and that of the model compound (−76.1°)[35] is positive (+60°). This indicates that the contribution by the helical main chain is dextrorotatory. Applying the relation found for poly(*tert*-butyl isocyanide), namely, an *M*-helix gives rise to a (+)- and a *P*-helix to a (−)-sign of optical rotation (see Section 9.3), it can be concluded that on polymerization, (*R*)-(*i*-C_3H_7)$CH_3CHN{=}C$ preferentially forms a left-handed helix, i.e., the polymerization proceeds diastereoselectively. For the series of optically active monomers (*R*)-R(CH_3)$CHN{=}C$ the degree of chiral induction was measured as a function of the bulkiness of R. It was observed that the enantiomeric excess (e.e.) of left-handed screws increases in the series R = C_2H_5 < n-C_6H_{13} < i-C_4H_9 < i-C_3H_7, i.e., it is correlated to the size of R. The highest e.e. was measured for R = i-C_3H_7 (62%).[35]

9.2.3. *Enantioselective Polymerization of Achiral Isocyanides*

Stereoselective polymerization of achiral isocyanides with the objective of forming polymers with an excess of one screw sense has been tried using chiral catalysts (nickel L-alaninate and nickel L-valinate) as well as using nickel catalysts in combination with a chiral solvent [e.g., (*S*)-*sec*-butanol], a chiral additive (e.g., 1-borneol, cinchonine), or a chiral ligand [e.g., (*R*,*R*)-DIOP, (*S*,*S*)-chiraphos, neomenthyldiphenylphosphine]. All these attempts were unsuccessful.[39] The chiral ligand or additive either expelled the isocyanide from the nickel center, in this way blocking the polymerization, or was unable to do so, in which case no chiral induction occurred. Screw-sense selective polymerization, however, could be achieved by using a catalyst obtained by adding an optically active initiator to Ni(C=NR)$_4$(ClO$_4$)$_2$.[39b] See also Section 9.4.3. Some results obtained in polymerization reactions of *tert*-butyl isocyanide using various optically active amines as initiators are presented in Table 9.4. The highest stereoselectivity is obtained when (*S*)- or (*R*)-1-phenylethylamine is used as the initiator, and when the catalytic complex has the very bulky 2-*tert*-butylphenyl isocyanide as a ligand (83% e.e.). Further increase in the bulkiness of this ligand has no positive effect as this impeded the approach of the initiator (Table 9.4, last entry).

A different, recently published procedure for obtaining optically active polymers is the selective inhibition of the growth of one of the two

TABLE 9.4. Enantioselective Polymerization of *tert*-Butyl Isocyanide by Ni(CNR)$_4$(ClO$_4$)$_2$ and Optically Active Initiator[a,b]

Initiator	e.e. (%)	Screw sense
(*S*)-(+)-$C_2H_5CH(CH_3)NH_2$	7	*P*
(*S*)-(−)-$C_6H_5CH(CH_3)NH_2$	61	*P*
(*S*)-(−)-$C_6H_{11}CH(CH_3)NH_2$	50	*P*
(*S*)-(−)-$C_2H_5CH(CH_3)NHCH_3$	0	*P* + *M*
L-prolinol	36	*M*
L-phenylalaninol	37	*M*
L-alanine methyl ester	47	*P*
(*S*)-(−)-$C_6H_5CH(CH_3)NH_2$[c]	83	*P*
(*S*)-(−)-$C_6H_5CH(CH_3)NH_2$[d]	22	*P*

[a] Data have been taken from Ref. 39b. Reaction conditions: neat; 1.0 mol% of catalyst and 1.0 mol% of initiator; T = 25 °C.
[b] R = t-C_4H_9.
[c] R = 2-(t-C_4H_9)C_6H_4.
[d] R = 2,6-di(i-C_3H_7)C_6H_3.

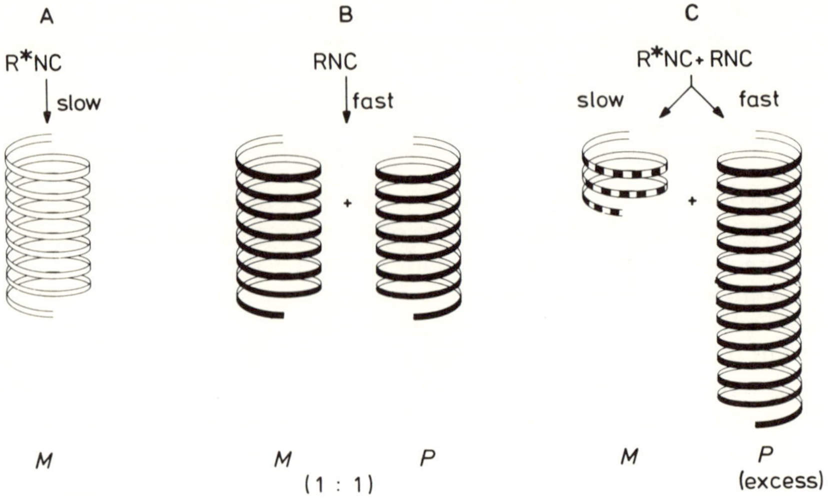

FIG. 9.1. Proposed mechanism for the enantioselective polymerization of achiral isocyanides in the presence of an optically active comonomer. Reproduced with permission from Ref. 41.

enantiomeric helices that arises from an achiral isocyanide.[(40,41)] In this procedure bulky, optically active isocyanides are used as inhibitors. Such bulky isocyanides are the esters of (*S*)-2-isocyanoisovaleric acid $(CH_3)_2CHCH(COOR)N{=}C$, in which R = CH_3, *i*-C_3H_7, *t*-C_4H_9.

In the presence of nickel(II) salts these ester isocyanides slowly polymerize to give homopolymers with predominantly left-handed screws (see Fig. 9.1A).

TABLE 9.5. Enantioselective Polymerization of Achiral Isocyanides in the Presence of (*S*)-*i*-$C_3H_7CH(COOR)N{=}C$[a,b]

R in $(RNC)_n$	$-[\alpha]_D^{20}$ (deg)	Screw sense	$\bar{M}_w$
C_6H_5	610	*P*	44,000
4-$CH_3OC_6H_4$	520	*P*	42,000
4-$CH_3C_6H_4$	550	*P*	36,000
2-$CH_3C_6H_4$	340	*P*	33,000
4-$(CH_3)_2NC_6H_4$	250	*P*	<5,000
4-ClC_6H_4	660	*P*	31,000
$C_6H_5CH_2$[c]	93	*P*	
n-C_8H_{17}[c]	58	*P*	35,000
i-C_3H_7[c]	105	*P*	45,000

[a] Data have been taken from Ref. 41. Reaction conditions: molar ratio achiral to chiral isocyanide 1:1; catalyst 1 mol% $NiCl_2$; neat; reaction temperature 20 °C.
[b] R = CH_3.
[c] R = *i*-C_3H_7.

When such an isocyanide is mixed with an achiral isocyanide and polymerized with nickel chloride, polymer samples are obtained which show high optical rotations (Table 9.5). These samples mainly contain the homopolymer of the achiral isocyanide. Interestingly, the CD spectra reveal that these homopolymers have right-handed screws (see Fig. 9.1C). This suggests that the polymerization is not a common copolymerization reaction, as in that case a left-handed screw sense would have been expected. (See also Section 9.4.3.)

9.3. STRUCTURE OF THE POLYMERS

Several structures of oligomers of isocyanides have been proposed in the literature.[20] The structure of the polymer of 1-phenylethyl isocyanide with a molecular weight in the range of 25,000–150,000 was proposed by Millich and Sinclair[26] to be $[CH_3CH(C_6H_5){-}N{=}C<]_n$ on the basis of its infrared and ultraviolet absorption spectrum. Yamamoto[17] ascribed a similar structure, $[R{-}N{=}C<]_n$, to polymers of alkyl isocyanides.

The structure, $[R{-}N{=}C<]_n$, was confirmed by Nolte[28] who noted the close resemblance between the infrared, ultraviolet, and ^1H NMR spectra of the polymers with those of the corresponding *N,N'*-dialkyl- and *N,N'*-diaryldiimines. The structure is in agreement with the tendency of isocyanides to react by α-addition. Although often called poly(isocyanide), according to the rules for the nomenclature of single stranded polymers proper names are poly(carbonimidoyl) or poly(iminomethylene).[42]

From, among others, space-filling molecular models Millich and Sinclair[26] concluded that the polymer molecules would have the configuration of a rather rigid helix. Since a helix complies with the definition of a chiral species, the polymer samples would actually be racemic mixtures of right- and left-handed helices. This suggestion of a highly ordered configuration was supported by the observation of an increase in optical rotation by one order of magnitude on polymerization of optically active 1-phenylethyl isocyanide.[43] The rigid helical structure was more strongly supported by the resolution of *tert*-butyl isocyanide into fractions with (+)- and (−)-sign of optical rotation.[44] The resolution was achieved using column chromatography. The support was a high-molecular-weight polymer of optically active (*S*)-*sec*-butyl isocyanide. After improving the resolution procedure, viz. by coating the optically active poly(*sec*-butyl isocyanide) on glass beads, a completely resolved polymer of *tert*-butyl isocyanide was obtained. Its optical rotation amounted to $[\alpha]_{578} = 56°$.[45]

Additional proof of a helical polymer structure came from circular dichroism (CD). The experimental CD spectrum of the completely resolved (+)-rotating poly(*tert*-butyl isocyanide) is shown in Fig. 9.2. In the region

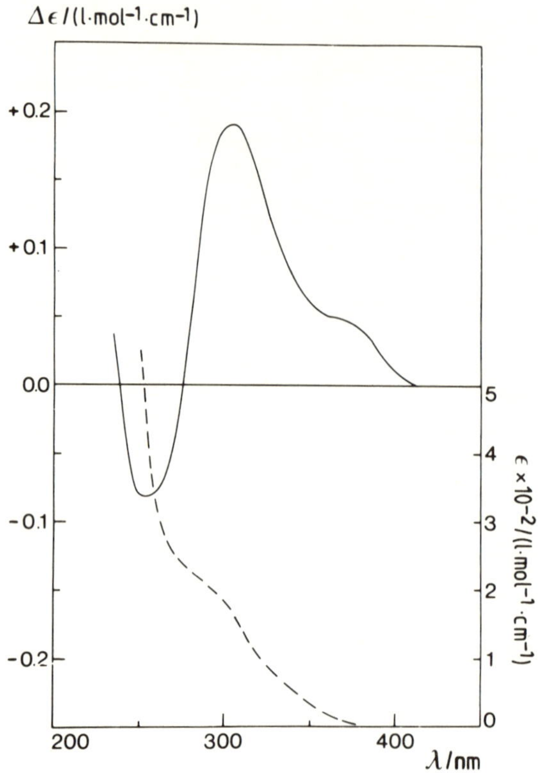

FIG. 9.2. UV (– – –) and CD (—) spectrum of completely resolved (*M*)-(+)-poly(*tert*-butyl isocyanide). Reproduced with permission from Ref. 32.

from 240 to 400 nm a so-called exciton couplet is present which is positive (S-shape) for the (+)-rotating enantiomer and negative (Z-shape) for the (−)-rotating enantiomer. This couplet is due to the $n \rightarrow \pi^*$ transition of the imino functions in the polymer main chain (this transition is visible in the UV spectrum as a weak shoulder at about 300 nm). According to exciton theory the CD spectrum can be described by a two-term equation. One term represents the so-called conservative contribution, the second the nonconservative contribution. These two contributions have been calculated by Hezemans[46] as a function of the number of repeating units per turn for a right-handed and a left-handed helical poly(*tert*-butyl isocyanide) molecule. The experimental curve was also analyzed in terms of conservative and nonconservative contributions. On comparing the calculated with the experimental spectra it was concluded that the (+)-rotating enantiomer had a left-handed screw and the (−)-rotating one a right-handed screw. The number of repeating units per turn was calculated to lie between 3.6 and 4.6.

Recently, more extensive CD calculations were carried out by Huige[47] using the polarizability theory of De Voe. These calculations revealed that the number of repeating units per turn in poly(*tert*-butyl isocyanide) is 3.81.

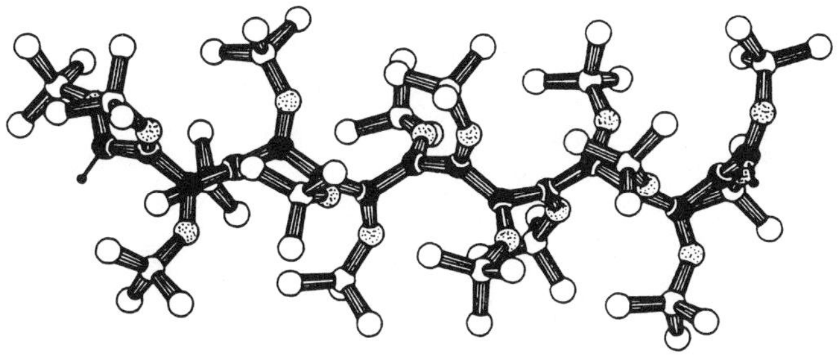

FIG. 9.3. Absolute minimum energy conformation of hexadeca(*tert*-butyl isocyanide). Reproduced from Ref. 47.

In addition, the following configurational parameters were derived: dihedral angle between two repeating units 77°, translation of these units along the helix axis 0.10 nm, and translation around the helix axis 94.5°.

Consistent force field (molecular mechanics) calculations[47] gave support to a helical structure for a long-chain poly(isocyanide) molecule. These calculations suggest that the end sections of the polymer chain are disordered. The dihedral angle N═C—C═N in hexa(*tert*-butyl isocyanide) was calculated to be 78.6 and the number of repeating units per helical turn 3.75. These data are in good agreement with the data obtained from the CD spectra. The calculations further revealed that reversals of screw sense are energetically unfavorable. This conclusion can also be drawn from the CD studies which showed that the intensities of the experimental and calculated CD spectra are of the same order of magnitude. According to the force field calculations, polymers of isopropyl isocyanide and ethyl isocyanide form helical structures as well. On the contrary, the chain of the polymer of methyl isocyanide was calculated to be disordered.[47]

A drawing of the secondary structure of hexadeca(*tert*-butyl isocyanide) as derived from the consistent force field calculations is presented in Fig. 9.3. A convenient schematic representation of a poly(iminomethylene) molecule is given in Fig. 9.4.

In principle, the side groups R of the monomeric units in a poly(iminomethylene) can adopt *syn*- or *anti*-configurations (Fig. 9.5).

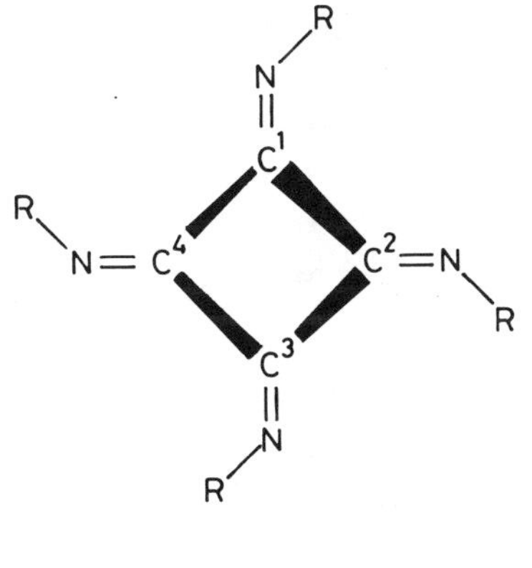

FIG. 9.4. View along the helical axis of a poly(isocyanide) molecule. Unit 5 is behind unit 1, etc.

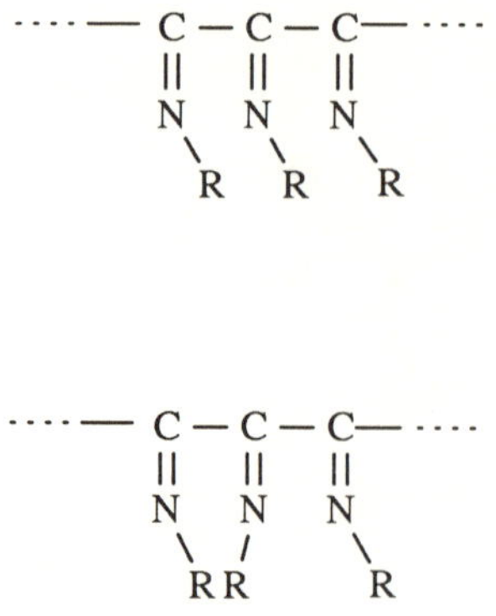

a

b

FIG. 9.5. Syn- (a) and *anti-* (b) configurations of the side groups R in a poly(isocyanide).

Molecular models[48] as well as molecular mechanics calculations[47] suggest that *syn–anti* structures in one polymer chain are not altogether possible for steric reasons. However, both Green[49] and Kamer[50] have recently reported that chemical shift dispersion occurs in the ^{1}H and ^{13}C NMR spectra of a number of poly(isocyanides). For instance, the polymer of *p*-tolyl isocyanide shows in its ^{1}H NMR spectrum two broad resonances for the CH_3 protons at 1.90 and 2.40 ppm. A similar feature is observed for the CH_3O protons in the corresponding methoxy substituted polymer, namely, signals are present at 3.25 and 3.75 ppm. Addition of strong acid results in a decrease in intensity of one of the CH_3 signals and an increase in the other. Protonation of the imino nitrogen probably allows the side chains of these polymers to relax to the thermodynamically most stable conformation, i.e., the all-*syn* or, what is equivalent, the all-*anti* configuration.

The tightly coiled helical structure of a poly(iminomethylene) suggests that the individual polymer molecules are rigid rods. This characteristic is supported by the solution properties of the polymers. The Mark–Houwink equations measured for poly(1-phenylethyl isocyanide) and poly(2-octyl isocyanide) in toluene at 30 °C are $[\eta] = 3.8 \times 10^{-7}\bar{M}_{w}^{1.30}$ [51] and $[\eta] = 1.4 \times 10^{-9}\bar{M}_{\text{light scat}}^{1.75}$,[32] respectively. The high value of the exponents in these equations points to stiff chains. On the contrary, poly(2-phenylethyl isocyanide) displays a much lower Mark–Houwink exponent, namely 0.68.[51] This and other data[49] suggest that the type of side chain has a large effect on the conformation of the polymer backbone. A similar conclusion was drawn from the molecular mechanics calculations.[47]

9.4. MECHANISM OF NICKEL(II) CATALYZED POLYMERIZATION

9.4.1. General

A mechanism has been proposed for the polymerization of isocyanides by nickel(II) salts[34,53] (Scheme 9.1). This mechanism is based on kinetic

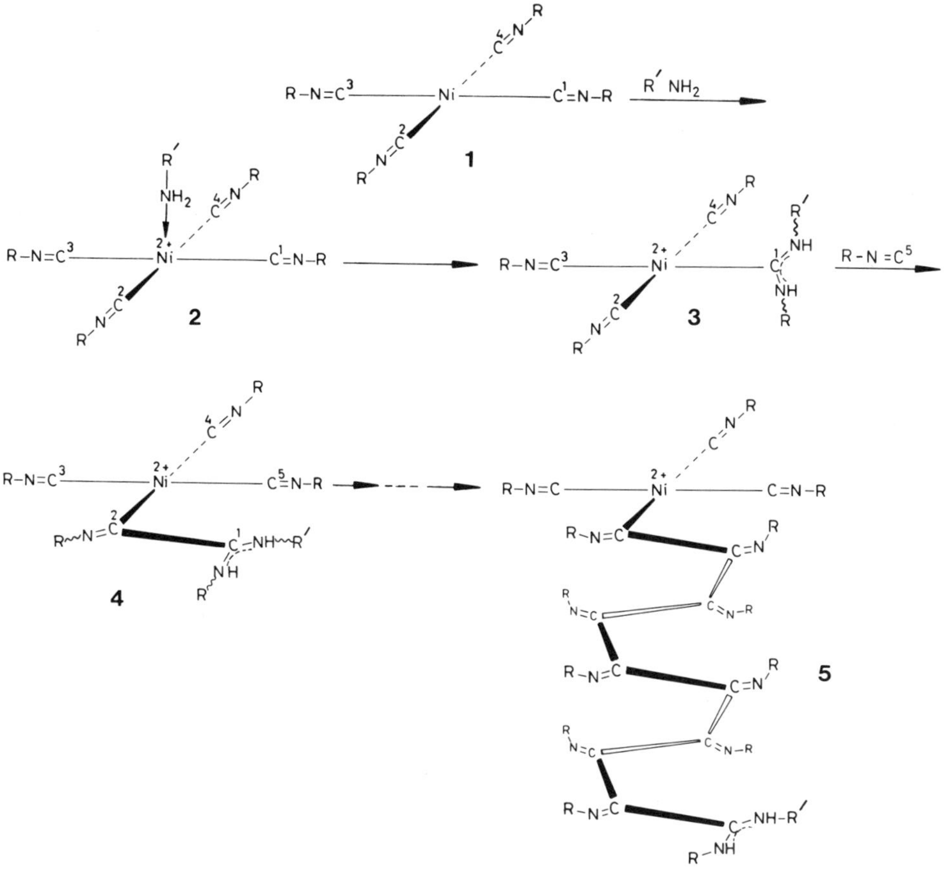

SCHEME 9.1.

measurements[29] and experiments with optically active monomers[32] and initiators.[39b,54,55]

The polymerization starts from a square-planar nickel complex, $Ni(CNR)_4^{2+}$, **1** (Scheme 9.1), which is formed when an excess of isocyanide is added to nickel chloride or nickel perchlorate. The structure of this complex was confirmed by an X-ray analysis.[55] A common reaction of a metal-coordinated isocyanide is nucleophilic attack on its terminal carbon atom. Therefore, a conceivable initiation step of the polymerization is attack by a nucleophile on one of the four coordinated isocyanide ligands of **1**. This nucleophile can be a solvent molecule (e.g., water or methanol), the counterion of Ni^{2+} (e.g., Cl^-), or an initiator that is added purposely, such as an amine as in Scheme 9.1. The nucleophile first coordinates to the nickel center, **2**, as was shown by low-temperature IR and UV experiments.[55] In a subsequent step it attacks a coordinated isocyanide. The resulting species, **3**, is a carbene complex which has been fully characterized.[54] The plane of the ligand $C^1(NHR^1)NHR$ in **3** is approximately perpendicular to the plane of the isocyanide carbons and nickel. For steric reasons there is no rotation around the Ni—C^1 bond. Carbon C^1 has gained in nucleophilicity allowing it to attack a neighboring isocyanide ligand. Such an attack is facilitated when a new isocyanide, C^5NR, is substituted for $C^1(NHR')NHR$. Attack can occur on either C^2 or C^4. In **4** it has occurred on C^2. When the sequence of attacks continues in the direction $C^1 \rightarrow C^2 \rightarrow C^3 \rightarrow C^4$ a left-handed helix is formed (**5**). In a similar way the sequence $C^1 \rightarrow C^4 \rightarrow C^3 \rightarrow C^2$ will lead to a right-handed helix. The polymerization stops when the active chain end accepts a proton.

The foregoing mechanism has been named a "*merry go round mechanism.*"[52] It accounts for the enantioselective and diastereoselective polymerization of achiral and chiral isocyanides, as will be discussed in Sections 9.4.2 and 9.4.3. It also explains why the tightly coiled helix of a poly(iminomethylene) is so easily formed: the monomers are preorganized in the nickel complex and are knitted together by reshuffling of bonds with a minimal amount of atomic movement. Since one turn of the helix consists of approximately four repeating units, each rotation around nickel contributes one turn to this helix. The entropy of activation was calculated to be $\Delta S^{\neq} = -54\,\mathrm{J\,mol^{-1}\,K^{-1}}$, which is a reasonable value for ligand substitutions in square-planar d^8 metal complexes.

9.4.2. *Diastereoselective Polymerization of Chiral Isocyanides*

The question has been addressed as to what determines the screw sense in the diastereoselective polymerization of chiral isocyanides[32] (Section 9.2.2). This polymerization may be a thermodynamically controlled process—i.e., the most favorable combination of side chain and helix is formed—or may be a kinetically controlled process at the catalytic center. To answer this question the models of Fig. 9.6 have been used.

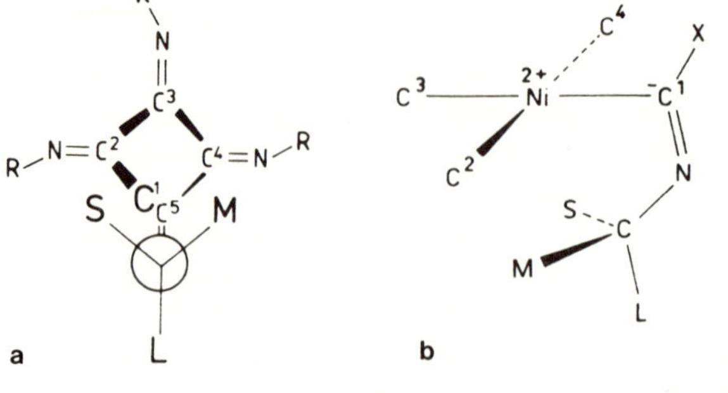

FIG. 9.6. Models for predicting the screw sense in the diastereoselective polymerization of isocyanides.

In Fig. 9.6a a projection of the polymer side chain along the C—N= bond is given. The substituents on the chiral carbon atom of this side chain are distinguished by their relative steric requirements: *S* for a small, *M* for a medium-sized, and *L* for a large substituent. The most severe steric hindrance for a substituent at the chiral carbon atom comes from C^1 as this atom is situated close to C^5. The next most severe hindrance comes from C^4. By placing *S, M,* and *L* in such a way that the least interference with C^1 and C^4 occurs, the most favorable combination of helix and chiral side chain can be derived. This reasoning has been applied to the polymers of Table 9.3. It was found that much disagreement existed between the experimentally determined screw sense and the predicted one.[32] Such disagreement did not occur when the model of Fig. 9.6b was applied. In this figure one of the first intermediates in the polymerization process is shown (see also Scheme 9.1). The substituents on the chiral carbon again are denoted by *S, M,* and *L* and are placed in such a way that the least steric hindrance occurs in the transition state of the first propagation step of the polymerization reaction. In Fig. 9.6b attack by C^1 will preferentially occur in the direction of C^4 as this is the sterically least hindered side. In this way a right-handed screw is generated. By similar reasonings the screw senses of the polymers in Table 9.3 have been predicted and were found to coincide with the experimental ones. From these findings it was concluded that the screw sense induction in the diastereoselective polymerization of isocyanides is a process kinetically controlled at the catalytic center.[32]

9.4.3. Enantioselective Polymerization of Achiral Isocyanides

In Section 9.2.3 it has been shown that enantioselective polymerization of achiral isocyanides can be achieved by the application of an optically active initiator. The screw sense that is induced by the chiral initiator can be predicted from the polymerization mechanism, by reasoning in a similar way as described for the diastereoselective polymerization of chiral isocyanides (Section 9.4.2).[39] The starting complex for the polymerization with (*S*)-(1-phenylethyl)amine as the initiator is shown in Fig. 9.7a. It is assumed that the substituents at the chiral carbon atom in this figure have become fixed due to interaction of the phenyl ring with the nickel center. This results in a preferential attack of C^1 on C^2 (H is the sterically least hindered side) and the formation of a right-handed helix in agreement with experiment (Table 9.4). Support for this induction mechanism is the fact that the enantiomeric excess considerably decreases when, in the starting complex of Fig. 9.7a the phenyl ring is replaced by a noncoordinating ethyl group (7% e.e.) or when, in the initiator one of the NH_2 protons is replaced by a methyl group (0% e.e.). In the latter case (Fig. 9.7b) there will be no preferential attack of C^1, as the *N*-methyl group and the methyl group

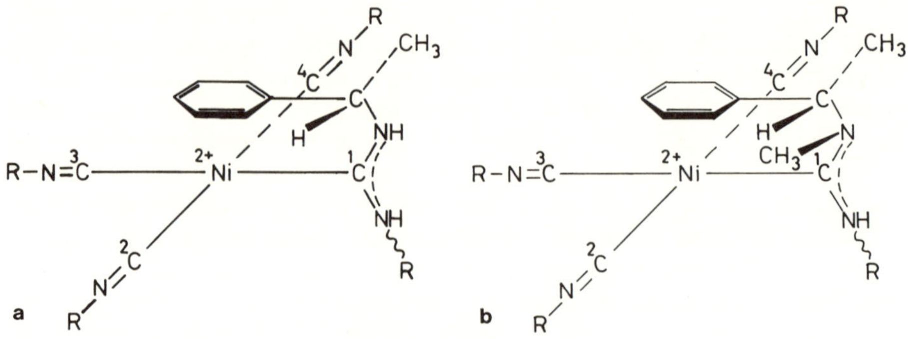

FIG. 9.7. Starting complex for the polymerization of isocyanides with (*S*)-1-phenylethylamine (a) and (*S*)-methyl(1-phenylethyl)amine (b) as the initiator.

at the chiral center point in opposite directions and now shield carbon atoms C^2 and C^4 to the same extent.

Another method of achieving enantioselective polymerization of an achiral isocyanide, also described in Section 9.2.3, is selective inhibition of the growth of one of the two enantiomeric helices. The mechanistic explanation is as follows.(41) The bulky chiral isocyanide (*S*)-2-isocyanovaleric acid is a slowly polymerizing monomer ($k_p \ll 10^{-7}\,s^{-1}$) and preferentially forms an *M*-screw (see Fig. 9.1A). The achiral monomers of Table 9.5 are fast polymerizing isocyanides ($k_p \approx 10^{-4}\,s^{-1}$) and form racemic mixtures of *P* and *M* screws (Fig. 9.1B). When the chiral isocyanide is copolymerized with the achiral one, the former has a preference for inclusion into *M*-helices and retards the formation of these helices from the latter one. The *P*-helices continue to grow on and eventually consume all the achiral monomer (Fig. 9.1C).

The proposed mechanism is supported by the fact that no chiral induction is observed when a combination of a fast polymerizing chiral and a fast polymerizing achiral monomer is used.(41) in line with the mechanism are also the results from gel permeation chromatography experiments. These show that the polymerization mixtures contain low-molecular-weight fractions with a high content of chiral monomer and high-molecular-weight fractions with a low content of this monomer.(41)

9.5. FUNCTIONALIZED POLYMERS

9.5.1. Polymers with Phosphine Functions

Because of the typical rigid rod configuration, all side groups R of polymers $[R{-}N{=}C<]_n$ are on the outside of the molecules. Approach to the terminal positions of these side groups should be relatively easy. Easy

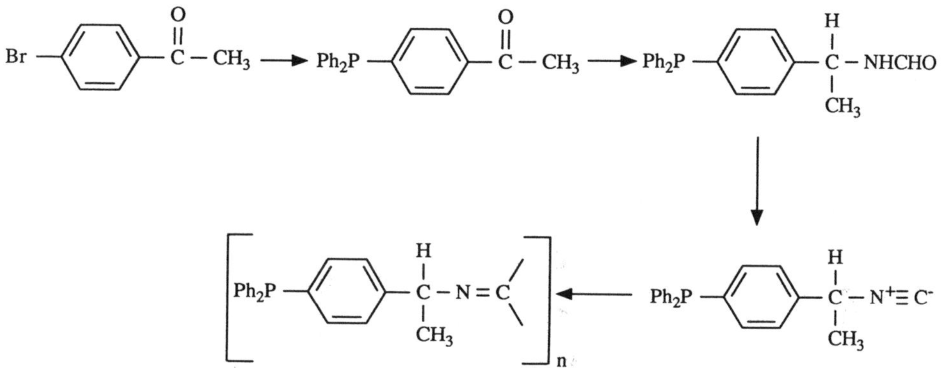

SCHEME 9.2.

approach might be advantageous when reactive groups such as catalytically active centers are anchored on to the polymers. Since phosphine functions are often used to anchor metal centers, polymers of phosphine-containing isocyanides have been prepared[56,57] (Scheme 9.2).

After protection of its carbonyl function, 4-bromoacetophenone was lithiated and the lithiated derivative reacted with chlorodiphenylphosphine. After removal of the protecting group by hydrolysis, the ketone was directly converted in a Leuckart reaction to *N*-formyl-1-[4-(diphenylphosphino)-phenyl]ethylamine. Dehydration of this formamide by perchloromethyl formate and triethylamine afforded the isocyanide. Each of the steps from 4-bromoacetophenone to isocyanide had a yield of 75–85%. The corresponding polymer was synthesized in 70% yield by treatment of an ethanolic solution of the isocyanide with nickel chloride hexahydrate for four days at room temperature. Addition of a solution of [RhCl(1,5-cyclooctadiene)$_2$] or [RhCl$(C_2H_4)_2$]$_2$ in THF to a solution of the phosphine polymer in the same solvent gave suspensions of polymer rhodium complexes which were insoluble, not only in THF, but also in benzene, alcohol, etc. The polymer rhodium complex was collected by centrifugation. A maximum loading of one rhodium to one phosphine could be reached. These rhodium-loaded polymers are catalytically active in the hydrogenation of cyclohexene. The catalyst prepared from the polymer and [RhCl$(C_2H_4)_2$]$_2$ showed maximum activity at a P/Rh ratio of 1.5–2.0. Despite the insolubility, the turnover per rhodium atom in THF at 30.0 °C is higher than that of the low-molecular-weight complexes RhCl$(PPh_3)_{1-3}$.

9.5.2. *Polymers with Imidazolyl Functions*

Proteolytic enzymes, such as chymotrypsin and elastase, hydrolyze esters and peptides in an enantioselective way. In the active centers of these

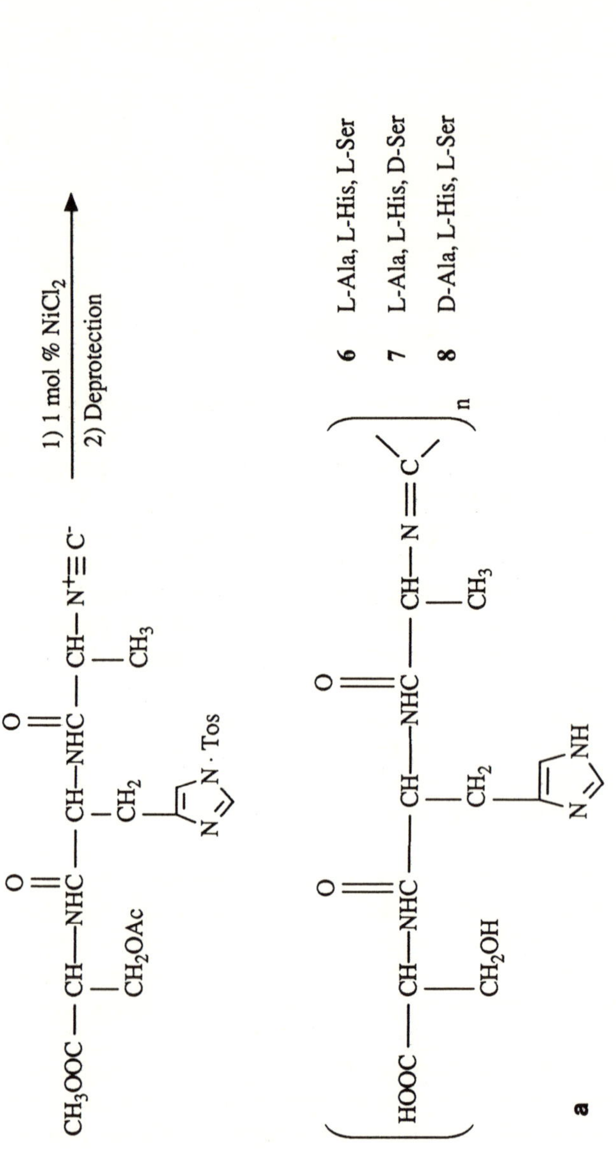

6 L-Ala, L-His, L-Ser

7 L-Ala, L-His, D-Ser

8 D-Ala, L-His, L-Ser

a

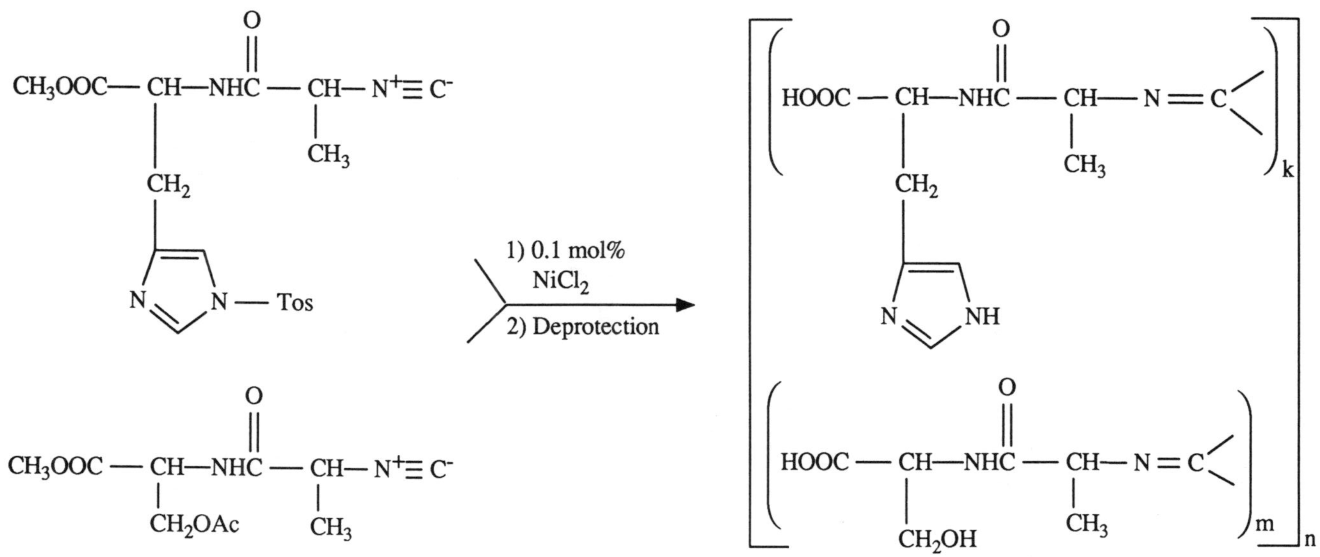

L-Ala, L-His; L-Ala, L-Ser

b

9 k=1, m=2
10 k=m=1
11 k=2, m=1

SCHEME 9.3. (a) Homopolymers from isocyanotripeptides; (b) copolymers from isocyanodipeptides.

enzymes imidazolyl, carboxylic acid, and hydroxymethyl functions are present. These catalytic functions act cooperatively in the cleavage of the ester or peptide bond. Optically active polymers and copolymers of isocyanides containing the above-mentioned catalytic functions have been synthesized and studied as protease models.[58–64] Starting compounds for the synthesis are the amino acids L- or D-alanine, L-histidine, and L- or D-serine. Two approaches have been followed:

1. Homopolymerization of isocyanides derived from tripeptides of alanine, histidine, and serine.
2. Copolymerization of isocyanides derived from dipeptides of alanine and serine and of alanine and histidine (Scheme 9.3).

Some physical data of the homopolymers and copolymers are presented in Table 9.6. The homopolymers have molecular weights in the range of $\bar{M}_v$ = 20,000–35,000. For the copolymers much higher values were found: 220,000–250,000. The pK_a values of the carboxylic acid groups in the polymers were determined by potentiometric titration. They appeared to be appreciably higher than in low-molecular-weight model compounds, such as histidine and serine [pK_a(COOH) 1.8 and 2.2, respectively]. Also, the pK_a(ImH^+) values of the polymers had increased considerably as compared to those of model compounds. This increase amounted to 1–3.5 pK units and was taken as evidence that the imidazolyl residues in the polymers are affected by the negative charges of the carboxylate ions. Copolymers **9–11** (Scheme 9.3) had two different kinds of imidazolyl groups, one with a pK_a value of about 7 and a second with a pK_a value of about 9.5. The latter value is unusually high and suggests that the corresponding imidazolyl groups are surrounded by one or more carboxylate ions and are strongly interacting with these ions.

Polymers **6–11** (Scheme 9.3) were tested as catalysts in the hydrolysis of achiral and chiral nitrophenyl esters.[64] The active species were found to be the neutral imidazolyl groups. The hydroxymethyl functions had no appreciable effect on catalysis. Generally, the homopolymers and copoly-

TABLE 9.6. Physical Data of Homopolymers and Copolymers Derived from Isocyanopeptides[a]

Polymer	$\bar{M}_v$	$[\alpha]_D^{20}$ (deg)	Screw sense	pK_a(COOH)	pK_a(ImH^+)
6	24,000	−8.0	*P*	3.8	7.1
7	20,000	+8.0	*P*	4.7	8.4
8	35,000	+30.2	—	5.8	8.7
9	220,000	−43	*M*	4.6	7.2, 9.6
10	250,000	−130	*M*	3.8	7.2, 9.7
11	220,000	−30	*M*	4.7	7.1, 9.6

[a] Data are taken from Refs. 62 and 63.

mers showed markedly higher activities than corresponding low-molecular-weight compounds. This enhancement in activity was ascribed to cooperative effects involving, among others, imidazolyl and carboxylate functions. The activities could be further enhanced by adding positively charged surfactants. These surfactants were supposed to arrange themselves around the negatively charged polymer molecules, forming a hydrophobic pseudophase. In the presence of surfactant the imidazolyl-containing polymers displayed enantioselectivity in the hydrolysis of chiral amino acid esters.[64]

9.5.3. Polymers with Crown Ethers

Ion transport across biological membranes frequently occurs by means of ion channels. Polymers of isocyanides containing crown ether rings have been synthesized as models of such channels.[65–67] The side chains in the molecules of a polymer of an isocyanide are arranged in four rigid stacks which run parallel to the axis of the polymer helix (see Section 9.3). When each side chain contains a crown ether ring, these rings are placed on top of each other and form four channels as shown schematically in Fig. 9.8. The synthetic route to a poly(isocyanide) with stacked 18-crown-6 rings is outlined in Scheme 9.4.[66] The molecular weight of the polymer amounted to $\bar{M}_v = 15{,}700$, which corresponds to a degree of polymerization of 43. Taking into account the 4 Å pitch of the polymer helix, the length of the channel can be calculated to be about 4.4 mm.

The cation binding properties of polymers with 15-crown-5, 18-crown-6, 21-crown-7, and 24-crown-8 rings have been reported.[66,67] Binding energies for alkali metal ions are in the range from $\Delta G° = -32$ to $-45\ \mathrm{kJ\,mol^{-1}}$ and are roughly $5\ \mathrm{kJ\,mol^{-1}}$ higher than those of low-molecular-weight model compounds. This difference was explained from the fact that in the polymers the metal ions can be sandwiched between consecutive crown ether rings, which favors binding.[66] In the model compounds such sandwiching is not possible.

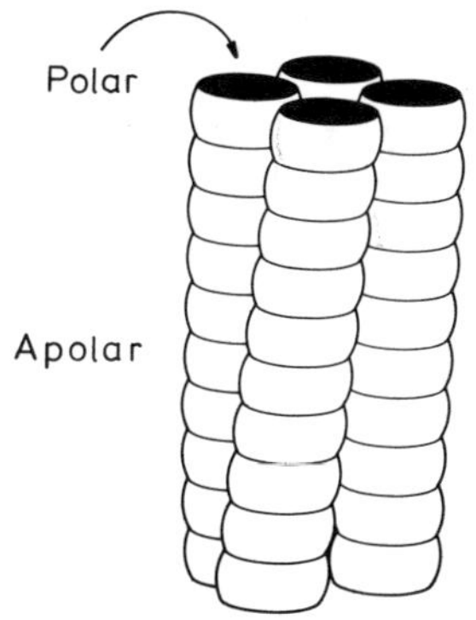

FIG. 9.8. Schematic representation of the channels formed by stacked crown ethers attached to a poly(isocyanide) backbone. Reproduced from Ref. 67.

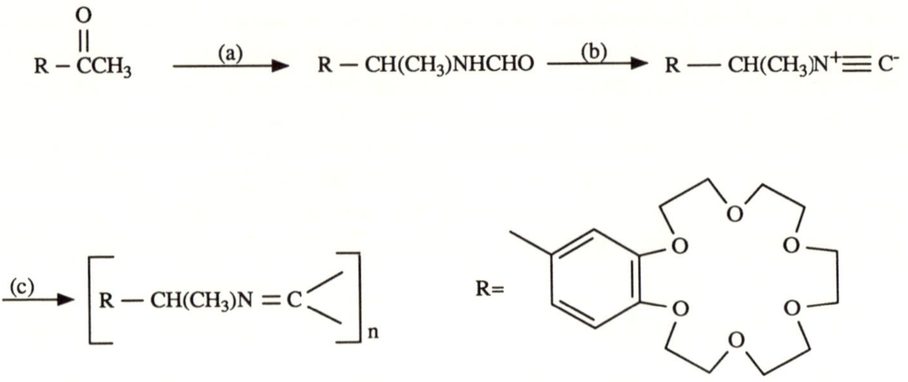

(a) Leuckart reaction ($NH_4^+HCO_2^-$)

(b) $POCl_3$ / Et_3N

(c) 1 mol % $NiCl_2.6H_2O$, neat

SCHEME 9.4.

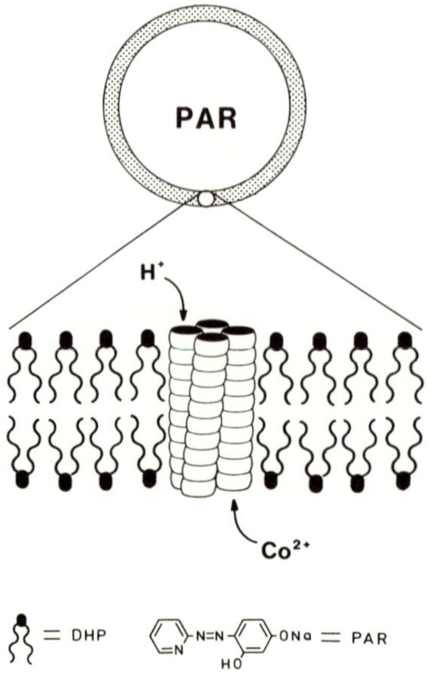

FIG. 9.9. Facilitated ion transport across bilayers of dihexadecyl phosphate vesicles by an artificial ion channel derived from a poly(isocyanide) containing crown ether rings. In the inner aqueous compartment of the vesicles the dye 4-(2-pyridylazo)resorcinol mono sodium salt (PAR) is present. Ion transport is measured by adding Co^{2+} ions to the vesicle dispersions and recording the increase in absorption of the cobalt–PAR complex in the UV-vis as a function of time. Reproduced with permission from Ref. 69.

The electrical properties of a polymer with 18-crown-6 rings have been studied by impedance spectroscopy and by using the polymer as a membrane in a concentration cell of the type K–amalgam/KCl–polymer complex/Hg.[68] These studies suggest that the polymer channels conduct cations. The same conclusion was drawn from experiments in which the channels had been incorporated into bilayers of dihexadecyl phosphate (DHP) vesicles.[67,69] The setup of the latter experiments is shown in Fig. 9.9. Vesicles without channels do not show ion transport across their bilayers, while vesicles with channels do.

9.5.4. Polymers with Pyridinium Functions

Several pyridinium containing polymers have been prepared according to Scheme 9.5, either through **13** or through **14**.[70] In aqueous solution the

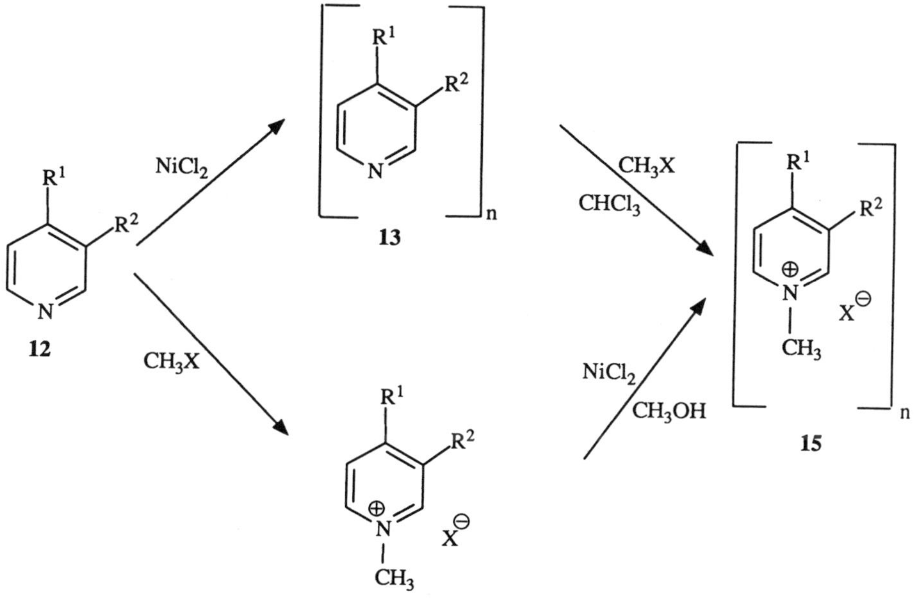

	R^1	R^2
a	H	$CH_2N^+{\equiv}C^-$
b	H	$CH(CH_3)N^+{\equiv}C^-$
c	$CH(CH_3)N^+{\equiv}C^-$	H
d	$CH_2CH_2N^+{\equiv}C^-$	H

SCHEME 9.5.

iodides **15** ($X^- = I^-$) show charge transfer absorption at 274–296 nm, which is approximately 20–40 nm higher than the charge transfer absorption of corresponding low-molecular-weight compounds. This shift is probably due to a low microenvironment polarity of the polymer rather than to an intramolecular proximity effect. More interesting in this respect are copolymers, e.g., those obtained after methylating the copolymers in which **12a** and **12d** (Scheme 9.5) are present in a ratio of 1:1 or 1:2. The UV/Vis absorption spectrum of these copolymers ($X^- = I^-$) has a band at the exceptionally high wavelength of 400–450 nm. These high-wavelength position could be due to the presence of some of the iodide ions in cavities formed by neighboring side chains of unequal length (Fig. 9.10).

Another copolymer was synthesized by treating a methanolic solution of 1-phenylethyl isocyanide and 1-(3-pyridyl)ethyl isocyanide (ratio 1:1) with 0.5 mol% of $NiCl_2 \cdot 6H_2O$ (Scheme 9.6). The pyridine moieties were quaternized by refluxing a solution in chloroform for 40 h with an eightfold excess of 2,4-dinitrobenzyl iodide. The copolymer has cavities (Fig. 9.11) in which a molecule of *N,N,N',N'*-tetramethyl-1,4-benzenediamine just fits as indicated by the formation of the radical cation of the latter species.

Two polymers have been prepared that contain in their side chain a pyridinium ring and, in addition, an electron-donor group.[71] The side groups are of the Dimroth/Reichardt type[72] (Scheme 9.7).

In the UV/Vis absorption spectrum the long-wavelength band of zwitterion polymer **18** (Scheme 9.7) shifts from 455 nm in methanol to 500 nm in the less polar solvent *tert*-butanol. These values do not differ appreciably from the low-molecular-weight model compound **19** (Scheme 9.7), indicating the absence of cooperative effects in the polymer.

Zwitterion **17** (Scheme 9.7) is a black-brown powder that is insoluble in

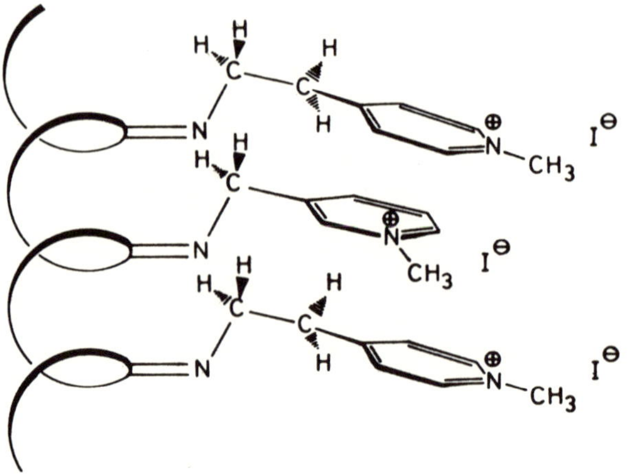

FIG. 9.10. Cavity in copolymer of **12a** and **12d**. Reproduced with permission from Ref. 70.

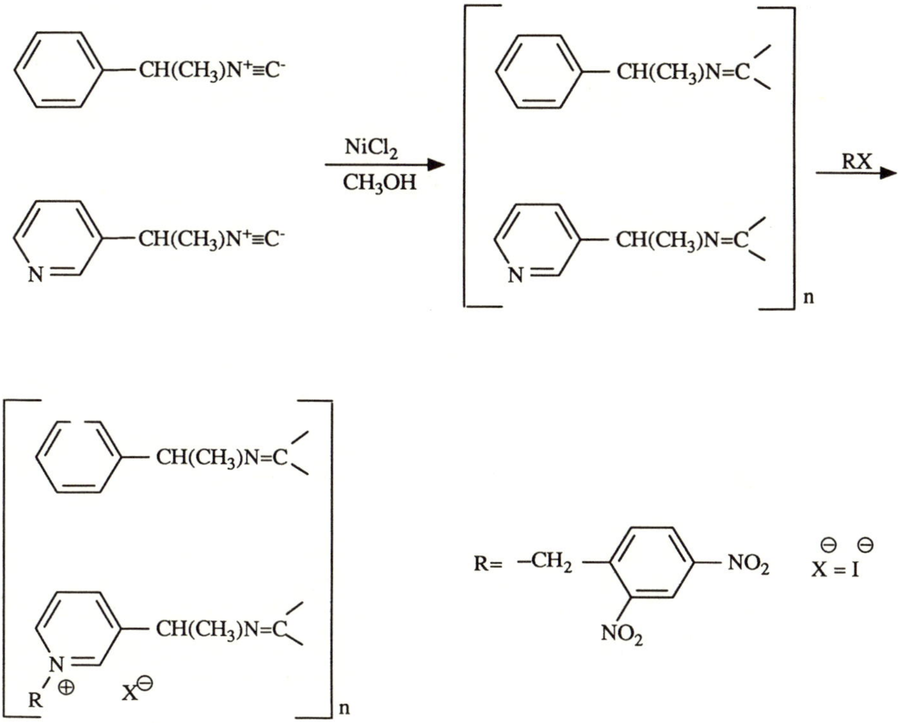

16

SCHEME 9.6.

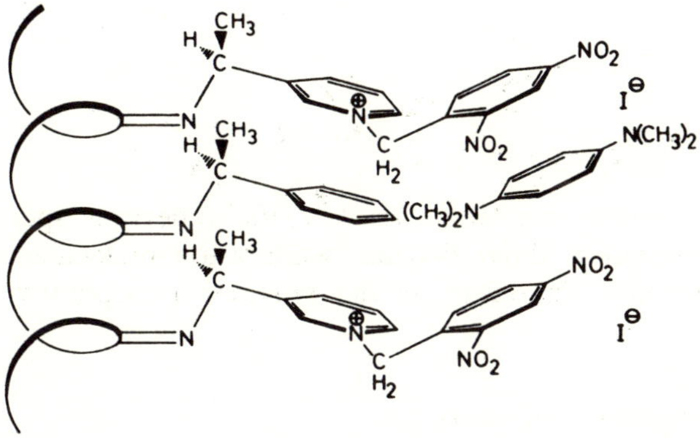

FIG. 9.11. N,N,N′,N′-tetramethyl-1,4-benzenediamine in a cavity of polymer **16**. Reproduced with permission from Ref. 70.

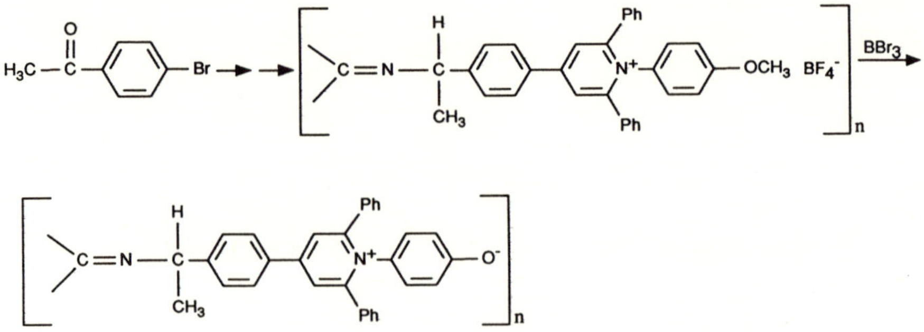

17

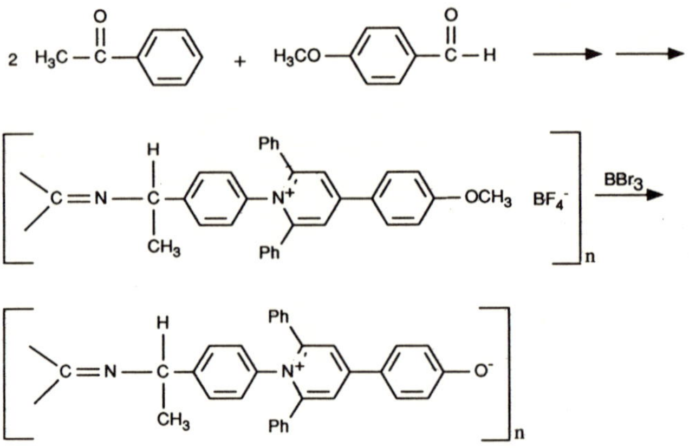

18

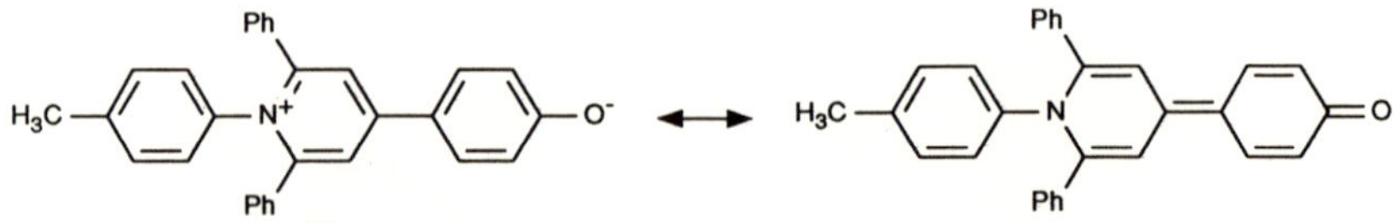

19

SCHEME 9.7.

most of the common solvents. Its UV/Vis reflection spectrum has an appreciable absorption above 600 nm, while a low-molecular-weight model compound does not. Therefore, in this polymer a cooperative effect might be present.[71]

9.5.5. *Amphiphilic Polymers*

Phospholipids and double-chain synthetic surfactants form closed bilayer structures in water, known as liposomes or vesicles.[73] These

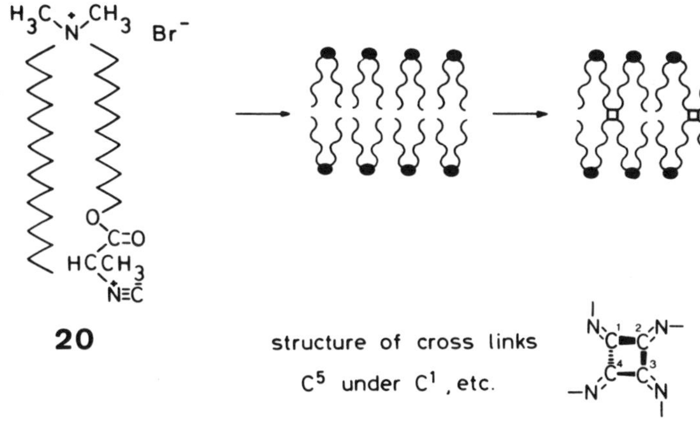

SCHEME 9.8.

aggregates have received great interest as models of biological membranes, as carriers of drugs, and as microreactors.[74] Vesicles have a limited lifetime, which obstructs their use in practical applications. To overcome this problem polymerized vesicles have been synthesized from amphiphiles containing a polymerizable function.[75] For this purpose also isocyano functions have been used.[76,77]

Amphiphile **20** (Scheme 9.8) was synthesized in four steps from dimethylhexadecylamine, 11-bromoundecanol, and L-alanine. On dispersal

FIG. 9.12. Electron micrograph of vesicles stabilized by polymerization of isocyano functions. Magnification × 100,000. Reproduced with permission from Ref. 76.

in water it formed closed vesicles with diameters ranging from 20 to 500 nm. These vesicles could be stabilized by polymerization of the isocyano functions in the bilayers with nickel capronate. The polymerized vesicles retained their structure as was shown by electron microscopy (Fig. 9.12), osmotic experiments, and fluorescence techniques. The degree of polymerization of the isocyano surfactants within the bilayers was estimated to be approximately 74.[77] Amphiphiles having their isocyano functions located in the middle of the hydrocarbon chains gave lower degrees of polymerization (11–50) on treatment with nickel capronate.[77]

9.6. SYNTHETIC PROCEDURES

9.6.1. Polymerization of tert-Butyl Isocyanide[38,45]

tert-Butyl isocyanide (21.8 g, 263 mmol) was polymerized heterogeneously by anhydrous nickel chloride (1.3 g, 10 mmol) without solvent at ambient temperature. After stirring for three days the volatile components were removed in vacuum. The resulting solid was dissolved in approximately 100 mL of chloroform, filtered, and concentrated to a volume of about 40 mL. This solution was added dropwise to a vigorously stirred mixture of 500 mL of methanol and 150 mL of water. The flocky yellow precipitate was collected on a glass filter, washed with 25% aqueous methanol, and dried in vacuum. The yield was 9.5 g (43%) of a bright yellow solid. The polymer did not melt when heated to 200 °C. IR (KBr): 1630, 1670 (shoulder) cm^{-1} (C═N); ^1H NMR ($CDCl_3$): δ 1.35 (s, t-C_4H_9); $\bar{M}_n = 860$ (VPO, toluene, 47 °C). With $NiCl_2 \cdot 6H_2O$ instead of anhydrous $NiCl_2$, the polymerization proceeds faster but the polymer is insoluble, probably because of a higher molecular weight.

9.6.2. Polymerization of (R,S)-2-Isocyanooctane[33,38]

To a stirred solution of 28 mg (0.12 mmol) $NiCl_2 \cdot 6H_2O$ in 10 mL of methanol were added at ambient temperature 6 mL (4.70 g, 33.8 mmol) of 2-isocyanooctane all at once. Immediately, an orange-red color appeared, followed by the precipitation of a cream solid. After 10 to 20 min the reaction mixture solidified, while evolution of heat occurred. Stirring was stopped and the reaction mixture was kept at ambient temperature for 5 h. Subsequently, the methanol was evaporated under water-pump pressure and the polymer dissolved in 50 mL of chloroform. This solution was added dropwise to 750 mL of vigorously stirred methanol–water (4:1, v/v). The precipitated polymer was collected by filtration, washed with methanol, and dried under reduced pressure at 50 °C over KOH. The yield was 4.35 g (92%) of a cream colored solid. The polymer did not melt when heated to 300 °C. IR (KBr): 1633 cm^{-1} (C═N); $[\eta] = 0.45$ $dL\,g^{-1}$ (toluene, 30.00 °C), which corresponds to a molecular weight of $\bar{M}_v = 65{,}000$ according to the relation: $[\eta] = 1.4 \times 10^{-9} \bar{M}_w^{1.75}$ (toluene, 30.00 °C).

9.6.3. *Diastereoselective Polymerization of (S)-1-(Ethoxycarbonyl)Ethyl Isocyanide*[32,41]

(*S*)-1-(Ethoxycarbonyl)ethyl isocyanide[32] ($[\alpha]_{578}^{21}$ + 16.7°, *c* 3.7, $CHCl_3$) was polymerized neat with 0.2 mol% of $NiCl_2 \cdot 6H_2O$ at ambient temperature. After five days the glassy mixture was dissolved in a small amount of chloroform and added dropwise to an excess of vigorously stirred methanol–water (4:1 v/v). The precipitated yellow polymer was collected by filtration, washed with methanol–water, and dried under reduced pressure at 50 °C over KOH. Yield 70%: $[\alpha]_{578}^{20}$ − 280° (*c* 0.36, $CHCl_3$). $[\eta] = 0.44\ \mathrm{dL\,g^{-1}}$ (toluene, 30.00 °C).

9.6.4. *Enantioselective Polymerization of tert-Butyl Isocyanide*[39b]

Preparation of the Catalyst. Tetrakis(*tert*-butyl isocyanide)nickel(II) perchlorate[78] (1.93 mmol) was suspended in 45 mL of dry methylene chloride under an atmosphere of dry nitrogen. To this dispersion was added at room temperature 1.93 mmol of (*S*)-1-phenylethylamine. The solvent was removed under vacuum and the complex was dried at 0.01 mm Hg. Yield 100% of yellow colored catalyst (for physical properties, see Ref. 54).

Polymerization. tert-Butyl isocyanide (758 mg, 9.1 mmol) was added to 0.091 mmol of the above-mentioned catalyst and stirred at ambient temperature for five days. The volatile components were removed under vacuum and the solid residue was dissolved in 2 mL of chloroform. The solution was added dropwise to 35 mL of vigorously stirred methanol–water (3:1 v/v). The pale yellow precipitate was collected by filtration and dried under vacuum at 50 °C. Yield 295 mg (39%) of polymer. $[\alpha]_D^{20}$ − 28.7° (*c* 0.1, $CHCl_3$). $\bar{M}_n = 2400$ (end group determination by ^{1}H NMR).

9.6.5. *Enantioselective Polymerization of 4-Methoxyphenyl Isocyanide*[41]

4-Methoxyphenyl isocyanide[79] (133 mg) was mixed with 144 mg of (*S*)-2-isocyanoisovaleric acid methyl ester[41] ($[\alpha]_D^{20}$ 5.5°, *c* 3.4, $CHCl_3$) and polymerized neat with 1 mol% of $NiCl_2 \cdot 6H_2O$. After 16 h, methanol was added to the reaction mixture. The yellow polymer was collected by filtration, washed with methanol, and dried under reduced pressure at 50 °C. Yield 158 mg of poly(4-methoxyphenyl isocyanide) (90% based on achiral monomer, 23 mol% incorporation of chiral monomer). $[\alpha]_D^{20}$ − 520° (*c* 0.03, $CHCl_3$); $\bar{M}_v = 42{,}000$.

Table 9.7 lists polymers of isocyanides, $[\!>\!C{=}N{-}R]_n$.

TABLE 9.7. Table of Polymers of Isocyanides, $[>C{=}N{-}R]_n$

R in $[>C{=}N{-}R]_n$	Ref.
Aliphatic Homopolymers	
CH_3	28, 80
CF_3	81
C_2H_5	28, 80, 82
n-C_3H_7	80
i-C_3H_7	27
$CH(CH_3)_2$	28
n-C_4H_9	28, 80, 82, 83
sec-C_4H_9	27, 82
(*S*)-*sec*-C_4H_9	44, 45, 46
(*R*)-*sec*-C_4H_9	44
i-C_4H_9	27
$C(CH_3)_3$	28, 44, 45, 46, 83, 84
n-C_5H_{11}	80
$CH(C_2H_5)_2$	83
$C(CH_3)_2C_2H_5$	45, 83, 84
(*R*)-$CH(CH_3)CH(CH_3)_2$	35
n-C_6H_{13}	25, 80
cyclo-C_6H_{11}	27, 28, 82
$CH_2C(CH_3)_3$	80
(*R*)-$CH(CH_3)CH_2CH(CH_3)_2$	35
n-C_7H_{15}	80
n-C_8H_{17}	80
$CH(CH_3)C_6H_{13}$	33
(*R*)-$CH(CH_3)C_6H_{13}$	35
n-C_9H_{19}	80
n-$C_{10}H_{21}$	80
4-(2,2,6,6-tetramethyl-*N*-oxyl-piperidinyl)	96
$CH_2C_6H_5$	80, 83
$CH(CH_3)C_6H_5$	25, 28, 44
(*S*)-$CH(CH_3)C_6H_5$	32, 43, 45
(*R*)-$CH(CH_3)C_6H_5$	43, 45
$CH(CH_3)[4\text{-}BrC_6H_4]$	56
$CH(CH_3)[4\text{-}C_6H_4P(C_6H_5)_2]$	56, 85
$C(CH_3)_2C_6H_5$	83
(*S*)-$CH(CH_3)CH_2C_6H_5$	32
$CH_2CH_2C_6H_5$	86
$CH{=}CHCH_3$	87
$CH{=}C(CH_3)_2$	87
$CH{=}CHCHC(CH_3)_3$	87
$CH{=}CHC_6H_5$	87
$CH{=}C(CH_3)C_6H_5$	87
$CH{=}CH[2,4,6\text{-}(CH_3)_3C_6H_2]$	87
$CH{=}CHC(CH_3){=}CHCH_3$	87
(*S*)-$CH(C_2H_5)CH{=}CH_2$	32
(*S*)-$CH(C_2H_5)C{\equiv}CH$	32
$CH(CH_3)COOCH_3$	88
(*S*)-$CH(CH_3)COOC_2H_5$	32
$CH(CH_2C_6H_5)COOCH_3$	88

TABLE 9.7. (Continued)

R in [>C=N—R]$_n$	Ref.
(*S*)-$CH(CH_3)CH_2OC(O)CH_3$	32
(*S*)-$CH[CH(CH_3)_2]CH_2OC(O)CH_3$	32
(*R*)-$CH(C_6H_5)CH_2OC(O)CH_3$	32
(*S*)-$CH_2CH(CH_3)P(C_6H_5)_2O$	32
$CH_2Si(CH_3)_3$	80
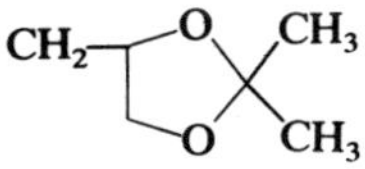	89
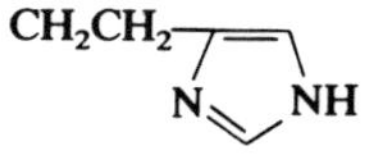	31, 90
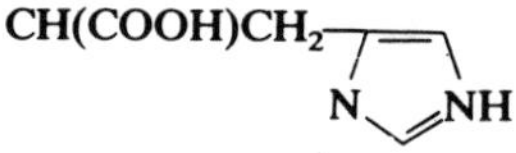	31, 90
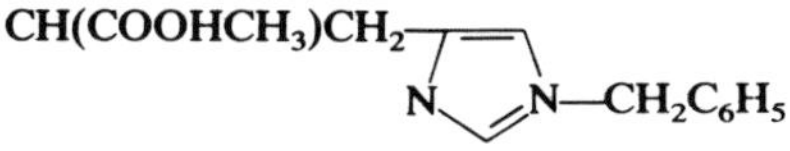	31
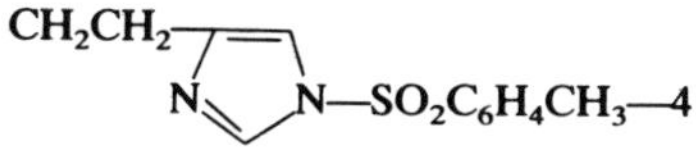	31, 91
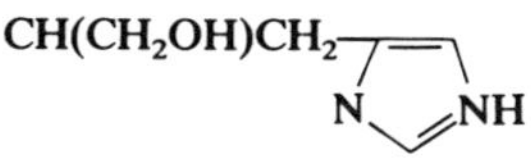	30
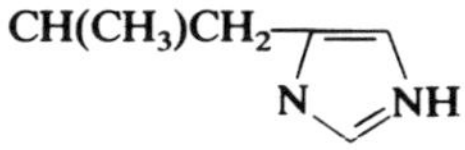	30
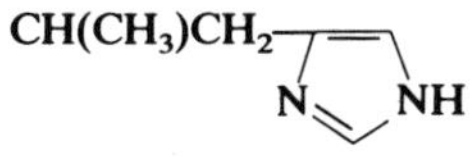	60
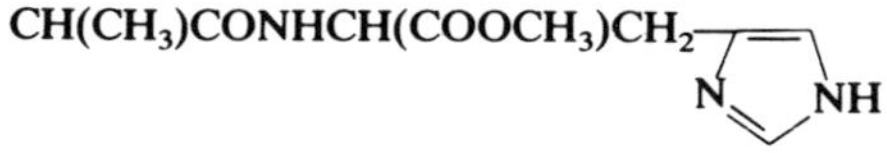	60
	60
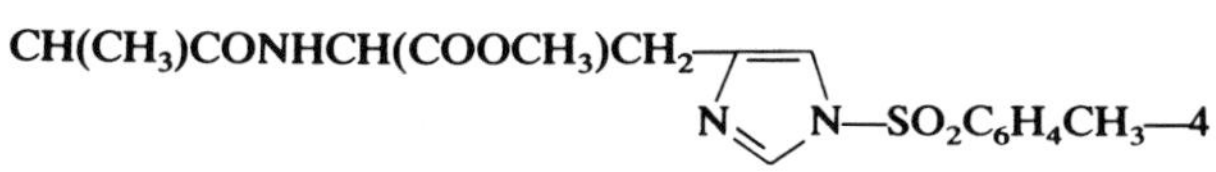	
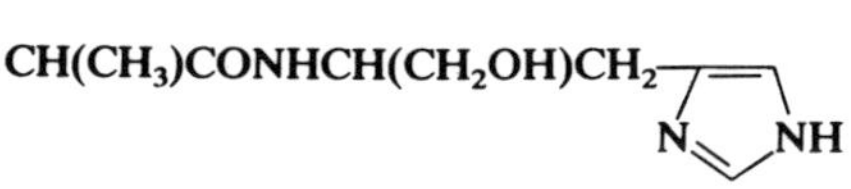	61

(*continued*)

TABLE 9.7. (Continued)

R in [>C=N—R]$_n$	Ref.
	37
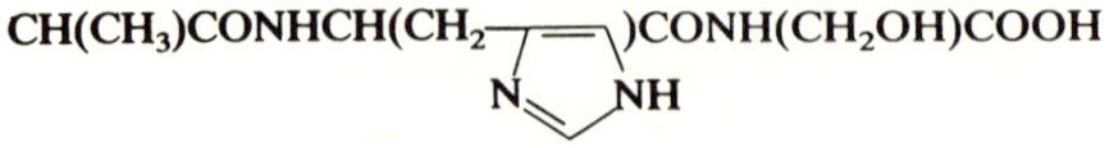	37
CH(CH$_3$)CONHCH(CH$_2$OCOCH$_3$)COOCH$_3$	37
CH(CH$_3$)CONHCH(CH$_2$OH)COOH	37
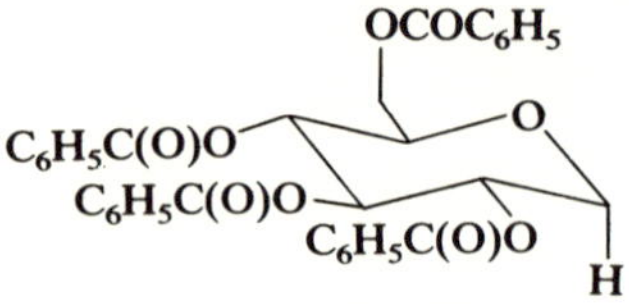	36
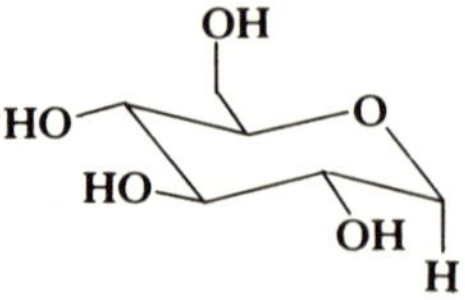	36
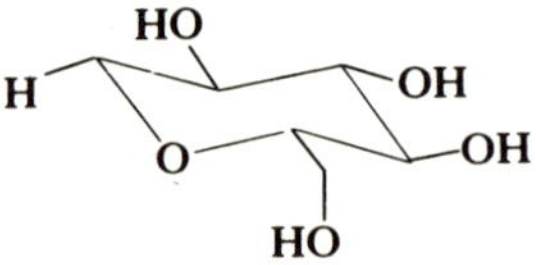	36
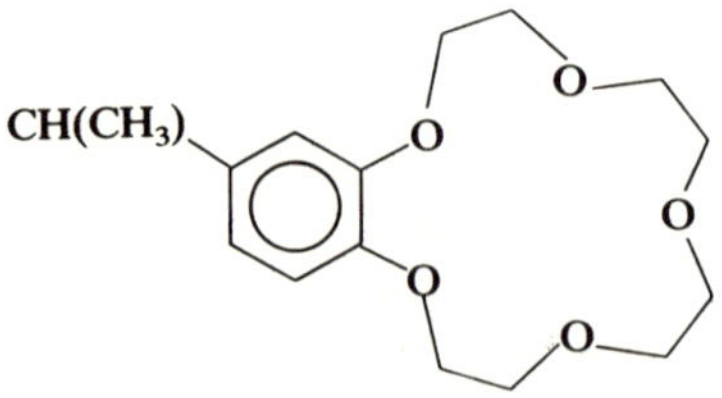	67
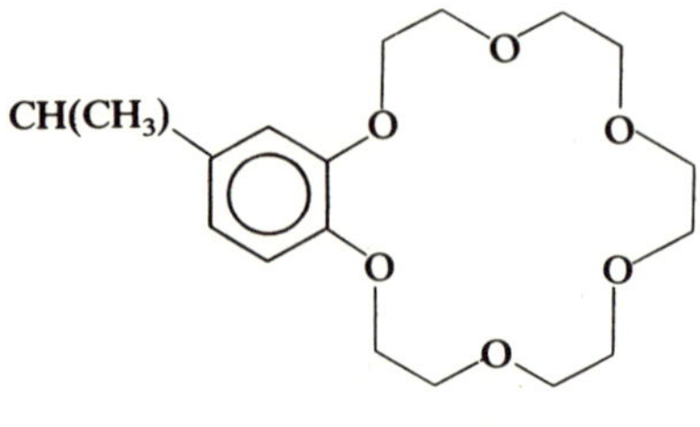	65, 66, 67

TABLE 9.7. (Continued)

R in [>C=N—R]$_n$	Ref.
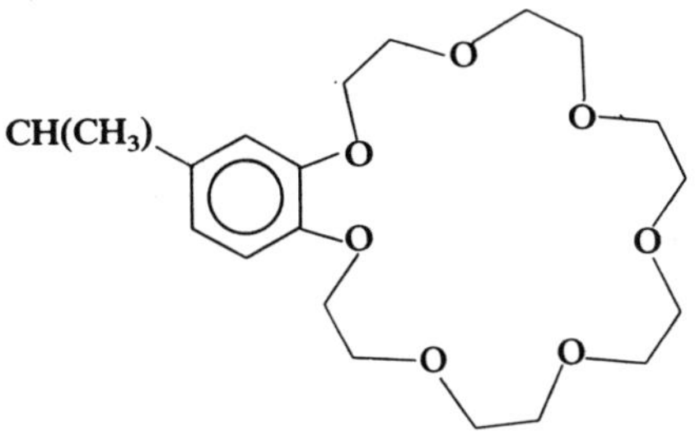	67
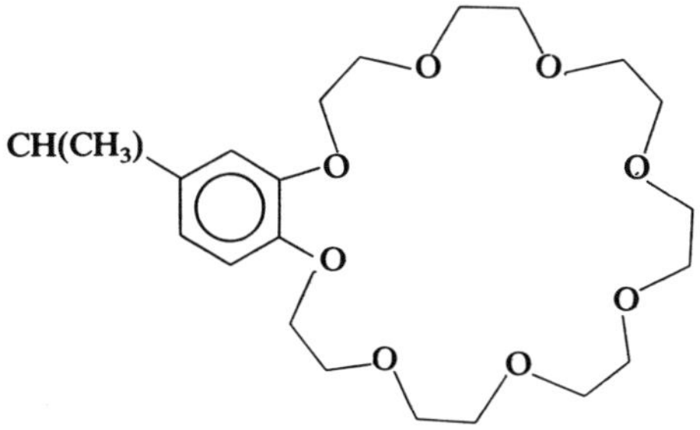	67
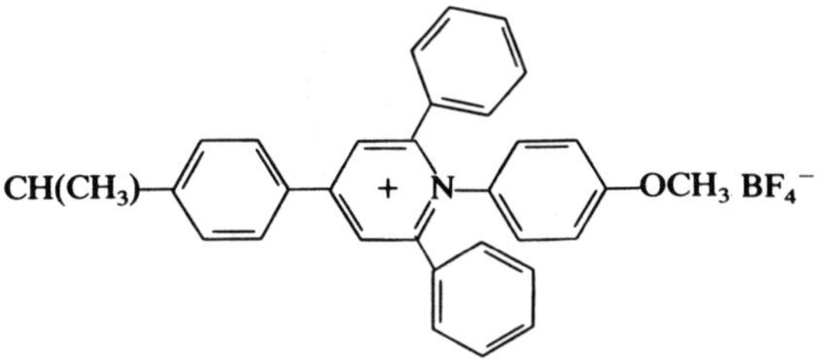	71
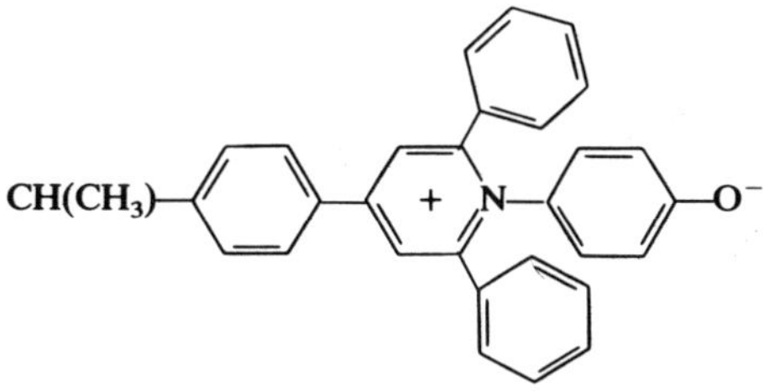	71

(continued)

TABLE 9.7. (Continued)

R in [>C=N—R]$_n$	Ref.
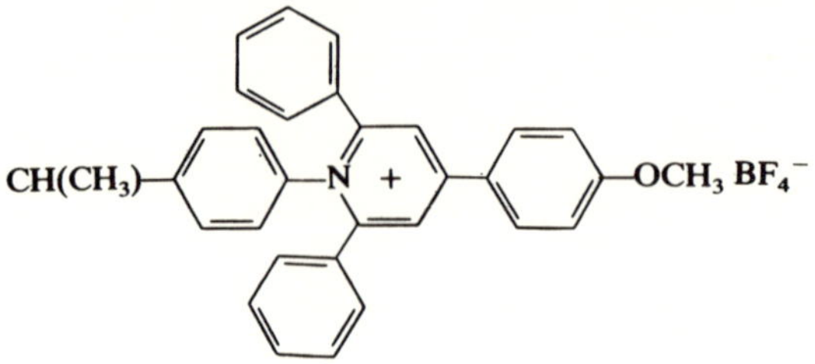	71
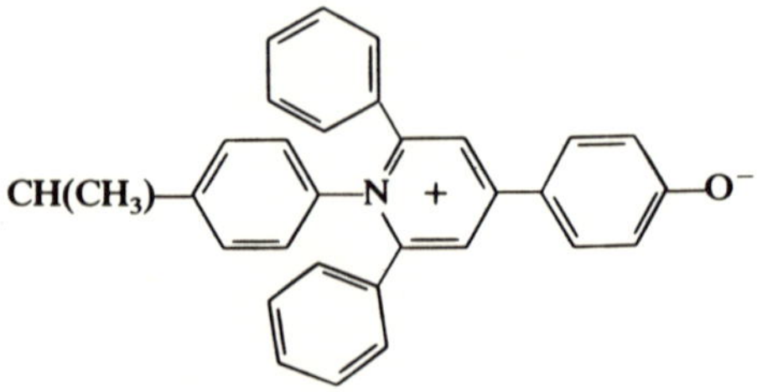	71
$CH(CH_3)COOC_{11}H_{22}N^+(CH_3)_2C_{16}H_{33}Br^-$	67, 76
$CH(i\text{-}C_3H_7)COOC_{11}H_{22}N^+(CH_3)_2C_{16}H_{33}Br^-$	67
$CH(n\text{-}C_6H_{13})COOC_{11}H_{22}N^+(CH_3)_2C_{20}H_{41}Br^-$	67
$CH(n\text{-}C_6H_{13})COOC_{11}H_{22}N^+(CH_3)_2C_{22}H_{45}Br^-$	67
$[CH(CH_3)COOC_{11}H_{22}]_2N^+(CH_3)_2Br^-$	67
(*S*)-$CH(COOCH_3)CH(CH_3)_2$	40, 41
(*S*)-$CH[COOCH(CH_3)_2]CH(CH_3)_2$	41
(*S*)-$CH[COOC(CH_3)_3]CH(CH_3)_2$	40, 41
(1*S*,2*S*)-$CH(COOCH_3)CH(CH_3)C_2H_5$	41
$CH(COOCH_3)CH_2C_6H_4OC(O)CH_3$-4	92
$CH(COOH)CH_2C_6H_4OH$—4	92
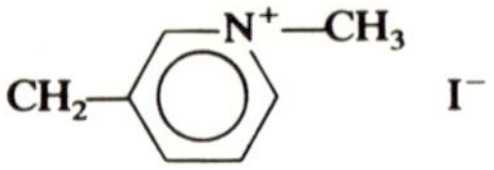	70
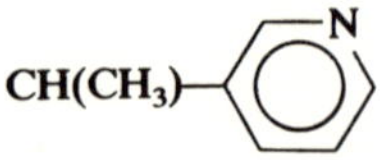	70
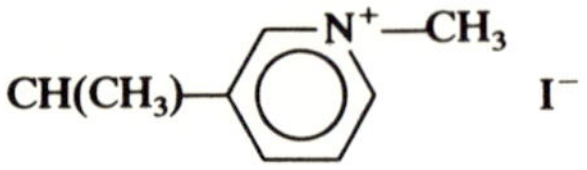	70
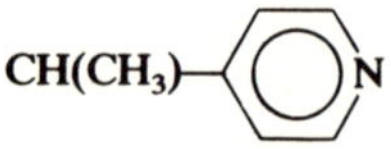	70

TABLE 9.7. (Continued)

R in [>C=N—R]$_n$	Ref.
	70
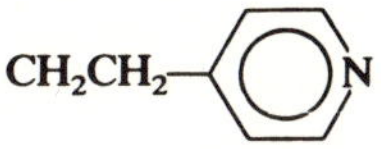	70
	70
Aromatic Homopolymers	
C_6H_5	27, 28, 80, 82, 83
$C_6H_4OCH_3$-4	27, 80, 82, 83
C_6H_4Cl-4	27, 80, 82
$C_6H_4CH_3$-4	27, 80, 82, 94
$C_6H_4N(CH_3)_2$-4	83
$C_6H_4CH_3$-2	26, 80
$C_6H_4C(CH_3)_3$-2	83
$C_6H_4C_6H_5$-2	83
$C_6H_4C_2H_5$-2	80
$C_6H_4OCH_3$-2	80
$C_6H_4CF_3$-2	80
$C_6H_4OCH_3$-3	80
$C_6H_3F_2$-2,6	83
$C_6H_3Cl_2$-2,6	83
$C_6H_3(CH_3)_2$-2,4	80
$C_6H_3(CH_3)_2$-3,5	80
C_6F_5	95
$C_6H_3OCH_3$-2—CH_3-4	83

REFERENCES

1. A. W. Hofmann, *Ann. Chem. Pharm.* **146**, 107–119 (1868).
2. A. W. Hofmann, *Ann. Chem. Pharm.* **144**, 114–120 (1867).
3. A. Gautier, *Ann. Chem. Pharm.* **146**, 119–124 (1868).
4. A. Gautier, *Ann. Chem. Pharm.* **142**, 289–294 (1867).
5. *Isonitrile Chemistry,* ed. by I. Ugi (Academic Press, New York, 1971).
6. *The Chemistry of the Cyano Group,* ed. by Z. Rappoport (Interscience, New York, 1970), Chap. 15, pp. 853–883.
7. *Nomenclature of Organic Chemistry, The Blue Book,* ed. by J. Rigaudy and S. P. Klesney (Pergamon Press, Oxford, 1979), Rule C 833.
8. I. Hagedorn and H. Tonjes, *Pharmazie* **11**, 409 (1956).
9. W. R. Hertler and E. J. Corey, *J. Org. Chem.* **23**, 1221–1222 (1958).
10. I. Ugi and R. Meyr, *Angew. Chem.* **70**, 702–703 (1958).

11. *Comprehensive Organic Chemistry*, ed. by D. Barton and W. D. Ollis (Pergamon Press, Oxford, 1979), Vol. 2, pp. 568–570.
12. J. Songstad, L. J. Stangeland, and T. Austad, *Acta Chem. Scand.* **24**, 355–356 (1970).
13. L. B. Engemyr, A. Martinsen, and J. Songstad, *Acta Chem. Scand.* **A28**, 255–266 (1974).
14. W. P. Weber and G. W. Gokel, *Tetrahedron Lett.* **17**, 1637–1640 (1972).
15. W. P. Weber, G. W. Gokel, and I. Ugi, *Angew. Chem.* **84**, 587 (1972).
16. Y. Yamamoto and N. Hagihara, *Bull. Chem. Soc. Jpn.* **39**, 1084–1085 (1966).
17. Y. Yamamoto, T. Takizawa, and N. Hagihara, *Nippon Kagaku Zasshi* (*J. Chem. Soc. Japan, Pure Chem. Section*) **87**, 1355 (1966).
18. S. Otsuka, A. Nakamura, and T. Yoshida, *J. Am. Chem. Soc.* **91**, 7196–7198 (1969).
19. The first observation by Millich of polymerization of an isocyanide occurred in 1962. He communicated this observation at the 16th Midwest Chemistry Conference of the Am. Chem. Soc., Kansas City, Mo, Nov. 19, 1964 and at the 150th National Meeting of the Am. Chem. Soc., Atlantic City, NJ, Sept. 1965. Millich is the author of several review articles in this field, e.g., Refs. 20 and 21.
20. F. Millich, *Chem. Rev.* **72**, 101–113 (1972).
21. F. Millich, *J. Polym. Sci., Macromol. Rev.* **15**, 207–253 (1980).
22. R. W. Stackman, *J. Macromol. Sci., Chem.* **A2**, 225–236 (1968).
23. T. Saegusa, N. Taka-Ishi, and Y. Ito, *J. Org. Chem.* **34**, 4040–4046 (1969).
24. T. Saegusa, Y. Ito, S. Kobayashi, and K. Hirota, *Tetrahedron Lett.*, 521–524 (1967).
25. F. Millich and R. G. Sinclair II, *J. Polym. Sci., Part A1* **6**, 1417–1430 (1968).
26. F. Millich and R. G. Sinclair II, *J. Polym. Sci., Part C* **22**, 33–43 (1968).
27. R. J. M. Nolte, Thesis, Utrecht, 1973.
28. R. J. M. Nolte, R. W. Stephany, and W. Drenth, *Recl. Trav. Chim. Pays-Bas* **92**, 83–91 (1973).
29. R. J. M. Nolte and W. Drenth, *Recl. Trav. Chim. Pays-Bas* **92**, 788–800 (1973).
30. J. M. van der Eijk, Thesis, Utrecht, 1980.
31. J. M. van der Eijk, R. J. M. Nolte, and W. Drenth, *Recl. Trav. Chim. Pays-Bas* **97**, 46–49 (1978).
32. A. J. M. van Beijnen, R. J. M. Nolte, A. J. Naaktgeboren, J. W. Zwikker, W. Drenth, and A. M. F. Hezemans, *Macromolecules* **16**, 1679–1689 (1983).
33. R. J. M. Nolte, A. J. M. van Beijnen, J. W. Zwikker, and W. Drenth, *Macromol. Synth.* **9**, 81–83 (1985).
34. A. J. M. van Beijnen, R. J. M. Nolte, J. W. Zwikker, and W. Drenth, *J. Mol. Catal.* **4**, 427–432 (1978).
35. A. J. M. van Beijnen, R. J. M. Nolte, W. Drenth, A. M. F. Hezemans, and P. J. F. M. van de Coolwijk, *Macromolecules* **13**, 1386–1391 (1980).
36. R. J. M. Nolte, J. A. J. van Zomeren, and J. W. Zwikker, *J. Org. Chem.* **43**, 1972–1975 (1978).
37. H. G. J. Visser, R. J. M. Nolte, J. W. Zwikker, and W. Drenth, *J. Org. Chem.* **50**, 3133–3137, 3138–3143 (1985).
38. A. J. M. van Beijnen, Thesis, Utrecht, 1980.
39. (a) R. J. M. Nolte and W. Drenth, unpublished results; (b) P. C. J. Kamer, R. J. M. Nolte, and W. Drenth, *J. Am. Chem. Soc.* **110**, 6818–6825 (1988).
40. T. Harada, M. C. Cleij, R. J. M. Nolte, A. M. F. Hezemans, and W. Drenth, *J. Chem. Soc., Chem. Commun.*, 726–727 (1984).
41. P. C. J. Kamer, M. C. Cleij, R. J. M. Nolte, T. Harada, A. M. F. Hezemans, and W. Drenth, *J. Am. Chem. Soc.* **110**, 1581–1587 (1988).
42. IUPAC, *Pure Appl. Chem.* **48**, 373–385 (1976).
43. F. Millich and G. K. Baker, *Macromolecules* **2**, 122–128 (1969).
44. R. J. M. Nolte, A. J. M. van Beijnen, and W. Drenth, *J. Am. Chem. Soc.* **96**, 5932–5933 (1974).

45. A. J. M. van Beijnen, R. J. M. Nolte, and W. Drenth, *Recl. Trav. Chim. Pays-Bas* **99**, 121–123 (1980).
46. A. J. M. van Beijnen, R. J. M. Nolte, W. Drenth, and A. M. F. Hezemans, *Tetrahedron* **32**, 2017–2019 (1976).
47. C. J. M. Huige, Thesis, Utrecht, 1985.
48. F. Millich, *Adv. Polym. Sci.* **19**, 117–141 (1975).
49. M. M. Green, R. A. Gross, F. C. Schilling, K. Zero, and Ch. Crosby III, *Macromolecules* **21**, 1839–1846 (1988).
50. P. C. J. Kamer, W. Drenth, and R. J. M. Nolte, *Polym. Prepr.* **30**, 418–419 (1989).
51. F. Millich, E. W. Hellmuth, and S. Y. Huang, *J. Polym. Sci., Polym. Chem. Ed.* **13**, 2143–2150 (1975).
52. R. J. M. Nolte, J. W. Zwikker, J. Reedijk, and W. Drenth, *J. Mol. Catal.* **4**, 423–426 (1978).
53. R. J. M. Nolte and W. Drenth, unpublished results.
54. P. C. J. Kamer, R. J. M. Nolte, and W. Drenth, *Recl. Trav. Chim. Pays-Bas* **107**, 175–181 (1988).
55. P. C. J. Kamer, R. J. M. Nolte, W. Drenth, H. L. L. M. Nijs, and J. A. Kanters, *J. Mol. Catal.* **49**, 21–32 (1988).
56. A. J. Naaktgeboren, R. J. M. Nolte, and W. Drenth, *Recl. Trav. Chim. Pays-Bas* **97**, 112–115 (1978).
57. A. J. Naaktgeboren, Thesis, Utrecht, 1981.
58. J. M. van der Eijk, R. J. M. Nolte, and W. Drenth, *Recl. Trav. Chim. Pays-Bas* **97**, 46–49 (1978).
59. J. M. van der Eijk, Ch. F. Gusdorf, R. J. M. Nolte, and W. Drenth, *Recl. Trav. Chim. Pays-Bas* **98**, 232–238 (1979).
60. J. M. van der Eijk, R. J. M. Nolte, W. Drenth, and A. M. F. Hezemans, *Macromolecules* **13**, 1391–1397 (1980).
61. J. M. van der Eijk, R. J. M. Nolte, V. E. M. Richters, and W. Drenth, *Recl. Trav. Chim. Pays-Bas* **100**, 222–226 (1981).
62. H. G. J. Visser, R. J. M. Nolte, J. W. Zwikker, and W. Drenth, *J. Org. Chem.* **50**, 3133–3137 (1985).
63. H. G. J. Visser, R. J. M. Nolte, J. W. Zwikker, and W. Drenth, *J. Org. Chem.* **50**, 3138–3143 (1985).
64. H. G. J. Visser, R. J. M. Nolte, and W. Drenth, *Macromolecules* **18**, 1818–1825 (1985).
65. A. J. M. van Beijnen, R. J. M. Nolte, J. W. Zwikker, and W. Drenth, *Recl. Trav. Chim. Pays-Bas* **101**, 409–410 (1982).
66. R. J. M. Nolte, A. J. M. van Beijnen, J. G. Neevel, J. W. Zwikker, A. J. Verkleij, and W. Drenth, *Isr. J. Chem.* **24**, 297–301 (1984).
67. M. F. M. Roks, Thesis, Utrecht, 1987.
68. J. H. van der Linden, J. Schoonman, R. J. M. Nolte, and W. Drenth, *Recl. Trav. Chim. Pays-Bas* **103**, 260–262 (1984).
69. U. F. Kragten, M. F. M. Roks, and R. J. M. Nolte, *J. Chem. Soc., Chem. Commun.*, 1275–1276 (1985).
70. E. Oostveen and W. Drenth, *Recl. Trav. Chim. Pays-Bas* **102**, 35–41 (1983).
71. A. J. M. van Beijnen, R. J. M. Nolte, and W. Drenth, *Recl. Trav. Chim. Pays-Bas* **105**, 255–261 (1986).
72. Ch. Reichardt and K. Dimroth, *Fortschr. Chem. Forsch.* **11**, 1–73 (1968).
73. J. H. Fendler, *Membrane Mimetic Chemistry* (Wiley, New York, 1982).
74. J. van Esch, M. F. M. Roks, and R. J. M. Nolte, *J. Am. Chem. Soc.* **108**, 6093–6094 (1986).
75. H. Ringsdorf, B. Schlarb, and J. Venzmer, *Angew. Chem., Int. Ed. Engl.* **27**, 113–158 (1988).

76. M. F. M. Roks, H. G. J. Visser, J. W. Zwikker, A. J. Verkleij, and R. J. M. Nolte, *J. Am. Chem. Soc.* **105**, 4507–4510 (1983).
77. M. F. M. Roks, R. S. Dezentjé, V. E. M. Kaats-Richters, W. Drenth, A. J. Verkleij, and R. J. M. Nolte, *Macromolecules* **20**, 920–929 (1987).
78. R. W. Stephany and W. Drenth, *Recl. Trav. Chim. Pays-Bas* **91**, 1453–1458 (1972).
79. G. Skorna and I. Ugi, *Angew. Chem.* **89**, 267–268 (1977).
80. R. B. King and M. J. Greene, *J. Polym. Sci., Polym. Chem. Ed.* **25**, 907–918 (1987).
81. R. E. Banks, R. N. Haszeldine, M. J. Stevenson, and B. G. Willoughby, *J. Chem. Soc., C,* 2119–2123 (1969).
82. Y. Yamamoto, T. Takizawa, and N. Hagihara, *Nippon Kagaku Zasshi* **87**, 1355–1359 (1966); Y. Yamamoto and N. Hagihara, *Nippon Kagaku Zasshi* **89**, 898–900 (1968).
83. P. C. J. Kamer, R. J. M. Nolte, and W. Drenth, *J. Am. Chem. Soc.* **110**, 6818–6825 (1988).
84. P. C. J. Kamer, R. J. M. Nolte, and W. Drenth, *J. Chem. Soc., Chem. Commun.*, 1789–1791 (1986).
85. A. J. Naaktgeboren, R. J. M. Nolte, and W. Drenth, *J. Am. Chem. Soc.* **102**, 3350–3354 (1980).
86. F. Millich and G.-M. Wang, Abstracts, *Midwest Chemistry Conference of the American Chemical Society,* Lincoln, *No. 617,* NB October 1970.
87. R. B. King and L. Borodinsky, *Macromolecules* **18**, 2117–2120 (1985).
88. J. Chenvanij, Thesis, University of Missouri—Kansas City, 1973.
89. M. M. Green, R. Gross, and A. Stassi, *Polym. Prepr.* **26**, 76–77 (1985).
90. J. M. van der Eijk, R. J. M. Nolte, V. E. M. Richters, and W. Drenth, *Biopolymers* **19**, 445–448 (1980).
91. J. M. van der Eijk, V. E. M. Richters, R. J. M. Nolte, and W. Drenth, *Recl. Trav. Chim. Pays-Bas* **103**, 46–50 (1984).
92. J. A. S. J. Razenberg, A. W. van der Made, J. W. H. Smeets, and R. J. M. Nolte, *J. Mol. Catal.* **31**, 271–287 (1985).
93. A. W. van der Made, J. W. H. Smeets, R. J. M. Nolte, and W. Drenth, *J. Chem. Soc., Chem. Commun.*, 1204–1206 (1983).
94. S. Iwatsuki and Y. Yamashita, *Kogyo Kaguka Zasshi* **70**, 1822–1825 (1967).
95. R. E. Banks, R. N. Haszeldine, and B. G. Willoughby, *J. Chem. Soc., Perkin Trans. 1,* 2451–2454 (1975).
96. E. J. Vlietstra, R. J. M. Nolte, J. W. Zwikker, W. Drenth, and E. W. Meijer, *Macromolecules* **23**, 946–948 (1990).

Index